Cytogenetics of Mammalian
Embryonic Development

Cytogenetics of Mammalian Embryonic Development

A.P. DYBAN

and

V.S. BARANOV

*Institute for Experimental Medicine
of the Academy of Medical Sciences
of the USSR, Leningrad*

English translation by V.S. Baranov

CLARENDON PRESS · OXFORD
1987

Oxford University Press, Walton Street, Oxford OX2 6DP

Oxford   New York   Toronto
Delhi   Bombay   Calcutta   Madras   Karachi
Petaling Jaya   Singapore   Hong Kong   Tokyo
Nairobi   Dar es Salaam   Cape Town
Melbourne   Auckland
and associated companies in
Beirut   Berlin   Ibadan   Nicosia

Oxford is a trade mark of Oxford University Press

Published in the United States
by Oxford University Press, New York

Russian edition first published by Nauka (Moscow) 1978
English edition first published by Oxford University Press 1987

British Library Cataloguing in Publication Data
Dyban, A.P.
Cytogenetics of mammalian embryonic development.
1. Embryology—Mammals
I. Title   II. Baranov, V.S.
599.03'34   QL959
ISBN 0-19-854584-3

Library of Congress Cataloging in Publication Data
Dyban, A.P. (Andreí Pavlovich)
Cytogenetics of mammalian embryonic development.
A revised and updated translation of the Russian text.
On t.p.: Institute for Experimental Medicine of the
Academy of Medical Sciences of the USSR, Leningrad.
Bibliography: p.
Includes index.
1. Embryology—Mammals.   2. Developmental genetics.
3. Developmental cytology.   I. Baranov, V.S.   II. Title.
QL959.D93   1987   599'.033   86-5187
ISBN 0-19-854584-3

Set by Colset Private Ltd, Singapore
Printed in Great Britain by
The Alden Press, Oxford

Preface to the English edition

The Russian version of this monograph, published in 1978, treated the developmental cytogenetics of mammals in a way which had no real equivalent either here in the Soviet Union or abroad. The book summarized numerous original cytogenetic investigations carried out in the Department of Embryology of the Institute of Experimental Medicine, Academy of Medical Sciences of the USSR. It encompassed the numerous key problems, the basic questions, and the major trends in the cytogenetics of mammalian embryogenesis. We tried to make Soviet readers more familiar with relevant work going on abroad, which is sometimes not very readily accessible. Hence our first edition covered most references to this foreign work up to the end of 1976.

We had some serious difficulties in preparing this monograph for the English edition. One major problem was the tremendous growth of experimental work in the field of developmental cytogenetics since 1977. Thorough incorporation of all this new information would have required complete revision of the text, resulting in a quite new, extensive monograph of the handbook type. We have made an effort to avoid this logical outcome, which would have substantially exceeded our mental and physical capabilities!

Another problem was that many comprehensive reviews and extensive monographs (mostly in English) have been published since 1978. We concluded that it would be unnecessary to include detailed accounts of these for the English-speaking reader, who can easily get the original publications.

A further complication results from the language barrier. Many of the theoretical accounts recently produced by our foreign colleagues are in line with basic conclusions already inferred from our own much earlier experimental studies. Unfortunately most of our experiments were published in Russian and thus passed unnoticed by our Western colleagues. Russian scientists, however, are always adequately informed on all scientific work published in English or in any other Western language. This imbalance of information exchange stimulated us to extend the description of experimental results obtained in our own department as well as by other Russian scientists working in developmental cytogenetics.

Trying to avoid major changes in the general structure of the original monograph, we nonetheless undertook a complete revision of some chapters where there has been substantial new information in recent years. These are Chapters 1 (Haploidy), 6 (Autosomal trisomy), 7 (Monosomy and nullisomy of autosomes), 12 (Influence of genes on early mammalian development), and 13 (Functional activity of chromosomes and mechanisms of early embryonic development). The text of the remaining chapters was updated. The chapter on genetic mosaics and chimaeras which appeared in the Russian edition is completely omitted, as this topic has been thoroughly discussed in several important monographs and major reviews. The English edition of the book does not include the Addendum, which was devoted to the description of standard methods of chromosome preparation at different stages of embryogenesis as well as to the standard karyotypes of common laboratory animals. Both are described satisfactorily in cytogenetic literature published in English.

It should be noted that the Russian edition of the book was prepared not only for narrow specialists in developmental cytogenetics but also for general biologists, and even for clinical geneticists. Hence we tried to avoid unnecessary complications, and some theoretical discussion was deliberately simplified. The English version of these chapters, although somewhat extended, has kept this same style of presentation, which we hope will make the monograph of value to many scientists outside the field of developmental cytogenetics.

Most of the aforementioned remarks concern Chapters 4, 5, 9, and 12. We realize that a good deal of recently published experimental work is not included in this monograph. Most of these works, especially monographs published in the last two or three years, are known to us only by their titles or as short abstracts, which do not allow proper citation in the text. Nevertheless, the list of publications is already overloaded, though still not complete.

We hope that English readers will find this monograph of sufficient use to justify the efforts we have made in preparing this new edition. We hope also that they will forgive any imperfections in presentation resulting from language difficulties, as well as from other obstacles in the way of efficient and rapid exchange of scientific information.

We shall appreciate any kind of constructive criticism and consider the goal of the book achieved if it is of use to the English readers.

Leningrad   A. P. D.
1986   V. S. B.

Contents

Acknowledgements

We wish to express our deep gratitude to I.I. Tichodeeva and L.A. Konopisceva, collaborators in the Department of Embryology, Institute for Experimental Medicine, for their patience and generous help with the typescript throughout its preparation, and wish to extend our thanks to O.V. Lazareva and N.V. Antonova for preparation of the book for print.

Introduction

The book is devoted to the borderline region of biological sciences related both to genetics and developmental biology. It should be remembered that the term 'cytogenetics' was coined for the science dealing with studies of normal chromosomal complement (karyotype chromosomal transformation, behaviour and interactions in mitoses and meioses) as well as with aberrations of chromosomal structure and number affecting the phenotype of particular cells and the organism as a whole.

In other words, cytogenetics is the study of genetic problems by means of cytological methods, primarily through the visualization of chromosomes displaying structural or numerical alterations which make it possible to elucidate their functional activity.

Hence, developmental cytogenetics might be conventionally admitted as a branch of cytogenetics exploring developmental problems in terms of chromosomal structure and functions related to gene action at different stages of ontogenesis. The cytogenetic approach might also be applied to the studies of gametogenesis, fertilization, embryogenesis, and postnatal stages.

This comprehensive definition of developmental cytogenetics enables one to trace the onset of this science back 50 years, to when the first attempts to study developmental peculiarities in mice with chromosomal aberrations were undertaken (Bonnevie 1934, 1940; Snell, Bodemann, and Hollander 1934; Brenneke 1937; Hertwig 1938). Since that time the main goal of this approach has been considered to be the search for the causative correlations between developmental peculiarities and chromosomal alterations which can help to pinpoint the gene loci responsible for the corresponding morphogenetic processes.

However, the absence of proper cytogenetical techniques and shortage of proper embryological information, especially concerning control mechanisms of morphogenetic reactions, made this constructive approach hardly accessible at that time.

Since before the Second World War, serious cytogenetic studies related to embryonic development have been carried out on *Drosophila*. Detailed knowledge of the polytene chromosome map of this species substantiated developmental analysis of X-chromosome deficiencies. The correlation between the size of missing segments and their phenotype manifestations has

1

been established. The early embryolethals usually corresponded to more widespread deficiencies and vice versa. The few exceptions from this general rule (i.e. early death and minor deficiency) bore witness to the presence of some genes crucial for early development in these particular chromosome regions (Poulson 1940, 1945). These observations led to the hypothesis of step-like action of genes in ontogenesis, according to which stages of embryonic death as well as of other developmental failures correspond in time to the onset of action of genes located in the missing chromosomal segment (Hadorn 1961).

It took many years to adapt the cytogenetic approach for studying embryonic development in mammals. This delay can primarily be attributed to the absence of adequate cytogenetic methods for detailed identification of mammalian chromosomes and their definite regions. For almost 50 years, mammalian chromosomes were studied exclusively in squash preparations, which did not even allow the exact number of chromosomes in humans and other mammalian species to be established with adequate precision.

It is not surprising therefore that after 50 years of research, almost nothing was known for sure on developmental effects of structural and numerical chromosomal aberrations in mammals. Meanwhile, useful information concerning developmental manifestations of gross genome imbalance (i.e. haploidy, triploidy, tetraploidy, and some even more complicated forms) had become available (Beatty 1957).

The onset of modern cytogenetics might be dated from the innovation of new techniques of metaphase chromosome spreading, based on major discoveries of 'hypotonic miracle' combined with other technical improvements (colchicine pretreatment, air-drying, etc.). The next fascinating achievements of cytogenetics were mainly due to the introduction of differential staining methods, which provided precise identification of metaphase chromosomes and their structural rearrangements. A historical essay with very peculiar presentation and intriguing details on this and many other cytogenetic studies in humans and other mammals has already been published (Hsu 1979), so there is no reason to go into further detail here.

It is noteworthy, however, that all methods successfully applied for chromosomal preparations from bone marrow cells or cell cultures are totally unsuitable for chromosome analysis of mammalian embryos.

Therefore, the experimental cytogenetics of mammalian embryogenesis especially its most intriguing part, concerned with oogenesis and preimplantation stages, lags at least 10 years behind clinical cytogenetics and might be dated back to 1966, when a convenient method of chromosomal preparations from oocytes and cleaving eggs was invented (Tarkowski 1966). Several reliable methods of chromosomal analysis of early post-implantation embryos were suggested soon after this (Wroblewska and Dyban 1969; Evans, Burtenshaw, and Ford 1972) and methods of meiotic chromosome study from both oocytes and spermatocytes were also significantly improved (Evans, Breckon, and Ford 1964; Dyban 1970b; Meredith 1969). Hence metaphase spreads from the oocytes and early mammalian embryos became amenable for detailed analysis from the early 1970s onwards.

In spite of all these technical achievements, a substantial gap still exists between experimental cytogenetics of mammalian development and developmental cytogenetics of human beings. There is a good deal of conclusive evidence on the major role of chromosomal aberrations in spontaneous abortions, inborn abnormalities, and well-known chromosomal diseases. At the same time, however, we still remain almost totally ignorant of the developmental effects triggered by chromosomal abnormalities in early human embryos.

The basic information on cytogenetic aspects of gametogenesis, fertilization, pre-implantation, and early post-implantation development can be gained from experiments with laboratory animals. Human embryos of equivalent stages are still rather rare and they are poorly studied both morphologically and cytogenetically.

Unfortunately, only a few laboratories are dealing with experimental cytogenetics of mammalian development at present. Our monograph summarizes the results of experimental studies carried out since 1966 on the developmental cytogenetics of laboratory mice and rats (plus a few studies in golden hamsters) in the Department of Embryology at the Institute of Experimental Medicine, Academy of Medical Sciences of the USSR (Leningrad). Comprehensive analysis of all, or most, available experimental works on cytogenetics of mammalian embryogenesis is also included.

Chapters 1–3 deal with peculiarities of mammalian embryogenesis induced by numerical deviations at the whole genome level (haploidy, triploidy, and tetraploidy, respectively). Chapters 4 and 5 summarize methodological prerequisites elaborated for the deliberate production of mouse embryos with numerical and structural aberrations of particular chromosomes. Developmental peculiarities of mammalian embryos (predominantly of mice) with trisomy, monosomy, or nullisomy of individual chromosomes as well as imbalance of gonosomes (X, Y chromosomes) are discussed in Chapters 6, 7, and 8, respectively. Chapter 9 is devoted to structural chromosomal aberrations. Relevant data on human cytogenetics are only briefly outlined in each chapter as there are many original publications, fundamental reviews, monographs, and textbooks on human cytogenetics, some of which are referred to in the text. Comparative analysis of the effects of chromosomal anomalies on embryos of humans and laboratory animals is presented in Chapter 11. Chapter 10 covers all issues concerned with a relatively new branch of developmental cytogenetics (behavioural peculiarities in mammals with chromosomal anomalies), which was initially launched in our collaborative studies with staff members of the Laboratory of Behavioural Genetics at the

Moscow State University. Although shortage of positive knowledge in current developmental cytogenetics significantly hampers theoretical considerations and even makes fundamental generalizations and theories premature, we found it possible to sum up all basic factors and offer some suggestions as to genetic (chromosomal) and epigenetic factors interacting in early mammalian embryos (Chapter 13). Our considerations at the chromosomal level are substantiated by available information on gene action in early mammalian embryos, reviewed in Chapter 12.

In conclusion, it should be stressed that in updating and revising the monograph we have been quite aware that developmental cytogenetics of mammalian species is still at the stage of fact-gathering. It is to be expected, therefore, that some of our generalizations or suggestions will actually turn out to be erroneous and will be refuted by new experimental evidence. Nevertheless, we have deliberately refused to confine ourselves to delivery of pure facts as we could not resist the temptation to make some modest theoretical generalizations and suggest some theoretical, hypothetical mechanisms. We hope the latter are ready for experimental verification or rejection and may stimulate further searches in this area.

1
Haploidy (androgenesis, gynogenesis, parthenogenesis)

1.1 Mechanisms of origin

Embryos with a haploid set of chromosomes may arise in the course of parthenogenesis, gynogenesis, or androgenesis. Let us dwell on each of these mechanisms in detail.

Parthenogenesis is defined as development of an egg containing only the female genome without the intervention of the male gamete (Beatty 1957; Kaufman 1978a). Gynogenesis or pseudogamy means the development of an egg fertilized spontaneously or experimentally by a genetically inactive spermatozoon (Beatty 1957; Graham 1974a). Only the female genome remains active both in gynogenetic and parthenogenetic embryos, so they might be considered as genetically similar. Non-genetic contribution of some unknown extrachromosomal factors by the fertilizing sperm should not be neglected however, and this is the main difference between gynogenesis and parthenogenesis. The development of an egg containing only a male genome, the female genome being completely extruded or genetically inactivated, is called androgenesis.

Parthenogenetic, gynogenetic, and androgenetic embryos may be haploid, diploid, or chromosomal mosaics, composed of cells of different ploidy as well as of aneuploid cell clones that may be revealed only in the course of detailed cytogenetical studies. Therefore the data on developmental peculiarities of androgenetic, gynogenetic, and parthenogenetic embryos should be taken with caution, if they are not confirmed by solid cytogenetic analysis.

Let us consider first the data on the developmental profiles of gynogenetic and androgenetic embryos in mammals. Cytogenetic studies of parthenogenetic embryos with special emphasis on developmental characteristics of haploid and diploid parthenogenetic embryos (parthenogenones) and the comparison of these genomes with the developmental capabilities of haploid or diploid androgenetic and gynogenetic embryos, will be considered thereafter.

1.2 Gynogenesis and androgenesis

These routes of abnormal development, known both in rats and in mice, are suggested to arise from ovum overripeness (Beatty 1957; Marston and Chang 1964). There are some indications of a very low incidence of androgenesis compared to gynogenesis in murine rodents (Beatty 1957; Kaufman 1983d). According to some recent data, however, spontaneous androgenesis in human embryogenesis is much more common.

1.2.1 Spontaneous androgenesis in humans

This route of embryonic development has been recently identified in studies of the rather common pathological syndrome known as hydatidiform mole. This syndrome is characterized by hydropic degeneration of chorionic villi, hyperplasia or dysplasia of the trophoblast, accompanied by the early death of the embryo or the fetus. Cytogenetic studies of these moles distinguish between partial and complete (true) diploid moles (Vassilakos, Riotton, and Kajii 1977; Szulman and Surti 1978, 1984). According to chromosomal analysis the partial moles were found to be triploid, while complete moles were diploid. The triploid specimens are characterized by the typical hydatidiform swelling and cystic degeneration of chorionic villi in only some patches of chorion and placenta. Many groups of chorionic villi seem quite normal. Focal trophoblastic hyperplasia is usually confined to the syncytiotrophoblast layer, which nonetheless shows no propensity to become malignant and transform into chorionepithelioma. These partial hydatidiform moles are distinguished by rather late embryonic death (after the ninth week) and are usually aborted during the third or fourth trimester. In some more rare cases the triploid embryos may reach term. They then display a definite syndrome of malformation (Makino, Sasaki, and Takishima 1964; Carr 1971a, b; Niebuhr 1974; Kajii and Niikawa 1977; see also Chapter 2).

True or complete hydatidiform moles are characterized by severe hydropic degeneration of all placental villi, giving a 'bunch of grapes' appearance. Chorionic villi do not contain capillaries, their connective stroma is swollen with many cistern-like cavities, lined by mesenchymal connective tissue. Trophoblastic hyperplasia and aplasia, predisposing to malignant transformation, are also very common features of true hydatidiform moles. The embryos of true moles die very early, before embryo–placental circulation is established. Hence, no embryonic remnants or nucleated blood cells can be seen in these moles. Detailed cytogenetical studies based on chromosomal polymorphism of both parents as well as on different methods of chromosome banding have unequivocally proved the paternal origin of both chromosomal sets in true hydatidiform moles. This means that the diploid complement of each cell of true moles contains exclusively paternal chromosomes (Szulman and Surti 1978; Kajii and Ohama 1977; Jacobs,

Wilson, Sprenkle, Rosenshem, and Migeon 1980; Wake, Takagi, and Sasaki 1978). These original findings have been recently confirmed by combined cytogenetical and biochemical studies. The latter included some marker enzymes such as phosphoglucomutase -1 (PGM-1) and HLA-haploid types (Jacobs *et al.* 1980). The overwhelming majority of true moles have an XX chromosome constitution and they are believed to result from fertilization of an 'empty egg' by a haploid sperm, with subsequent duplication of its genome to give a 46 XX complement. In some rare cases complete moles were shown to be Y-chromosome positive: these may arise from penetration of an egg by two sperms followed by ejection of the female pronucleus (Ohama, Kajii, Okamoto, Fukuda, Imaizunni, Tsukuhara, Kobagashi, and Hagiwara 1981). The time of maternal haploid chromosomal set elimination, its exact mechanism, as well as the factors predisposing for maternal chromosome loss remain entirely unknown as yet.

Homozygosity for some recessive paternal genes responsible for cell pro-liferation has been implicated as one of the possible reasons for malignant transformation of true mole cells (Kajii and Ohama 1977). These genes are unlikely to be sex-linked, however, as both XX and XY true moles are prone to malignant transformation. Whether such transformation is actually caused by some malfunction of paternal chromosomes or is due to some other factors apart from androgenesis still remains totally unclear.

Thus, diploid androgenesis in humans is quite compatible with pre-implantation and early post-implantation development. All of these embryos are inevitably resorbed during the third to fourth weeks of development, while the chorionic part of the embryonic sac survives for some time after embryonic death. The cells of the chorion often reveal a complex of pathological changes and may often undergo malignant transformation at more advanced stages of embryogenesis. No evidence in favour of the exist-ence of haploid androgenetic embryos in humans has been obtained so far. Diploid androgenesis in humans affects embryonic survival and development somewhat differently than in murine rodents, as no cases of true hydatid-iform moles with two sets of paternal chromosomes and omitted female genome have ever been reported in rodents.

The existence of testicular teratomas arising from germ cells has been repeatedly reported in murine rodents (Stevens 1967). They might be con-sidered as an example of androgenesis (Graham 1977). They start as groups of undifferentiated germ cells inside testis tubules, which gradually trans-form into unorganized ectoderm and endoderm epithelia surrounded by mesenchymal cells. The groups of numerous neuroepithelial cells sometimes resemble nerve tube rudiments but no typical trophectodermal or giant cells appear and no morphogenetic processes equivalent to any normal steps of embryogenesis have ever been found.

Thus, androgenetic germ cells with an XY chromosome set possess all the properties necessary for control of tissue differentiation, but the full embryonic programme encoded by the male genome can be realized only inside the ooplasm. We will come back to this question later (see Section 1.3.1.2).

1.2.2 Experimental production of gynogenetic and androgenetic embryos in mice

Several experimental procedures are already available for the deliberate production of both gynogenetic and androgenetic embryos in mice. The same techniques with some minor modifications may also be applied to early embryos of other mammals.

Androgenetic and gynogenetic haploid embryos in mice may be obtained by mechanical removal of one (male or female) pronucleus from the fertilized egg (Modlinski 1975, 1980; Markert and Petters 1977; Hoppe and Illmensee 1977; Surani and Barton 1983) or by bisection of the zygote into halves, each containing a pronucleus, using a dissecting microscope with a glass needle operated by hand (Tarkowski 1977; Tarkowski and Rossant 1976). Other methods for pronucleus removal include extrusion of one pronucleus with the aid of osmotic shock (Opas 1977); or centrifugation of zona-free mouse zygotes in a gradient density, resulting in two parts, a cytoplast and a karyoplast (nucleated fragment). Some of the latter contain both pronuclei, while others possess only a single pronucleus, thus being androgenetic or gynogenetic (Dyban, Waisman, and Golinsky 1983). All the above-mentioned procedures result in one-cell gynogenetic or androgenetic haploids. If cultured in cytochalasin B or D supplemented medium, which specifically blocks cell division, these one-cell haploids do not cleave but their DNA content doubles. This technique has been widely used for the production of homozygous diploid androgenetic and gynogenetic embryos (Hoppe and Illmensee 1977; Markert and Petters 1977; Modlinski 1980; Surani and Barton 1983). An even more efficient procedure for the production of gynogenetic heterozygous diploid embryos has been offered recently (Borsuk 1982). It consists of two successive steps. First, mouse zygotes are treated with cytochalasin B (or D), which blocks emission of the second polar body and results in zygotes with three pronuclei (one male and two female). (This step is now adopted for the efficient production of triploid embryos—see Chapter 2.) Second, the male pronucleus is withdrawn mechanically, the two female pronuclei being left intact. The resulting product will give rise to a heterozygous gynogenetic diploid embryo capable of further development (see Section 1.3.4).

After dispermic fertilization, withdrawal of female pronucleus will result in heterozygous diploid androgenetic embryos. These and many other possible interferences in the ploidy of zygotic genomes are thoroughly discussed elsewhere (Markert and Seidel 1981; Seidel 1983), though many of these theoretically possible routes have not been tested experimentally as yet.

1.2.3 Embryonic development in gynogenetic and androgenetic haploidy and diploidy

The data on developmental capacities of androgenetic and gynogenetic embryos in mice are still rather contradictory. A proper consideration of available facts is significantly hampered by the diversity of procedures used to produce these embryos. Some of these procedures are quite damaging and may significantly reduce the survival capacity of the resulting gynogenetic and androgenetic embryos. For instance, pronucleus expulsion after osmotic shock is especially harmful for the resulting haploid zygotes, which usually show very poor development *in vitro* and are eliminated in a few cleavages. However, if transplanted into the oviducts, some of these zygotes proceed to the morula stage. Of 22 haploid embryos induced by osmotic shock, four embryos recovered from the oviduct on the fifth day were at the morula stage, with cell counts between 12 and 62. Two of them were successfully karyotyped and were found to be gynogenetic in origin (Opas 1977). Thus, expulsion of the female pronucleus may be more damaging for subsequent development than absence of the male one.

Withdrawal of the female pronucleus according to Modlinski's original technique is also highly traumatic. This is probably because of peculiarities of male and female pronuclear origin and the microfilamentous links formed between each pronucleus and the oolemma membrane. Microfilaments of the male pronucleus are formed relatively late and do not seem to come into intimate contact with sub-oolemma regions of the egg, being rather loosely dispersed in the ooplasm. The female pronucleus results from disjunction of the meiotic chromosomes of the ovum. During the second meiotic division the mitotic spindle rotates so that its main axis becomes radially oriented and the group of chromosomes left after emission of the second polar body gives rise to the female pronucleus. This remains connected with the oolemma membrane by numerous filaments. Mitotic fibres of the female pronucleus are still evident after completion of the second meiotic division. The destruction of these links between pronucleus and oolemma membrane during female pronucleus removal results in much more severe damage to the oolemma and sub-oolemma regions than removal of the male pronucleus, which is still rather loosely connected with ooplasm structures. These morphological peculiarities might be responsible, at least in part, for the poor development of androgenetic haploids (removal of female pronucleus) compared to gynogenetic haploids (removal of male pronucleus). Such an explanation might be admitted as quite plausible for haploid embryos obtained by Modlinski's original technique, which did not involve any artificial derangement of microfilaments and microtubuli before pronucleus withdrawal. The more sophisticated microsurgical technique used at present includes as its initial step the incubation of the zygotes in the presence of cytochalasin B and colcemid (or nocadozol), to destroy microfilament and microtubular systems. This treatment significantly improves the survival capacities of microsurgically operated eggs. In initial studies, 37 micro-surgically obtained haploid embryos proceeded up to morula and blastocyst stages and all of them were gynogenetic, while androgenic haploids never developed beyond four to six cells (Modlinski 1975). However, when the eggs were first incubated in a medium containing cytochalasin B, the survival capacities of the operated eggs were significantly improved. Most of the cleaving haploid embryos, both gynogenetic and androgenetic, survived to the six- to eight-cell stage with at least some of them (eight of 74) forming blastocysts. Whether these blastocysts were produced by gynogenetic or androgenetic haploids remained unspecified (Modlinski 1980).

The least traumatic technicque, invented by A. Tarkowski, is based on bisection of a fertilized one-cell egg into sister halves, with the maternal and paternal pronuclei being separated in the two halves (Tarkowski 1977; Tarkowski and Rossant 1976). Most of the operated eggs survived. Andro-genetic haploids carrying a Y rather than an X chromosome do not survive beyond the four-cell stage, while some of androgenetic haploids bearing an X-chromosome, as well as many gynogenetic haploids, proceed in their development to compacted morulae and sometimes even to blastocysts. These results were confirmed in our laboratory. In microsurgically bisected eggs, two halves could develop to compacted morulae or even to blastocysts, though the mean cell count in developing halves was less than normal (Archangelskaya 1985).

The experimental data on the crucial lethal phases in the development of androgynetic and gynogenetic haploid embryos in mice are still very fragmentary and need further investigation.

It might fairly be concluded at the present stage of our knowledge that haploid embryos in mice are eliminated during implantation (Tarkowski 1977; Modlinski 1980). Androgenetic haploids show lowered developmental capabilities compared to their gynogenetic counterparts, especially during *in vitro* culture. Nevertheless, at least some of them may be recovered at late blastocyst stage and thus their developmental profile resembles that of gyno-genetic haploids. Androgenetic haploid embryos with a male pronucleus carrying a Y-chromosome do not survive beyond the four-cell stage (Tarkowski 1977), thus confirming the crucial role of X-chromosome genes in the control of initial stages of embryogenesis in mice (see Chapter 8). Artificial diploidization of the previously haploid genome substantially improves the developmental capacities of both gynogenetic and androgenetic embryos. The same is also true for parthenogenetic embryos (see Section 1.3.4). The embryos with doubled haploid genome at the first cleavage division show better survival than usual haploid embryos but somewhat worse than diploid embryos produced by blocking second polar body emission. It should be remembered, however, that the former diploids are genetically homozygous while the latter are genetically heterozygous. Moreover, because one cleavage division is actually omitted, the embryos of the former group usually have less blastomeres than normal diploids. Although the mean cell count in homozygous diploid blastocysts was almost 40 per cent lower than in haploid ones, the former arise much more frequently (Modlinski 1980). Homozygous gynogenetic diploid embryos have a lower mean cell count than control diploids, and this may have an unfavourable effect on inner cell mass formation and further progress of these embryos. Of special interest in this respect are the results obtained from gynogenetic heterozygous diploid embryos produced by microsurgical removal of the male pronucleus from an originally digynous triploid zygote. These embryos displayed normal development up to implantation, with 34 cells as morulae and 38–84 cells as blastocysts, and did not differ morphologically from the control embryos of the same age (Borsuk 1982). Moreover, at least some of these gynogenetic diploid embryos could implant, though no details of their further development are presented in the interesting work of Borsuk (1982). This conclusion has been experimentally supported by Surani and Barton (1983), who used Borsuk's method for the production of heterozygous gynogenetic diploid embryos. There are also some indications in favour of much better survival of heterozygous gynogenetic diploids both before and after implantation compared to homozygous gynogenetic diploids. Further studies are needed to clarify whether the survival peculiarities of homo- and heterozygous diploids depend on the state of specific alleles or are due to some other still unknown cause.

Conclusive evidence from several different laboratories indicated that both gynogenetic and androgenetic diploid embryos show normal preimplantation development (Modlinski 1975, 1980; Markert and Petters 1977; Surani and Barton 1983). Information on their developmental capabilities after implantation is still rather meagre and contradictory. Though much is still left for further study, there have been some serious experimental achievements in this field in recent years. Gynogenetic diploids produced by microsurgical withdrawal of the male pronucleus from initially triploid zygotes successfully proceeded up to early neurula stage or even to more advanced stages. Some of them remained alive up to the tenth day. A small and poorly formed yolk sac and some minor abnormalities were combined with severe retardation and progressive degeneration. One gynogenetic embryo recovered alive on the eleventh day had 25 somites but revealed some degenerative changes (Surani and Barton 1983). Similar results have been obtained in pronuclear exchange experiments. Both diploid gynogenetic and androgenetic embryos with their pronuclei coming from two different parents (high level of heterozygosity) proceed quite normally *in vitro* up to the blastocyst stage. Afterwards, if transplanted to the uteri of pseudopregnant females, androgenetic diploids were capable of inducing the formation of decidua that contained some trophoblastic derivatives. Gynogenetic diploids could successfully implant and form egg cylinders, and some developed up to early somite stage (McGrath and Solter 1984).

Developmental capacities of homozygous androgenetic diploids (removal of both female pronuclei from initially triploid zygotes) and gynogenetic diploids (removal of one male and one female pronucleus) with subsequent diploidization of the remaining genome by cytochalasin B treatment were rather poor. Gynogenetic homozygous diploids could implant, but most of them died soon thereafter. Out of 95 transplanted into the uteri of pseudopregnant females, only one severely retarded embryo has been recovered on the tenth day. Not a single androgenetic homozygous diploid embryo has been found after implantation, though at least some of them could induce a decidual reaction.

Thus diploid heterozygous androgenetic embryos are eliminated predominantly soon after implantation, while heterozygous gynogenetic embryos sometimes reach somite stage but inevitably die soon thereafter (Surani and Barton 1983; Surani, Barton, and Norris 1984; McGrath and Solter 1984). All the aforementioned data are consistent with a statement that both gynogenesis and androgenesis are not compatible with advanced postimplantation survival of mouse embryos (Surani et al. 1984; McGrath and Solter 1984). Some early studies announcing successful birth of both androgenetic and gynogenetic mice (Hoppe and Illmensee 1977) should now be considered as technical artefacts due to incomplete withdrawal of pronuclear tissue (McGrath and Solter 1984).

1.3 Parthenogenesis

Much original experimental work outlined in the numerous comprehensive reviews (Beatty 1957; Graham 1974a; Kaufman 1975; Tarkowski 1975; Whittingham 1980) and in the recently published monograph (Kaufman 1983a) has been devoted to parthenogenesis in mice. Most important for developmental genetics are the data indicating that parthenogenesis does not always imply haploidy. Four major types of parthenogenones may be generated from artificially activated eggs (Whittingham 1980):

1. 'Genetically uniform haploids'—the oocytes have completed their second maturation division with one haploid set of chromosomes, forming a single pronucleus, while the other set is being discarded into the second polar body.

2. Mosaic haploids with two pronuclei—the oocytes accomplish the first maturation division but omit the second, and cleave into halves, giving rise to two haploid blastomeres. This route is designed as 'immediate cleavage' (Braden and Austin 1954).

3. Heterozygous diploids with two pronuclei—activated oocytes without

second polar body emission, so that the two haploid sets of chromosomes form two haploid pronuclei.

4. Heterozygous diploids with one pronucleus—a single pronucleus develops from metaphase II (M-II) chromosomes; after DNA replication and successive cleavage division two diploid blastomeres are formed.

Routes of parthenogenetic development depend on the specificity of the activation agent and its dose as well as on the post-ovulatory age of the activated ovum, the osmotic pressure and the temperature of the culture medium. Deliberate changes of these conditions may have a substantial effect on the incidence of parthenogenones and on the route of parthenogenetic development (see Whittingham 1980). Under standard activation procedure genetically uniform haploids (one pronucleus, extruded second polar body) usually prevail over the other feasible types of parthenogenone. Experimental blockage of second polar body omission may result in a high proportion of parthenogenones with one diploid or two haploid pronuclei. Increasing dosages of activation agents may induce a higher output of mosaic haploids and both types of diploid parthenogenones.

Thus, parthenogenetic development by itself does not prove the existence of haploidy. Detailed cytological and especially chromosomal studies are needed to verify the exact route of nucleus progression and so to assess the exact ploidy of the parthenogenones. Chromosomal studies, however, should not be limited to one-cell parthenogenones only, as the selection of haploid embryos immediately after activation does not guarantee the stability of the haploid genome throughout subsequent development. The loss of some chromosomes and transformation into haploid–diploid mosaics is a rather common feature of initially haploid specimens. Therefore data on the developmental effects of haploidy should be treated with caution unless the morphological studies are substantiated by cytogenetic analysis. Up to now, however, such a stage-by-stage chromosomal analysis has not been performed, and this complication substantially hampers discrimination of diploid and haploid parthenogenones by their developmental effects (see Section 1.3.3).

Let us now consider developmental effects in spontaneous and induced parthenogenones, with special reference to the cytogenetic peculiarities of these specimens.

1.3.1 Spontaneous parthenogenesis

This section deals with both spontaneously activated ovulated eggs of different mammals and parthenogenetic development of follicular oocytes giving rise to ovarian teratomas.

1.3.1.1 Parthenogenetic embryos in different mammals. Information on the frequency of spontaneous haploidy in mammals is still very scarce. Cases of parthenogenetic development of ovulated eggs have been reported for human beings and other mammals (Slye, Holmes, and Wells 1920; Harman and Kirgris 1938; Krafka 1939). These early studies, however, lacked an adequate cytological and cytogenetic analysis. Whether these exceptional cases were true haploid parthenogenones or represented fragmentation of unfertilized oocytes remains still unknown. Not a single case of haploidy has ever been observed in induced or spontaneous abortuses in human (Boué and Boué 1974, 1978; Boué and Lazar 1975; Boué, Phillippe, Giroud, and Boué 1976; Sankaranarayanan 1979; Lauritsen 1982), but cytogenetic studies in human conceptuses are routinely performed at rather advanced stages of embryogenesis, never earlier than about 3–4 weeks.

Haploids of spontaneous origin in humans are presumably therefore eliminated during the initial few days of development. This suggestion has recently been substantiated by cytogenetic studies of pre-implantation human embryos fertilized *in vitro*. Out of 11 cleaving specimens composed of seven to eight blastomeres, two embryos were found to be haploid and another had trisomy D. The haploidy of one was proved by nuclear photometry, while the second embryo displayed three metaphase plates with haploid chromosome sets and concomitant nullisomy of chromosome 15 in each metaphase. Detailed parental chromosomal analysis confirmed the maternal origin of a single chromosome set in this unique embryo (Angell, Aitken, Van Look, Lumsden, and Templeton 1983). Whether it was a parthenogenetic embryo, according to the authors' suggestion or whether it was a gynogenetic haploid, remains obscure. Thus *in vitro* fertilization may result in a rather high incidence of spontaneous haploidy in humans, while the actual impact of spontaneous haploidy in embryonic wastage is still completely unknown.

Spontaneous parthenogenesis in laboratory mammals is encountered exceptionally rarely (see Tables 1–3). This is true for pre-implantation embryos of rats, rabbits, and pigs. In the golden hamster 10 per cent of ovulated eggs undergo spontaneous activation if not fertilized, reaching a 'two-cell stage'. The initial stages of parthenogenetic development were suggested for these aged eggs, but their haploidy was not confirmed cytogenetically (Szollosi 1975).

Detailed analysis of spontaneous parthenogenesis in mice has been carried out on lacmoid-stained total preparations of the eggs recovered from mice of the C57BL strain or F_1 hybrids (CBA × C57BL) with spontaneous or hormonal induced ovulation (Noniashvili 1985). Not a single case of spontaneous activation has been registered in spontaneously ovulated mice, while after gonadotropin-induced superovulation about 2 per cent of the eggs completed meioses and commenced parthenogenetic development, which never continued beyond the two-cell stage. A few haploids have been recovered in cytogenetic studies of three- to four-day-old mouse embryos. Among 1027

Table 1. Incidence of spontaneous chromosomal aberrations at different stages of embryonic development in mice (see Sections 5.1.2, 5.2; also Nijhoff and de Boer 1981)

References	Strains of mice	Embryonic age (days post coitum)	Karyotype of studied embryos									
			Diploid		Heteroploid								
			total	%	total	%	$1n$	$3n$	$4n$	$2n+1$	$2n-1$	Mosaics	Fragments
Before implantation													
Vickers 1969	PDE	3.5	301	97.4	8	2.6	1	1		2			2
Yamamoto et al. 1971	?	3.5	70	87.5	10	12.5				4	2		4
Donahue 1972c	?	1	328	96.4	10	3.6		4	1		1		4
Hansmann and Rohrborn 1973	F$_1$(103×C3H)	3	21	75.0	7	25.0							7
Gosden 1973	CBA/H-T6	3.5	60	95.2	3	4.8		3					
Kaufman 1973	CFLP	1	184	95.4	9	4.6				1	8		
Baranov 1976b	CBA	4	203	95.3	10	4.7		8	1	1			
Martin-Deleon and Boice 1983	ICR	1	280	92.1	24	7.9		3		11	9		1
After implantation													
Baranov and Dyban 1972a	C3H, A	8	176	97.3	5	2.7		3				2	
Yamamoto et al. 1973	CF-1	10.5	145	97.4	4	2.7		2				2	

Table 2. Incidence of spontaneous chromosomal aberrations at different stages of embryonic development in laboratory rats

Embryonic age (days post coitum)	Number of embryos	Karyotype of studied embryos									Reference
		Diploid	Heteroploid								
			total	%	3n	4n	2n+1	2n−1	Mosaics	Fragments	
Before implantation											
5	251	229	22	8.4	12	1	4	3	2		Dyban et al. 1971
4.5	230	226	4	1.7	3				1		Mikamo and Hamaguchi 1975
1	134	125	9	6.7	3		4			2	Chebotar 1978
After implantation											
12–13	410	404	6		3				3		Butcher and Fugo 1967
11	50[a]	44	6	12	5	1					Dyban et al. 1971a

[a] Only empty extra-embryonic membranes or malformed embryonic specimens.

Table 3. Incidence of spontaneous chromosomal aberrations at early stages of embryonic development in rabbits and in domestic pigs

Embryonic age (days post coitum)	Number of embryos	Karyotype of studied embryos										References
		Diploid	Heteroploid									
			Total	%	3n	4n	Mixoploid	2n+1	2n−1	Mosaics	Fragments	
Rabbits												
6	47	42	5	12	5							Shaver and Carr 1967
6	58	57	1	1.7						1		Shaver and Carr 1969
6	73	68	5	6.9		2	2					Shaver and Carr 1969
6	72	71	1	1.4	1			1				Widmeyer and Shaver 1972
6	75	66	9	11.2					2	3	4	Hofsaess and Meacham 1971
6	125	124	1	0.8			1					Martin-Deleon, Shaver, and Gammal 1973
5–6	46	46	—									Fujimoto *et al.* 1974
Domestic pigs												
10	88	79	9	10	4	3				2		McFeely 1967
10	169	169	—									Dolch and Chrisman 1981

specimens, seven haploid embryos were found in silver mice (Beatty 1957; Fischberg and Beatty 1952). According to other cumulative data (see Table 1) one haploid embryo has been recovered from over 1000 three to four-day-old mouse embryos of known karyotype. One more haploid specimen at the morula stage has been picked up among 849 pre-implantation embryos in mice with different Robertsonian translocations (Baranov 1983a; also see Chapter 7).

If considered as true spontaneous parthenogenones, these rare cases might be taken as evidence of very low (0.1–0.6 per cent) levels of spontaneous haploid parthenogenesis in laboratory mice.

No conclusive data on the incidence of spontaneous diploid parthenogenesis in mice or other mammals has been obtained so far.

1.3.1.2 Spontaneous parthenogenesis and ovarian teratomas. Spontaneous teratomas are well known in humans and other mammals, though the exact mechanisms of their origin remain chiefly unknown (Pierce 1967; Stevens 1967). These and other problems concerned with spontaneous parthenogenesis and ovarian teratomas in mammals have been vastly illuminated by the numerous experimental studies in mice of LT/Sv strain distinguished by an exceptionally high frequency of spontaneous ovarian teratomas (Stevens and Varnum 1974; Stevens 1975). Some ovarian follicles of these mice from 24–30 days of age contain parthenogenetically activated eggs. These start cleavage divisions, and development proceeds to the blastocyst stage and sometimes up to early egg cylinder stage. Afterwards development becomes disorganized and they transform into typical teratomas, composed of non-differentiated embryocarcinomal cells, plus neural, epithelial, muscle, connective tissue, mesenchymal derivatives, rudiments of retina, lens, and other organs (Stevens and Varnum 1974; Dyban 1981). By the ninetieth day of post-natal life almost 50 per cent of LT/Sv and practically all F_1(LT × BJ) females display ovarian teratomas (Stevens and Varnum 1974; Stevens 1975; Steven, Varnum, and Eicher 1977). Two distinct populations of oocytes were detected in the growing pre-antral follicules of these mice. The oocytes of one population are distinguished by their rather large size (about 70 μm in diameter) with substantial reduction in the number of follicular cell layers. The oocytes of this particular population are inclined to start parthenogenetic development (Eppig 1978; Dyban 1982). Suppression of granulosa cell proliferation in these follicles has been demonstrated by means of ^{3}H-thymidine histoautoradiography (Dyban 1982). Thus, desynchronization of oocyte growth and proliferative activity of follicle cells was typical for many follicles in LT/Sv mice (Dyban 1982).

Indeed, not only intrafollicular oocytes but also about 10–20 per cent of all ovulated eggs in LT/Sv mice are spontaneously activated in the oviducts and start cleavage (Stevens and Varnum 1974; Eppig 1978). *In vitro*, half of these eggs successfully complete cleavage and give rise to normal looking blastocysts (Eppig 1978, 1981), some of which were shown to be capable of inducing a decidual reaction in the uterus. All are resorbed, though some reach egg cylinder or early somite stages before dying (Stevens and Varnum 1974; Stevens 1975). Both the decidual reaction and implantation itself prove the existence of a quite unusual hormonal balance in the virgin LT/Sv mice, whose uteri are able to respond properly to the presence of cleaving eggs, i.e. to the stimulus generated by trophectodermal cells. The exact mechanism of this interesting phenomenon remains obscure, though some correlation between ovarian teratomas and hormonal balance of LT/Sv female mice might be suspected.

The actual routes of parthenogenetic development of both intrafollicular and ovulated oocytes in LT/Sv mice remain mostly unknown. Direct measurements of nuclear volume in one-cell stage LT/Sv parthenogenones carried out quite recently (Anderson, Hoppe, and Lee 1984) indicate the existence of a diploid content from the very start of development. Chromosomal studies of 74 metaphases from 36 parthenogenones at different stages of pre-implantation development (from the two-cell stage to expanded blastocysts) gave full support to these results. However, no conclusive cytogenetic analysis has been made in one-cell parthenogenones. It is still possible, therefore, that initially haploid parthenogenones undergo spontaneous diploidization during subsequent development. The latter statement is consistent with some early findings postulating the presence of haploid, diploid, and polyploid metaphases in the blastomeres of parthenogenetic blastocysts of LT/Sv mice (Stevens and Varnum 1974).

Taking into account that the extrusion of both polar bodies indicates the accomplishment of both meiotic divisions, the presence of a diploid set in the one-cell parthenogenones must indicate chromosome endoreduplication in the haploid pronucleus. This mechanism is favoured by the presence of diplochromosome figures during the first cleavage division of some parthenogenones (Kaufman 1983a). According to another suggestion, LT/Sv parthenogenones are generated by oocytes that have already accomplished an additional (unscheduled) round of pre-meiotic reduplication of their DNA content (Anderson et al. 1984).

Phase-contrast examination of lacmoid-stained whole-mount preparations of one-cell LT/Sv embryos undertaken in our department (Noniashvili 1985) has shown that activated LT/Sv eggs are a rather heterogeneous group. Most of them possessed only one pronucleus and a quite distinct second polar body and so were true haploids. Some of them, however, lacked a second polar body and contained two distinct pronuclei (blockage of the second meiotic division). Some displayed only one pronucleus but had no second polar body, so most probably arose through complete blockage of the second maturation division. Thus, spontaneous parthenogenesis in LT/Sv mice might proceed through various routes, and in this respect resembles ovarian teratomas in man.

1.3.2 Experimental production of parthenogenetic embryos

The first attempts to induce experimental parthenogenesis in mammals date back to the pioneering studies on rabbits by Pincus and his collaborators (Pincus and Enzmann 1937; Pincus 1939a,b). The results of these encouraging studies were rather contradictory and open to criticism. Successful attempts at parthenogenetic activation in different mammalian species were later confirmed in numerous publications (Amoroso and Parkes 1947; Edwards 1954a,b, 1957a,b; Chang 1954). Parthenogenetic activation in these works was achieved by a number of procedures, such as temperature shock to unfertilized tubal eggs (Chang 1954; Beatty 1957), genetic inactivation of chromosomal sets in the male and female gametes induced by X-rays or ultraviolet rays (Edwards 1957a,b), as well as by some radiomimetic agents and by colchicine (Edwards 1954a,b). According to some of these early data parthenogenetic development proceeds as far as the blastocyst stage in rabbits (Pincus and Shapiro 1940) and mink (Chang 1957). The techniques employed in these studies are likely to produce diploid parthenogenesis, while haploid embryos presumably cease their development after a few cleavage divisions, most probably due to the toxic effects of the applied activating agents.

More efficient and reliable techniques for inducing parthenogenesis in mammals were invented later (Tarkowski, Witkowska, and Nowicka 1970; Graham 1970; Tarkowski 1971). They provide a high rate of parthenogenetic activation, and permit detailed studies of developmental patterns in mammalian parthenogenones of different types. Extensive information on mammalian parthenogenesis has been provided by several fundamental reviews (Tarkowski 1971, 1975; Graham 1974a; Kaufman 1975, 1978a, 1981, 1983a; Whittingham 1980). Different methods used for production of parthenogenetic embryos are outlined in these reviews, particularly Kaufman (1978a), Whittingham (1980), and Kaufman (1983a).

The methods that have recently been employed for obtaining parthenogenetic embryos may be divided into two basic groups depending on whether the oocytes are activated directly *in situ* within the oviducts or isolated from the reproductive tract, cultured, and stimulated *in vitro*. Both approaches are equally reliable and are widely used for the deliberate production of parthenogenones of a desired karyotype, including haploid.

Many different agents are capable of inducing oocyte activation. Most of these agents are rather noxious for the ovulated eggs and subsequent parthenogenones. It should be noted, however, that the harmful effects of the activation treatments could easily be missed in the 4–6 h interval when the results of activation are usually checked. Visual examination of these eggs under a phase contrast or dissecting microscope provides an adequate identification of activated, non-activated, and fragmented eggs. But it is now clear that activated eggs (parthenogenones) with one normal pronucleus inside are not always capable of further development. Too harsh treatment as well as the use of highly toxic agents can efficiently induce parthenogenesis, but it does not proceed beyond the one-cell stage or arrests after a few cleavage divisions. Thus, not only the incidence of activated eggs but their capacity for further development should always be kept in mind in attempting to compare the efficiency of different activation techniques.

With all of these limitations in mind, the most reliable data on the developmental capacities of parthenogenones are obtained with just a few experimental procedures. These are the following: *in situ* electrical stimulation of oviducts (Tarkowski *et al.* 1970; Witkowska 1937a,b), heat shock (Komar 1973; Balakier and Tarkowski 1976; see also Dyban and Noniashvili 1984), hyaluronidase treatment (Graham 1970; Kaufman 1973a; Graham and Deussen 1974), treatment with standard culture medium without Mg^{2+} and Ca^{2+} or both (Surani, Barton, and Kaufman 1977; Kaufman 1978a), and ethyl alcohol. All of these techniques are described comprehensively in a review by Kaufman (1978a) as well as in his recent book (Kaufman 1983a). Let us deal in detail with the ethyl alcohol procedure, which is quite new and is therefore not included in the aforementioned reviews.

Although the egg activating properties of ethyl alcohol and ether treatments have been known for a long time (Austin and Braden 1954b; Braden and Austin 1954), it is only recently that we have successfully applied the former for the induction of parthogenetic development in mice (Dyban and Khozhai 1980). The procedure has been modified for *in vitro* activation of mouse eggs (Kaufman 1982a, 1983a; Cuthbertson 1983). Ethyl alcohol is now generally accepted as one of the most efficient agents for parthenogenetic activation of mammalian eggs and is widely used for this purpose in a number of laboratories (Cuthbertson 1983; Kaufman 1982a, 1983b; Surani and Barton 1983; Waksmundzka 1984). Successful activation of mouse eggs has also been achieved with benzyl alcohol (Cuthbertson 1983). The activating capacities of other alcohols have not yet been studied. Egg activation induced by ethyl alcohol can be achieved either *in situ*, i.e. inside the maternal body, or *in vitro*. Let us consider each procedure in detail.

It was originally discovered in our laboratory that the injection of 0.35 ml 25 per cent ethanol intraperitoneally 7–9 h after ovulation induces activation of 30–70 per cent of eggs (Dyban and Khozhai 1980). Most of these parthenogenones extruded their second polar body and contained only one pronucleus, so they were true haploids initially. Stage-by-stage studies of these activated eggs showed that they developed normally up to the late morula stage (16–20 cells). Approximately 12 per cent of day 4 parthenogenones were quite normal blastocysts with a trophectoderm layer and a well developed inner cell mass (Dyban and Khozhai 1980; Khozhai 1981). Some of them were able to implant, but even the most advanced parthenogenones always degenerated at the early egg cylinder stage (Khozhai 1981). Parthenogenetic activation in our experiments was achieved only if ethyl alcohol was applied intraperitoneally but not by mouth. These findings have been recently confirmed (Kaufman 1982a, 1983b), but according to the latter author, the activation effect of ethyl alcohol could also be extended to intragastric administration of the alcohol.

This report stimulated us to undertake an extensive study with more than 2300 mouse eggs to compare the egg activation capacities of ethyl alcohol with respect to the route of administration, i.e. intraperitoneally or by mouth (Noniashvili 1985; Dyban and Noniashvili 1985a,b). The results of our studies summed up in Table 4 clearly show that ethyl alcohol readily induces parthenogenetic activation if applied intraperitoneally. Subtoxic doses administered by means of a stomach tube do not appear to activate, but, high doses (close to the LD_{50}) applied through the stomach tube can activate parthenogenetic development in at least 14 per cent of ovulated eggs. Subtoxic doses of ethyl alcohol in combination with ether anesthesia, which is known to potentiate the action of ethyl alcohol and thus prolongs its narcotic effect, do not induce parthenogenetic activation of ovulated eggs. These results imply that it is not the general narcotic effect of ethyl alcohol *per se* but rather its specific local action on the ovulated eggs that is responsible for its activation properties. Perhaps when injected intraperitoneally

Table 4. The effects of ethyl alcohol, administered by mouth or intraperitoneally, on ovulated eggs of mice. F_1(CBA × C57BL)

Group	Ethyl alcohol treatment per os Concentration (%)	per os Volume (ml)	i.p. Concentration (%)	i.p. Volume (ml)	Egg age (h after HCG)	Total	Activated	(%)	Fragmented	(%)
1			10	1.0	17	76	12	15.6	—	—
2			25	0.35	17	289	84	29.0	—	—
3	10	1.0			17–19	690	3	0.4	129	18.7
4	10	1.0			22	281	—	—	57	20.2
5	25	0.35			17–19	56	1	1.8	30	53.6
6[a]	10	1.0			17	320	5	1.6	43	13.4
7[a]	12	1.0			13.5	149	—	—	30	20.1
8[a]	25	0.35			17	442	2	0.5	95	21.5
9[b]	25	0.5			19	55	8	14.3	5	0.9

[a] Ethyl alcohol + slight ether anaesthesia.
[b] Dose close to LD_{50} for females.

ethyl alcohol penetrates directly into the oviducts, where its concentration gradually increases and eventually reaches the level necessary for egg activation. Administered orally, ethyl alcohol gets to the oviduct through the bloodstream and thus can hardly reach the level of activation. It can not be excluded that the blood level of ethyl alcohol in the mice of our strain was lower than in the mice used by Kaufman (1983a,b), though the doses of ethyl alcohol applied in both studies were the same.

In the course of our recent studies we have also accumulated a good deal of experimental evidence on the activation properties of ethyl alcohol applied *in vitro* to ovulated mouse eggs. Over 4650 ethyl alcohol *in vitro* activated eggs have been studied (Dyban and Noniashvili 1985a,b; Noniashvili 1984).

An efficient activation of ovulated eggs can be achieved by immersing the excised oviducts containing freshly ovulated eggs in the medium with ethyl alcohol. The results of this *in situ* activation are similar to activation of isolated eggs. The time interval between ovulation and ethyl alcohol treatment is the most crucial step of this technique. According to our data ethyl alcohol treatment 17–19 h after HCG injection or 7–8 h after spontaneous ovulation gives the best results. Temperature of the medium is another crucial point of this technique (Dyban and Noniashvili 1985a,b; Noniashvili 1984). There is a clear-cut reverse correlation between temperature level and ethyl alcohol concentration. Relatively high temperature in combination with moderate ethyl alcohol doses may have just the same activation effect as reduced temperature and high concentration of ethyl alcohol. Varying both experimental conditions one may create optimal conditions for highly efficient activation (up to 90 per cent) and concomitant prolonged survival of parthenogenetic eggs. (See Table 5.)

The data assembled in Table 4 deal with the eggs of F_1(CBA × C57BL) female mice. Eggs from other strains of mice or from different crosses may not respond to the activation stimulus of ethyl alcohol in the same manner. Ethyl alcohol *in vitro* predominantly induces haploid parthenogenesis, i.e. embryos with one haploid pronucleus and extruded second polar body—genetically uniform haploids (see above). Only a rather small proportion of the eggs display mosaic haploidy (route 2)—due to immediate cleavage of oocytes—or transform into diploid parthenogenones (the embryos with two haploid or one diploid pronucleus).

These data are in line with similar observations of other authors (Kaufman 1982a, 1983a,b; Cuthbertson 1983) on *in vitro* activation of mouse eggs by ethanol treatment.

Survival and post-implantation development of parthenogenetic embryos induced by ethyl alcohol treatment of ovulated eggs is not inferior to that of parthenogenones induced by a medium deficient in Ca^{2+} and Mg^{2+}, the treatment that has been considered as the least damaging activation procedure until recently (Kaufman 1978a).

Table 5. The incidence of eggs activated *in vitro* by ethyl alcohol treatment under different temperature regimes of the culture medium

Series	Group	Activation conditions		Eggs examined		
		E.A. concentration (%)	Temperature of medium °C	Total	No. activated	Per cent activated
I[a]	1	—	21.0	387	99	22.9 ± 2.1
	2	7.0	17.0	199	68	34.0 ± 3.4
	3	7.0	21.0	597	448	75.2 ± 1.8
II[b]	4	7.0	24.5	52	47	90.3 ± 4.0
	5	4.0	17.0	126	31	24.6 ± 3.8
	6	4.0	26.3	32	13	40.6 ± 8.7
	7	4.0	30.3	74	39	52.7 ± 5.8
	8	4.0	33.2	85	62	72.9 ± 4.8
	9	4.0	37.0	94	69	73.4 ± 4.6
	10	2.0	37.0	78	22	25.6 ± 4.9
	11	—	37.0	109	8	7.3 ± 2.4
	Total			1843		

[a] Medium with ethyl alcohol + eggs with cumulus flushed out of oviducts.
[b] Medium with ethyl alcohol + incised oviducts with eggs inside.

Thus, ethyl alcohol should be considered as one of the few agents with pronounced ability for efficient activation of ovulated eggs, the developmental capacities of which are not much affected by the treatment.

1.3.3 Cytogenetic studies of different types of parthenogenones

Few data are available on this question as yet. Most are limited to cytogenetic studies of parthenogenones during the initial steps of their development, i.e. during metaphase of the first cleavage. The relevant information concerning parthenogenones of more advanced stages is still very fragmentary. Few experiments have been devoted to parthenogenones at the morula–blastocyst stages, and there has been even less cytogenetic study of post-implantation embryos. The data on differential staining of metaphase chromosomes of parthenogenones are still quite preliminary.

Let us look in some detail at the cytogenetic analysis of the first cleavage division in haploid parthenogenones. Since the first cytogenetic study of haploid parthenogenones, undertaken 10 years ago, a rather high level of aneuploidy has been known to occur in these specimens (Graham and Deussen 1974). Further cytogenetic studies confirmed this finding. Metaphase chromosomes of the first cleavage division were counted in 270 parthenogenetic embryos induced by ethyl alcohol activation of ovulated eggs and also in 89 parthenogenones induced by hyaluronidase treatment or by some mechanical action (Kaufman 1982a, 1983a,b,c). The incidence of aneuploidy in the parthenogenones of the former (ethyl alcohol) group was found to be in the range 13.6–18.8 per cent. Only genetically uniform haploids were used for cytogenetic analysis in this study. There was only one aneuploid embryo in both control groups (mechanical activation and hyaluronidase treatment). It was therefore concluded that the high level of aneuploidy in the former group was due to the direct and specific action of ethyl alcohol on chromosomal disjunction at the second meiotic division.

We have carried out a detailed cytogenetic study of ethyl alcohol or heat shock activated mouse eggs, in which the initially haploid genome was diploidized by cytochalasin B treatment. Chromosome non-disjunction at the first meiotic division, i.e. before activation, will result in 42 or 38 chromosomes in total in both pronuclei. Non-disjunction after completion of the second meiotic division, i.e. after activation and probably due to the activation stimulus, will result in two pronuclei with aneuploid but complementary numbers of chromosomes, giving in total 40 (see Table 6). With these theoretical considerations in mind, we have undertaken direct cytogenetic studies of chromosome complements in diploid parthenogenones (Dyban and Noniashvili, 1985; Dyban 1984). The eggs were activated *in vitro* by either ethyl alcohol or heat shock and were cultured thereafter for 4–6 h in medium with cytochalasin B or D to induce diploidization, as described elsewhere (Balakier and Tarkowski 1976). The embryos with two pronuclei but without

Table 6. Chromosomal non-disjunction at the first meiotic division estimated by chromosome count in the cleaving diploidized parthenogenones

Group	Activation	Total	Euploids (2n = 40)	Aneuploids Total	38	39	41	42
		Metaphases of the first cleavage division studies			Nombre Fondamental (NF)			
I	Ethyl alcohol	281	263	18	2	15	—	1
	Heat shock	276	267	9	4	3	1	1
II	Ethyl alcohol	104	59	45	19	2	3	21

Group I — ovulated oocytes from mice with normal karyotype.
Group II — ovulated oocytes from mice doubly heterozygous for different Robertsonian translocations sharing chromosome 19 in common Rb(9.19) 163H/Rb(5.19)1Wh.
NF (Nombre Fondamental) — number of chromosomal arms.

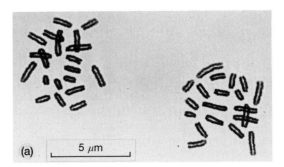

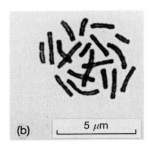

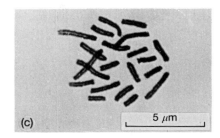

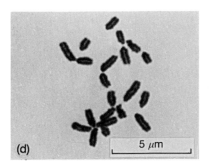

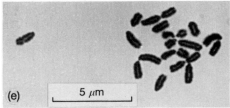

Fig. 1. Diploidized parthenogenetic mouse embryo after oocyte activation with ethyl alcohol. Chromosome complements at the first cleavage division.

(a) Normal haploid sets of chromosomes in both groups. (b), (c) Malsegregation of chromosomes at the second meiotic division to give 19 and 21 chromosomes respectively. (d), (e) Malsegregation of one chromosome during the first meiotic division: both groups with 21 chromosomes.

a second polar body predominated in these studies. They were selected under a stereomicroscope, transferred into cytochalasin-free medium and cultured in the presence of a small amount of colcemid for the next 14–17 h (up to metaphase of the first cleavage division). A minimal dose of colcemid was

Table 7. Ethyl alcohol effects on chromosomal disjunction at the second meiotic division in mouse oocytes – F_1(CBA × C57BL)

Activation procedure		Metaphases of the first cleavage division studied					No. of parthenogenones with chromosomal non-disjunction
Ethyl alcohol level in medium	Tempera-ture of activating medium (°C)	Total	Chromosome number				
			20 + 20	21 + 19	22 + 18	24 + 16	
8.8	+ 26.0	21	6	9	6	—	15(21)[a]
8.8	+ 20.0	9	6	2	1	—	3(9)
7.0	+ 26.0	12	3	7	2	—	9(12)
7.0	+ 20.0	172	137	32	3	—	35(172)
4.0	+ 26.0	20	11	7	1	1	9(20)
2.0	+ 26.0	6	6	—	—	—	0(6)

[a] Total number of parthenogenones in brackets.

used, that had little effect on chromosome spiralization but was sufficient to prevent chromosome coalescence of the two pronuclei on a common metaphase plate. We were thus able to discriminate which chromosomal groups belonged to which pronucleus (Fig. 1).

Cumulative data obtained with ethyl alcohol or heat shock treatments are presented in Tables 7 and 8 respectively. The total incidence of chromosomal non-disjunction at the first meiotic division was similar for both groups. Out of 251 ethyl alcohol activated eggs only one with 42 chromosomes and two with 38 chromosomes were recorded. Hence the relative non-disjunction rate was about 0.8 per cent (if we double the number of hyperhaploid eggs) or 1.2 per cent (if we sum the numbers of hyperhaploid and hypohaploid eggs). These estimates corroborate well with the very low incidence of chromosomal non-disjunction in oogenesis of young mice determined by chromosomal counts in non-activated M-II oocytes (see Chapter 5; also Hansmann 1983).

The presence of 15 embryos with 39 chromosomes in the ethyl alcohol activated group and three similar embryos in the heat-activated group should also be mentioned (see Tables 7 and 8 respectively). One cannot, however, exclude chromosomal loss in the course of chromosome preparations, and hence these specimens were neglected in our calculations of non-disjunction rate in the first meiotic division. Meanwhile, the existence of this type of non-disjunction with either 20 dyads + one half dyad (one chromatid plus) or 19 dyads + one half dyad (one chromosome missing) has recently been suggested (Hansmann and El-Nahass 1979; see also Hansmann 1983 for review). This particular type of non-disjunction predicts the origin of two types of parthenogenones with uneven chromosome sets 39 and 41. Not a single case

Table 8. Heat shock effects on chromosomal disjunction at the second meiotic division in oocytes of mice F_1(CBA × C57BL)

Tempera- ture of activating medium (°C)	Metaphases of the first cleavage division studied						Number of embryos with abnormal chromosome disjunction
	Total	With chromosomal number					
		20 + 20	21 + 19	22 + 18	23 + 17	24 + 16	
+ 42.0	23	2	8	4	4	5	21(23)[a]
+ 41.5	27	12	4	3	5	3	15(27)
+ 41.0	103	93	8	1	1	—	10(103)
+ 40.5	25	19	5	1	—	—	6(25)
+ 39.5	31	27	4	—	—	—	4(31)
+ 39.0	40	32	7	—	1	—	8(40)

[a] Total number parthenogenones in brackets.

of 41 chromosomes has been identified in our studies so far and this problem requires further study (Fig. 1).

Thus, cytogenetic studies of diploidized one-cell parthenogenones provide an estimate of meiotic non-disjunction during the first meiotic division. Mitotic chromosomes of one-cell specimens are easily scored compared to meiotic chromosomes of ovulated eggs. Moreover, mitotic chromosomes are quite suitable for differential staining (banding) and thus afford an exact identification of non-disjunction products.

The chromosome number in the one-cell parthenogenones activated by heat shock (Table 8) or by ethyl alcohol treatment (Table 7) was recorded separately for each pronucleus. The number of specimens with different chromosome scores in the two pronuclei was somewhat higher in the ethyl alcohol group (30.9 per cent) than in the heat shock group (26.1 per cent). The sum of chromosome numbers in both pronuclei of the same parthenogenone was always diploid ($2n$ = 40), so these figures can be interpreted as strict evidence of the actual level of chromosome non-disjunction at the second meiotic division.

Mutagenic (clastogenic) effects of ethyl alcohol treatment on meiotic chromosomes in mice are temperature-dependent, i.e. the meiotic spindle seems to be more vulnerable to ethyl alcohol effects at elevated temperatures (Table 7).

Heat shock treatment alone can induce chromosomal non-disjunction during second meiotic metaphase. Out of 241 ovulated eggs subjected to heat shock treatment, 64 (26.1 per cent) possessed pronuclei with different chromosome counts. Complementary hypo- and hyperhaploid pronuclei were found in almost all parthenogenones of the different experimental series. For instance, the transient treatment with + 42.0°C resulted in complementary non-disjunction in 21 out of 23 specimens (chromosomal sets

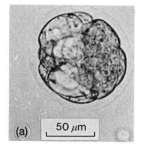

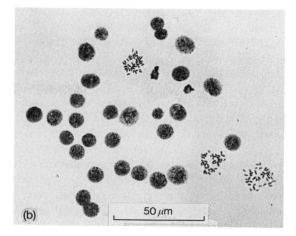

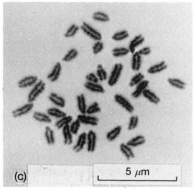

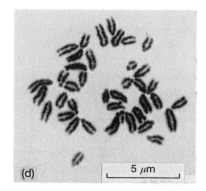

Fig. 2. Parthenogenetic mouse embryo, 96 h, cultured *in vitro* after ethyl alcohol oocyte activation. Nascent blastocyst.

(a) Phase contrast, general view; (b) air-dried preparation, 32 cells; (c,d) metaphase plates, 40 chromosomes in each (Dyban and Noniashvili 1985a,b).

22 + 18; 23 + 17; 24 + 16). The proportion of affected specimens drastically decreased with reduction of temperature. Heat treatment at + 41 °C resulted in a 10 per cent non-disjunction rate only (10 out of 103) and embryos with non-disjunction of only 1 chromosome prevailed (19 + 21 sets).

Moreover, it seems that the critical threshold of alcohol treatment is somewhat higher for the egg as a whole than for meiotic spindle fibres, i.e. activation of parthenogenetic development may be achieved at a somewhat higher temperature than induction of chromosomal loss from meiotic II metaphases.

Chromosomal non-disjunction from meiotic II metaphases can be achieved by 7-min treatment of ovulated eggs at 39°C (Dyban and Noniashvili 1984). These fluctuations of body temperature are often encountered in clinical practice. One can therefore not neglect the possible damaging effects of elevated body temperature on chromosome non-disjunction in female gametes. This problem needs further study.

Not only ethyl alcohol (Kaufman 1982a, 1983a,b; Dyban and Noniashvili 1984), and heat shock (Dyban and Noniashvili 1984), but also hyaluronidase treatment of ovulated eggs (Graham and Deussen 1974) can induce both oocyte activation and meiotic chromosome non-disjunction. Clastogenic effects of other oocyte activating agents also seems highly probable.

The consequences of meiotic chromosome non-disjunction are very different in haploid and diploid parthenogenones. Haploid parthenogenones generated from oocytes that have suffered meiotic non-disjunction will inevitably be nullisomic or disomic for some particular chromosome. If second polar body extrusion is suppressed, both haploid genomes with complementary types of aneuploidy will be fused again during metaphase of the first cleavage division (Tarkowski 1971, 1975), so the resulting diploid genome of the parthenogenone will again be balanced.

Thus chromosomal imbalance resulting from non-disjunction provoked by activating agents applied to ovulated eggs might be very harmful for subsequent progression of haploid parthenogenones. Unfortunately, no adequate cytogenetic analysis (including chromosome identification by banding) of parthenogenones at different stages of embryogenesis has yet been carried out. However, some information relevant mostly to the ploidy of parthenogenones at the morula and blastocyst stages is available.

Since the initial cytogenetic studies of parthenogenetic embryos, it has been known that some blastomeres of haploid embryos readily undergo spontaneous polyploidization, resulting in hypodiploid mosaics (Tarkowski et al. 1970; Witkowska 1973a). Subsequent investigations confirmed these findings (Kaufman and Sachs 1975; Kaufman 1975, 1978b, 1983a,b). It is known that a haploid genome by itself is not stable and may readily transform into a balanced diploid genome (Tarkowski 1975, 1977). Spontaneous diploidization of an initially haploid genome has been repeatedly observed in parthenogenones and also in androgenetic or gynogenetic haploids produced by direct microsurgical bisection of fertilized eggs (Tarkowski 1977). Out of six blastocysts generated from these 'halves', three displayed haploid metaphases exclusively, while the other three were composed of both haploid and diploid blastomeres (Tarkowski and Rossant 1976). It might be inferred from this unique observation that instability of a haploid genome is not provoked by parthenogenetic development itself. Spontaneous diploidization occurs both in androgenetic and gynogenetic haploids, i.e. when the two parental genomes are separated and cannot fuse to form a normal diploid nucleus. We shall come back to this problem later (see Section 1.4).

The mode of diploidization of the initially haploid genome remains obscure. The appearance of blastomeres with two nuclei at the initial stages of diploidization has been suggested (karyotomy without cytotomy; Tarkowski et al. 1970; Witkowska 1973a). Chromosome endoreduplication has been suggested as another feasible route for diploidization (Kaufman 1983a). During prolonged culture of haploid parthenogenones (93–100 h), both diploid and tetraploid blastomeres appear. Thus some specimens transform into haploid–diploid and some others into diploid–tetraploid mosaics (Kaufman and Sachs 1975).

Tetraploidy by itself may reflect the onset of trophoblast cell polyploidization, which is known to occur from the 25–30 cell stage in normal mouse embryos (Dyban, Samoshkina, Patkin, and Chebotar 1976). The exact stage of spontaneous diploidization and polyploidization in haploid embryos remains unknown, but it seems to be asynchronous and is time-specific for each parthenogenone.

Detailed cytogenetic analysis of parthenogenetic embryos during the second and the third cleavages is still missing. Available data on this subject are limited to cytogenetic analysis of day 4 parthenogenones transplanted at the one-cell stage into the oviducts of pseudopregnant females. Out of 17 16–20-cell parthenogenones of the first experimental group, 14 specimens displayed haploidy, two haploidy and diploidy, and one possessed only diploid metaphases. Out of 138 parthenogenones of the second group, 102 (72.3 per cent) were haploids, 35 (24.8 per cent) haploid–diploid mosaics, and the remaining four pure diploids (Kaufman, Robertson, Handyside, and Evans 1983; Kaufman 1983a). The low frequency of diploidy in these experiments was probably due to improvements in the activation technique. The preponderance of haploid mitoses in haploid–diploid mosaics was probably due to the enhanced mitotic activity of the haploid blastomeres (Tarkowski 1975; Tarkowski et al. 1970; Witkowska 1973a).

In spite of this high mitotic activity, haploid blastomeres seem to be selectively inferior to diploid ones, especially under in vitro culture conditions. Numerous attempts to produce haploid stem cell lines from haploid parthenogenetic blastocysts have resulted only in diploid cell cultures, often with concomitant deletions in the distal arm of the X-chromosome (Kaufman et al. 1983; Kaufman 1983a). Both spontaneous diploidization and structural rearrangements are a clear-cut indication of the instability of the haploid genome.

Structural rearrangements of chromosomes in 8–12 cell haploid parthenogenones are not limited to X chromosomes exclusively but extend to the autosomes as well, with numerous isochromosomes, dicentric chromosomes, centric fragments, and chromatid gaps (Dyban and Noniashvili 1985a,b). Isochromosomes and abnormal giant chromosomes are also often encountered in parthenogenetic embryos (Kaufman 1982a, 1983a,b). The blastomeres with structural or numerical chromosomal imbalance may give rise to genetically abnormal cell clones with drastically reduced viability and thus may substantially affect the survival capacity of haploid parthenogenones.

Cytogenetic studies of presumably haploid parthenogenones during post-implantation development have been carrried out only by Kaufman (1983a). These genetically mosaic parthenogenones arose from immediately cleaving eggs (see Section 1.1). No cytogenetic studies of genetically uniform parthenogenones after implantation have been reported as yet.

Immediate cleavage of ovulated eggs has been induced by transient treatment in Ca^{2+}, Mg^{2+}-free medium with subsequent *in vitro* culture of the activated eggs up to the blastocyst stage. These presumably haploid blastocysts were then transferred to the uteri of day-3 pseudopregnant females. All recipient females were ovariectomized thereafter and the pregnancy was maintained by progesterone and estradiol injections. At autopsy on the ninth day 48 implantation sites out of 96 transplanted blastocysts were found. Most contained no embryonic tissues, but 10 decidua contained parthenogenones at the early egg cylinder stage, corresponding to a 6.5–7.5-day normal embryo. Three of these were degenerate and had no mitotic figures. Cytogenetic studies of the remaining seven parthenogenones revealed true haploidy in two and haploid–diploid mosaicism in five. The total number of mitotic plates scored in these seven parthenogenones was low (16 haploid and 36 diploid metaphases). The small number of metaphases scored and the absence of chromosome banding do not permit a proper estimate of the actual level of aneuploidy and do not reveal structural aberrations or numerical changes such as pseudodiploid or pseudohaploid karyotype. However, the results of this preliminary work (Kaufman 1978b) indicate that at least mosaic forms of haploidy are sometimes compatible with early post-implantation survival up to the early egg cylinder stage, i.e. germ layer formation, differentiation of extraembryonic ectoderm, and entoderm. A few more advanced, presumably haploid, parthenogenones were found at an early somite stage (Kaufman 1981, 1983a) but these unique specimens were not studied cytogenetically.

The differentiation capacity of haploid cells can readily be studied in experimental teratomas generated from haploid parthenogenones trans-planted into ectopic sites. For this purpose haploid blastocysts generated from hyaluronidase-activated eggs were transferred under the testis capsules of syngeneic mice. Teratomas derived from these parthenogenetic grafts were similar morphologically to those from normal diploid embryos and included nerve tissue, different types of epithelium, cartilage, bone, skeletal muscle, hemopoietic cells, and adipose tissue. All these derivatives lacked teratocarcinoma or yolk sac embryocarcinoma stem cells and hence did not survive prolonged cloning. Cytogenetic analysis of six cell lines derived from initially haploid teratoma cells revealed a diploid karyotype in four, hypertetraploidy in one, and diplo-tetraploid mosaicism in one (Iles and Evans 1977; McBurney, Bramwell, Deussen, and Graham 1975). Cytophotometric studies showed a substantial proportion (80 per cent) of cells with a DNA content below 2c and above 1c. Selective advantage of diploidy over haploidy was suggested to be a functional prerequisite for abnormal replication and non-disjunction of mitotic chromosomes, giving rise to aneuploid (hypodiploid) cell clones (Graham *et al.* 1975; Graham 1977). Thus, advanced differentiation of parthenogenetic allografts transplanted ectopically does not reflect the actual developmental capacities of haploid cells, since most of them spontaneously multiply their karyotype (Graham 1977; Graham, McBurney, and Iles 1975).

The chromosomal complement of haploid parthenogenones is evidently prone to spontaneous numerical and structural changes. Initially haploid chromosome complements readily transform into diploid ones by the egg cylinder stage. More studies are needed to clarify the karyotype of parthenogenetic cells at more advanced stages of embryogenesis still compatible with their survival.

1.3.4 Developmental peculiarities of parthenogenones

This section reviews developmental capacities of parthenogenetic embryos with special emphasis on pre-implantation and post-implantation development of haploids.

Parthenogenetic embryos can successfully accomplish pre-implantation development to give normal-looking blastocysts (Tarkowski *et al.* 1970; Witkowska 1973a; Graham 1974a; Tarkowski 1975). Some of them can implant and proceed through the early post-implantation stages (Witkowska 1973b; Kaufman, Barton, and Surani 1977; Kaufman 1978b, 1981).

1.3.4.1 Pre-implantation stages. Developmental capacities of parthenogenetic embryos both *in vivo* and *in vitro* are inferior to those of normally fertilized eggs from the one-cell stage onwards. Developmental variations between different types of parthenogenones (see Section 1.4) are also rather conspicuous (Kaufman and Sachs 1975). The survival time of diploid parthenogenones always surpasses that of haploid specimens, and mosaic haploids usually develop better and can reach more advanced stages of embryogenesis than genetically uniform ones (Tarkowski 1975; Kaufman and Sachs, 1975; Kaufman 1983a,b).

Diploid and haploid parthenogenones generated through electrical stimulation of ovulated eggs contain almost equal blastomere numbers on the fourth day, while day-5 haploid parthenogenones have almost twice as many cells as diploid ones (Witkowska 1973a). These data seem to indicate a substantially higher cleavage rate of haploid parthenogenones, but the reason for this discrepancy is complex.

Haploid parthenogenones start their first cleavage approximately 12.5 h after activation, that is 4.5–5.5 h before diploid parthenogenones or normally fertilized eggs (Kaufman 1973b, 1983a). Haploid parthenogenones activated by ethyl alcohol treatment start their first cleavage at 15.5 h, i.e. almost 3 h later than the hyaluronidase-treated group (Kaufman 1983a).

The first round of DNA replication in pronuclei of mouse parthenogenones is detected 4–6 h after activation and lasts for about 4 h. DNA replication in at least some parthenogenones is postponed for about 8–12 h after activation (Graham and Deussen 1974; Abramczuk and Sawicki 1975). Hence, the unusually short interval between activation and the first cleavage division in haploid parthenogenones is not due to any specific deviation in initiation of DNA synthesis.

The first cleavage division (from early prophase to late telophase) lasts for about 116 min (from 109 to 120 min in different strains of mice) both in normally fertilized eggs and in diploid parthenogenones, while it extends up to 155 min in genetically uniform haploids (Kaufman 1983a,b).

The total duration of the one-cell stage correlates with embryonic ploidy, being about 3–4 hours less in haploid than in diploid eggs.

Mosaic haploids generated by immediate cleavage reach the two-blastomere stage about 4–5 h after activation, i.e. almost 10–12 h earlier than parthenogenones of other types or normally fertilized eggs. If the cleavage rate for subsequent pre-implantation stages remained the same, 'mosaic' haploids would contain twice as many cells as diploid parthenogenones. But this is not actually the case: after 96–100 h in culture diploid parthenogenones, mosaic haploids, and genetically uniform haploids were composed of 21.9 ± 2.4 cells, 17.9 ± 2.3 cells, and 12.3 ± 0.7 cells, respectively (Kaufman and Sachs 1975).

Thus development of all parthenogenones proceeds somewhat more slowly than that of normally fertilized eggs. The rate of cleavage of haploid parthenogenones is especially low. Kaufman (1981) found that haploid parthenogenones resemble diploid ones in their cleavage rate up to the eight-cell stage, while their subsequent development up to the early morula stage is substantially retarded (Graham et al. 1975; Kaufman 1983a,b).

In spite of a significant increase in blastomere number, mosaic haploids are late in compaction and blastulation. Compaction of normally fertilized eggs occurs at the eight-cell stage, i.e. after the third cleavage division (Ducibella 1976). The same is also true for diploid parthenogenones. On the other hand genetically uniform haploids and especially mosaic haploids display a substantial delay in compaction, which may not occur until the 16-cell stage, i.e. one cleavage later than in controls (Kaufman 1981, 1983a).

One more specific feature of haploid parthenogenones is decompaction, first discovered *in vitro* both in genetically mosaic and genetically uniform parthenogenones (Kaufman 1983a). Because of decompaction, subsequent recompaction and atypical cavitation, abnormal-looking haploid morulae and 'early' blastocysts usually contain more blastomeres than diploid embryos at a similar stage. Thus, an increased blastomere count *per se* in some parthenogenetic embryos at the morula and blastocyst stages on the fifth day of development does not show increased cleavage rate but indicates a delay in compaction and cavitation. The former is known to be an indispensable prerequisite for the normal transition from morula to blastocyst stage (Ducibella 1977; Ducibella and Anderson 1975).

Cavitation processes do not depend on blastomere number (Tarkowski and Wroblewska 1967), but correlate with the biological age of the cleaving egg, as well as with rounds of DNA replication and nucleo-cytoplasmic relationships (Smith and McLaren 1977). Substantial delay of cavitation in parthenogenones, and especially in haploid embryos, even with an elevated blastomere count, is in good agreement with these findings. [We shall come back to this subject in the final chapter, dealing with chromosomal activity during early development in mammals (see Chapter 13).]

Asynchronous cleavage is typical for all haploid parthenogenones and may result in pronounced heterogeneity of the cell population, comprising morulae and blastocysts (Witkowska 1973a; Tarkowski 1975; Graham 1974a). Immunofluorescent studies of parthenogenetic morulae as well as of haploid blastocysts with unusually high blastomere numbers revealed the presence of cells lagging far behind for one or even two cell divisions compared to the rest of the blastomeres (Handyside 1980; Kaufman 1981, 1983a,b). Scanning electron microscope examination of haploid blastocysts revealed the existence of both small and large cells of different shape in the trophectoderm layer (Kaufman 1983), which were never detected in control blastocysts (Ducibella 1977; Bergstrom 1978). Moreover, the inner cell mass (ICM) of haploid embryos had a significantly reduced cell number, and some inner cells remained uncommitted and were capable of transforming into trophoblast giant cells. Thus, the actual number of inner cell mass components giving rise to all embryonic tissues as well as to most of the extra-embryonic membranes is much less in haploid parthenogenones than in control embryos of the same stage (Kaufman 1981, 1983a,b). Though looking normal, the haploid blastocysts are distinguished by some severe abnormalities crucial for implantation and subsequent development.

1.3.4.2 Post-implantation stages. Post-implantation development of parthenogenetic embryos is studied much less thoroughly than pre-implantation development. This is partly because of the rather poor survival rate of parthenogenetic specimens.

At implantation, parthenogenetic embryos represent a rather heterogeneous population. Some of them, most severely affected, are unable to implant and cannot even induce a decidual reaction. More advanced parthenogenones induce decidualization and transform into empty tropho-blastic vesicles. A few haploid embryos survive implantation and proceed to further development.

About 25–30 per cent of electrical shock stimulated eggs can induce a decidual reaction and reach an egg cylinder stage by six to seven days. Embryonic mortality increased drastically by the eighth day, with only a few embryos still alive on the ninth to tenth day as advanced egg cylinders (Tarkowski *et al*. 1970). One morphologically normal eight-somite partheno-genone has been recovered on the tenth day (Witkowska 1973b).

Many more parthenogenones, including haploid ones, proceeded to the early egg cylinder stage in artificially delayed implantation experiments (Kaufman *et al*. 1977; Kaufman 1978b, 1981).

A small ectoplacental cone with concomitant delay in extra-embryonic ectoderm differentiation was a common morphological trait for most parthenogenetic haploids (Kaufman 1978b). In at least one case, a pre-sumably haploid embryo remained alive up to the ninth day. This unique specimen looked like a normal nine-day embryo with seven somite pairs, and properly developed ectoplacental cone, amnion, allantois, and yolk sac (Kaufman 1983a,b). Complete diploidization of an initially haploid genome has been suggested though not proved cytogenetically in this case.

The developmental capacity of diploid parthenogenones is undoubtedly much better than that of haploid ones both before and after implantation. Under natural conditions, i.e. without artificial delay in implantation by means of ovariectomy, diploid parthenogenones readily reach egg cylinder or even neurula stage (7 to 8 days; Tarkowski *et al*. 1970; Witkowska 1973b; Graham *et al*. 1975). This developmental profile was also typical for haploid parthenogenones diploidized by cytochalasin B treatment at the one-cell stage after heat shock or ethyl alcohol activation (Balakier and Tarkowski 1976; Cuthbertson 1983). These were more capable of advanced post-implantation development than eggs activated in other ways. Thus post-implantation survival of parthenogenones depends partly on the specificity of activating agents, with ethyl alcohol treatment being the most suitable and most delicate procedure for egg activation (Dyban and Noniashvili 1985a,b).

Parthenogenones which attain a proper cell number during pre-implant-ation delay in ovariectomized females routinely reach 10–11 days of normal embryogenesis and some of them recovered at the 25–30 somite stage possessed quite normal-looking forelimb buds (Kaufman *et al*. 1977; Kaufman 1978b, 1981, 1983b). No morphological abnormalities could be detected by outside inspection of these rare specimens, but some displayed moderated retardation in their extra-embryonic membrane differentiation (Kaufman 1981, 1983a,b). Not a single parthenogenetic embryo has been recovered after the eleventh day of development. The actual causes of their abrupt elimination remain unknown.

1.4 Causes of death in parthenogenetic, gynogenetic, and androgenetic embryos

All attempts to achieve normal development of mouse parthenogenones beyond active organogenesis have so far failed. Developmental profiles of parthenogenetic embryos of other mammalian species are even less clear, but none survive post-natally. Embryolethal stages of androgenetic and gynogenetic parthenogenones are still not known for sure, but their developmental profiles probably resemble those of haploid and diploid parthenogenones.

Many thousands of parthenogenones have been studied in different laboratories and quite a few have survived into advanced post-implantation development. Experimental data on the development of androgenetic and gynogenetic embryos are much fewer even at pre-implantation stages. More thorough studies of both androgenesis and gynogenesis in mice may help to distinguish intrinsic pathomorphological mechanisms underlying their poor survival.

The death of parthenogenones is not limited to one particular stage. On the contrary it may extend from the onset of cleavage through early post-implantation development. Some activated eggs do not start cleavage at all or cannot survive beyond the two-cell stage. Some other parthenogenones are eliminated during cleavage, compaction, or blastulation, while most die at implantation. This may be due to asynchrony between their developmental age and that of the uterus (Graham 1974a; Graham et al. 1975).

According to a more recent hypothesis (Kaufman 1981, 1983a), compaction and cavitation in parthenogenetic embryos occurs in a rather heterogeneous blastomere population, some of which lag in their cell cycles, remain uncommitted and form abnormally small ICM (Inner Cell Mass), unable to support effective morphogenetic processes thereafter. This hypothesis seems to be quite suitable for haploid parthenogenones, most of which display cleavage disturbances.

Thus embryonic death during cleavage and implantation is primarily due to defects in cleavage rate and subsequent abnormal blastulation.

Some haploid parthenogenones proceed to the egg cylinder stage, while others succeed in reaching the limb-bud stage. The reason for their death remains obscure.

It has been already pointed out (see Section 1.3.1.2) that after transplantation into ectopic sites parthenogenetic blastocysts form extensive outgrowths, transforming into teratomas (Iles et al. 1975; Graham et al. 1975; Graham 1977). Thus the death of parthenogenones cannot be attributed to cell lethals. This statement gains further support from experimental chimaeras made by aggregation of normal diploid and parthenogenetic embryos (Stevens et al. 1977; Surani et al. 1977). Parthenogenetic cells are encountered in different organs and embryonic anlagen as well as in various tissues of adult chimaeras. Moreover, some parthenogenetic cells give rise to functionally active germ cells (Stevens *et al.* 1977; Stevens 1978).

Developmental capacities of parthenogenetically activated eggs can be substantially improved by microsurgical injection of age-equivalent stem cells from normal diploid embryos. Almost complete absence of genetic mosaicism in injected parthenogenones developed up to the 18–25 somite stage was an unexpected finding of these studies. According to Kaufman, Evans, Robertson, and Bradley (1984), enhanced survival of the injected parthenogenones was probably due to the substantial increase of ICM repopulated by injected normal cells.

Two main hypotheses are suggested to explain developmental failures of post-implantation parthenogenones as well as gynogenetic and androgenetic embryos. According to one, cytoplasm of activated eggs is functionally inadequate for normal development. If transplanted into a previously enucleated zygote, a parthenogenetic nucleus may support normal development and give rise to normal looking post-natal mice (Hoppe and Illmensee 1982). All attempts to reproduce these results have so far been unsuccessful. It has been repeatedly demonstrated, however, that a male pronucleus transplanted into a haploid parthenogenone provides quite normal prenatal development of the 'reconstructed' zygote. On the other hand, gynogenetic embryos generated through microsurgical transplantation of a female pronucleus proceed only to the late egg cylinder stage, so their survival period is comparable to that of ordinary parthenogenones (Surani *et al.* 1984). Microsurgical transplantation of both paternal and maternal pronuclei into previously enucleated parthenogenetic eggs resulted in normal development up to birth. On the contrary, both pronuclei of parthenogenetic origin, if transferred into the ooplasm of normally fertilized eggs with previously withdrawn parental genome, cannot support embryonic development beyond implantation (Mann and Lovell-Badge 1984). All of these data indicate some kind of functional insufficiency of parthenogenetic nuclei.

Experiments with microsurgically achieved pronuclear exchange gave similar results. Both heterozygous androgenetic and gynogenetic diploids perished soon after implantation and never survived beyond a 25-somite stage. On the other hand, artificially produced combinations of one male and one female pronucleus readily support normal embryonic development and are quite compatible with live newborns (McGrath and Solter 1984; Surani *et al.* 1984). It has been speculated that the presence of both male and female pronuclei is indispensable for normal embryonic development and some hypothetical complementary interaction between parental genomes has been suggested. We shall come back to this problem in the final chapter of this book.

1.5 Conclusions

Haploidy may result from parthenogenetic, gynogenetic, or androgenetic routes of development. The methods for deliberate production of parthenogenetic, gynogenetic, and androgenetic embryos are well elaborated. Spontaneous haploidy is rarely encountered in any mammalian species studied cytogenetically. One strain of mice is known with an unusually high incidence of spontaneous parthenogenesis. Spontaneously activated eggs can generate either parthenogenetic embryos released into the oviduct or spontaneous teratomas.

Two types of haploid and two types of diploid parthenogenones can be produced artificially. Developmental capacities of haploid parthenogenones are always inferior to those of diploid parthenogenones from the very onset of embryogenesis. Unstable karyotypes with spontaneous diploidization, chromosomal non-disjunction, and structural rearrangements are common traits of haploid parthenogenones, both androgenetic and gynogenetic.

Haploidy in mice is quite compatible with accomplishment of pre-implantation morphogenesis and blastulation in parthenogenetic, gynogenetic, and androgenetic embryos. Survival capacities of androgenetic haploids are more limited than those of gynogenetic or parthenogenetic haploids. Diploidization of genome content substantially improves survival capabilities of parthenogenetic, androgenetic, and gynogenetic embryos. Embryolethal stages of diploid parthenogenones, androgenetic, and gynogenetic embryos coincide well and correspond to neurulation and active organogenesis. Functional insufficiency of the nuclei due to the absence of proper complementation between male and female chromosomes from the initial stages of embryogenesis is suspected as a feasible cause of the inevitable death of parthenogenetic, gynogenetic, and androgenetic embryos.

2
Triploidy

2.1 Mechanisms of origin

Triploidy in mammals and other animals may result from digyny or diandry (see Fig. 3).

2.1.1 Digyny

The extra chromosome set of digynous triploids has a maternal origin and may come from fertilization of an egg with a non-reduced (diploid) chromosome number by a normal (haploid) spermatozoon. Three basic routes known to evoke digynous triploidy in mammals are distinguished (Beatty 1957; Austin 1960). Type I digyny results from a meiotic block operative at the anaphase stage of the first maturation division with homologous bivalents being already detached and two meiotic spindles formed. Subsequently, two second polar bodies each carrying haploid sets of chromosomes are extruded, with the remaining two sets being left in the ooplasm. After normal fertilization the latter are transformed into two female pronuclei. Thus, the resulting embryo should have two sets of chromosomes from its mother and one from its father (Austin 1960, 1961, 1969). Type II digyny stems from failure at anaphase of the second meiotic division, which prevents extrusion of the second polar body. This egg, carrying one diploid or two separate haploid pronuclei, gives rise if fertilized to a triploid zygote (Thibault 1959). Type III digyny relates to fertilization of unusual giant diploid oocytes (Austin 1969), resulting from meiotic division of both binuclear and mononuclear giant eggs, occurring in mammals (Austin 1961). The latter seem to be produced either by nuclear division unaccompanied by cytoplasmic division of oogonia, or by cytoplasmic fusion of two adjacent oogonia at initial or advanced stages of oogenesis (Austin 1969). Initially, binucleate eggs usually become mononucleate, with both their nuclei fused together before maturation divisions commence.

Tetraploid mononuclear giant eggs have been observed in mice (Pesonen 1946; Austin and Braden 1954a) and polynuclear oocytes have been found in various mammalian species (mice, rats, hamsters, goats, and humans) (Austin 1961, 1969). Not a single case of proven type III digyny was reported

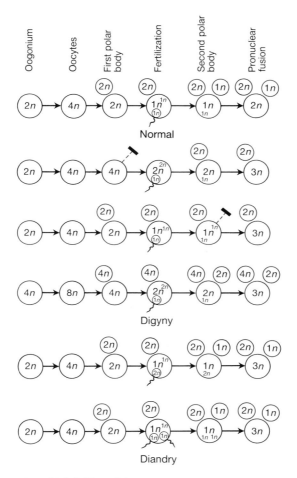

Fig. 3. Basic routes of triploidy origin.

in the literature until recently. These giant eggs were assumed to die in the initial stages of development. Full viability of giant diploid oocytes was recently demonstrated in Chinese hamsters (Funaki and Mikamo 1980). After completion of the two meiotic divisions they participate in fertilization, giving rise to digynic triploids. Their pre-implantation development seems to be less affected than that of any other type of triploid (see Section 2.4.1). Thus the possibility of a much greater impact of type-III digyny on the origin of triploid embryos in various mammals including humans is substantially reinforced. Nevertheless, digyny of types I and II is still considered as the most common for all mammals. These two types of digyny can be

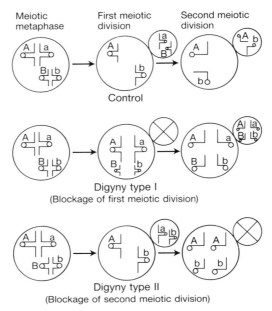

Fig. 4. Distribution of two marker chromosomes with blockage of the first or second meiotic divisions.

distinguished with certainty only by means of chromosomal or biochemical markers. Peculiarities of segregation for two different marker chromosomes due to the failure of meiotic divisions of the egg are shown in Fig. 4. If the emission of the first polar body is blocked, the resulting egg will contain both marker chromosomes of the mother. The block to second polar body extrusion will result in the egg carrying double copies of only one marker chromosome. Type-II digyny can be assessed biochemically only if there is no crossing over in the chromosome region carrying the marker genes, otherwise both types of digynic egg will be biochemically identical. Thus the assessment of the mode of origin of digyny offers substantial difficulties. Only a few cases of proven type I (Edwards, Yuncken, Rushton, Richards, and Mittwoch 1967; Jonasson *et al.* 1972; Niebuhr, Sparrevohn, Hinnigsen, and Mikkelson 1972) and one presumptive case of type II (Penrose and Delhanty 1961) digyny are reported so far in the literature.

The hypothesis has been put forward that the routes of digyny are genetically predetermined at least in laboratory mice (Braden 1957, 1958). This remains to be confirmed by novel cytogenetic and biochemical methods. It also remains unknown whether type II digyny predominates in mice, rats, and hamsters (Beatty 1957; Braden and Austin 1954; Austin 1969).

2.1.2 Diandry

The extra chromosome set in these triploids has a paternal origin and may result from the fertilization of normal (haploid) oocytes by diploid (non-reduced) spermatozoa or from dispermic fertilization. Both types of diandry are relevant to mammals though the exact contribution of each type to the origin of triploidy remains obscure for most species. Diploid sperm may arise through failure of the first or second meiotic divisions and is known to occur quite often in humans (Sumner 1971; Carothers and Beatty 1975) and in rabbits (Beatty and Fechheimer 1972). Diploid spermatozoa may make a substantial contribution to the origin of triploidy in humans (Beatty 1974). This suggestion was initially inferred from a good correlation of triploidy rates in spontaneously aborted human fetuses with the roughly estimated number of diploid spermatozoa in human sperm samples (Beatty 1978).

This type of diandry seems to be indicated by chromosome marker analysis of some human triploids (Ushida and Lin 1972). This conclusion is not firm, as the same pattern of marker chromosomes may result from dispermic fertilization (Nikawa and Tadashi 1974). No diploid sperm have so far been found in mice, in spite of thorough analysis of numerous cytogenetic data (Stolla and Gropp 1974; Beatty, Lim, and Coulter 1975). In some strains of mice quite a few diploid sperms are encountered. These unique spermatozoa have reduced motility and their participation at fertilization seems hardly possible (Krzanowska 1974). Thus this type of diandry is not of major importance for the mode of origin of triploidy in mice.

Simultaneous penetration of the ovum membranes by more than one spermatozoon has been described in many species including rats, mice, hamsters, and pigs (Piko 1958; Braden and Austin 1954; Beatty 1957; Thibault 1959). In case of dispermy it may give rise to diandric triploids. The incidence of dispermic fertilization does not completely coincide with that of diandric triploids as many eggs penetrated by several spermatozoa are eliminated before cleavage starts (Austin 1969). The reason for this premature death remains unknown.

2.1.3 Relative importance of digyny and diandry

The relative proportion of diandric and digynic mechanisms responsible for production of triploid embryos varies among different species of mammals. It appears from some early evidence that at least 60 per cent of human triploids are due to digyny and only 40 per cent to diandry (Niebuhr 1974; Beatty 1974). According to more recent data based on a method of maximum likelihood analysis (Jacobs et al. 1978) over three-quarters of triploids are diandric and only the remaining quarter is digynic.

Thus the data on the actual impact of diandry and digyny in the origin of triploidy in humans is still controversial. More definite results on this subject have been obtained in laboratory animals. For instance, in rabbits both spontaneous triploidy and triploidy arising after 8 h delayed insemination are primarily digynic (Carothers and Beatty 1975; Beatty and Coulter 1978), but in rats the diandric mechanism primarily due to dispermic fertilization is predominant (Bomsel-Helmreich 1965). In laboratory mice, like in rabbits, the eggs seem to be more prone to digyny, due to the failure of second polar body emission rather than to polyspermic fertilization. This is shown by cytological studies of freshly fertilized eggs (Braden and Austin 1954) as well as by chromosomal analysis of pre-implantation triploids, none of which was found to have XYY gonosome constitution (Vickers 1969; Wroblewska 1971; Baranov 1976b). More conclusive data on this matter were obtained in crosses of male mice homozygous for marker chromosome translocations to female mice without visible chromosomal rearrangements (Wroblewska 1971; Baranov 1976b). All 26 triploids studied cytogenetically at early post-implantation stages of embryogenesis were assessed to be digynic by their origin as there was only one marker male chromosome in their karyotype. Unfortunately, experiments of this kind are still very scarce at early pre-implantation stages of embryogenesis. Four cases of triploidy were recorded in female mice crossed to males homozygous for the marker translocations of Robertsonian type, Rb(9.19)163H. Two male pronuclei and only one female pronucleus were present in some of these triploids (Donahue 1972c). Direct evidence of a diandric mechanism in pre-implantation triploids was obtained in our recent studies of mice with chromosomal translocations. Three of 18 triploids recovered on the third and the fourth days had two distinct male marker chromosomes, indicating a paternal origin for the extra chromosome set (Baranov 1983a).

It remains to be seen whether the relative importance of the diandric route is significantly inferior to the digynic one for the production of triploidy in laboratory mice.

Some discrepancies in experimental data on the origin of triploidy in mice may be due to the different viability of androgenic and gynogenic triploids (see Section 2.4).

Polyspermic fertilization is common for golden hamsters but not a single triploid embryo of XYY gonosome constitution was ever recovered in this species (Yamamoto and Ingalls 1972). Probably some of the diandric triploids are eliminated soon after fertilization, owing not primarily to chromosome imbalance but to the failure of proper fertilization and oocyte activation (Austin 1969).

2.2 Frequency of spontaneous triploidy

Triploidy is one of the most frequent spontaneous genome aberrations in embryos of different mammalian species, including humans. The frequency of triploidy in all human conceptuses has been estimated to be about

1 per cent (Boué and Boué 1970; Boué *et al.* 1976; Carr 1971a,b). Among aborted fetuses triploidy accounts for about 15–18 per cent of all abnormal cases (Carr 1972; Bochkov, Zakharov, and Ivanov 1984; Kuliev 1974; Lauritsen 1982).

The frequency of spontaneous triploidy in animals varies among species and depends on the genetic background of the particular strain (Tables 1–3), as is well illustrated in laboratory mice (Beatty 1957; Vickers 1969; Baranov 1976a,b; Martin-Deleon and Boice 1983). The 0.3 per cent incidence of triploidy in mice of A/Fa and C57BL/Fa strains is almost eight times less than the 2.5 per cent recorded in silver strain mice (Beatty and Fischberg 1949, 1951b; Fischberg and Beatty 1952), and still lower than the 4–5 per cent recorded for mice of C57BL or CBA-T6T6 strains (Kaufman 1973b; Gosden 1973). The incidence of triploidy in one-cell mouse embryos of ICR strain was about 1 per cent (Martin-Deleon and Boice 1983) and about 2 per cent in post-implantation embryos of PDE, C3H, A, and CBA strains (Vickers 1969; Baranov 1976a,b). Mating between strains usually gives a significantly higher proportion of triploidy than intrastrain matings (Beatty 1957).

The incidence of triploidy in laboratory rats does not exceed 1 per cent according to Butcher and Fugo (1967) and Mikamo and Hamaguchi (1975), but was found in about 4.7 per cent of all pre-implantation embryos by Dyban, Udalova, Akimova, and Chebotar (1971; see also Table 2).

The frequency of triploidy in rabbit blastocysts is very low. According to some early observations (Table 3) only one case of triploidy was revealed among 611 rabbit embryos before implantation (five to six days of gestation). Combined data of more than 1000 rabbit embryos of the same age yielded only six cases of triploidy (Beatty and Coulter 1978).

The frequency of triploidy in domestic pigs was found to approach the same level as in laboratory mice (McFeely 1967), but in more recent work by Dolch and Chrisman (1981) it was reported that 169 ten-day-old pig embryos had a normal diploid chromosome number. Thus the normal background of spontaneous triploidy in the pig, though probably rather low, still remains unknown.

2.3 Experimental production of triploidy

Table 9 summarizes the results of numerous publications on the experimental production of triploidy in different mammalian species.

Both digynic and dispermic triploidy can be induced experimentally by various techniques including colchicine, colcemid or cytochalasin B treatments, temperature shock, and artificial delay in fertilization. In mice, rabbits, rats, pigs, and hamsters the frequency of induced triploidy varies widely depending on the method applied and the stage at which the embryos are examined. Some inducing agents were found to be highly toxic for

Table 9. Experimental production of triploid embryos in mice, rats, rabbits, pigs, and hamsters in early experimental studies (see Niemierko and Opas 1978 for details of modern techniques)

Treatment	Time of treatment	Day of pregnancy	Number of triploid embryos (%)	Mechanism	References
Mice					
Local heating of oviducts (45+45.5°C, 5–10 min)	2.5 h after mating	3.5	11	digyny	Fischberg and Beatty 1952
Delayed fertilization	5–11 h after ovulation	3.5	2.9	digyny	Vickers 1969
Superovulation		3.5 6.5	32 24	digyny	Takagi 1971 Takagi 1971
Delayed fertilization + artificial insemination	12–13 h after ovulation	1	7.5	digyny diandry	Marston and Chang 1964
Colchicine intrauterine	Before mating	1 3.5	50 10	digyny	Edwards 1958b, 1954a, b
Colchicine i.p. + artificial insemination	0.5–3.5 h after ovulation	1	6.3–8	digyny	McGaughey and Chang 1969
Cytochalasin B	2.5–3.5 h after mating	1	50	digyny	Niemierko 1975
Rats					
Local heating of oviducts	2.5 h after mating	1	14	diandry	Austin and Braden 1954b
Total hyperthermy	During 3 h after mating	8–15	10	diandry	Piko and Bomsel-Helmreich, 1960
Delayed fertilization	6–10 h after ovulation	1	8.8	diandry	Austin and Braden 1953; Braden 1958

Treatment	Timing		%	Type	Reference
Delayed fertilization	10 hr after ovulation	1	7	diandry	Piko 1958
Colchicine i.p.	2 h after mating	1	90	digyny	Austin and Braden 1954b
Colchicine i.p. + delayed mating	1.5–3.5 h after fertilization	8–15	4.5	digyny diandry	Bomsel-Helmreich 1967
48 h delay in ovulation by pentobarbital	Fourth day of oestrus cycle	1	5.8	diandry	Mikamo and Hamaguchi 1975
Colchicine i.p.	2.5 h after mating	8–15	4.8	diandry	Piko and Bomsel-Helmreich 1960
Rabbits					
Delayed fertilization	8–10 h after ovulation	1	13	digyny diandry	Austin and Braden 1953
Delayed fertilization + hyperthermy	10 h after ovulation	10	10	diandry	Piko and Bomsel-Helmreich 1960
Delayed fertilization	10 h after ovulation	1	45	digyny	Thibault 1959
Colchicine	During fertilization	1	90–97	digyny	Bomsel-Helmreich 1965
		6	71	digyny	
Superovulation		6	2		Fujimoto *et al.* 1974
Domestic pigs					
Delayed fertilization	6–8 h after ovulation	17	3	—	Bomsel-Helmreich 1967
Delayed fertilization	6–20 h after ovulation	1–2	20	digyny	Thibault 1959
			11	diandry	
Hamster					
Delayed fertilization	5–6 h after ovulation	1	30	digyny	Chang and Fernandez-Caro 1958

ovulated eggs and caused premature death of triploids. One of these agents is colchicine, which yields a very high number of triploids but is very toxic to the ovulated eggs. Intraperitoneal injection of colchicine shortly after mating blocks second polar body extrusion in about 90 per cent of ovulated eggs, though a few are capable of further development (Austin and Braden 1954b). In laboratory mice treated with colchicine and colcemid, 19 per cent or 50 per cent respectively of all eggs had no second polar body but were otherwise apparently normal. Most of the eggs after colchicine pre-treatment were eliminated during cleavage, with only 10 per cent being recovered on the fourth day of gestation. Some other unfavourable side effects of colchicine pre-treatment include aneuploidy due to chromosome non-disjunction (Edwards 1958b, 1961) and asynchrony of pronuclear progression (McGaughey and Chang 1969).

Toxic effects of colchicine were to some extent less evident in rabbit embryos subjected to this treatment *in vitro*, then thoroughly rinsed in fresh culture medium, and transferred to the uteri of pseudopregnant does. This treatment resulted in a very high frequency (up to 97 per cent) of triploid embryos due to the suppression of the second polar body. The development of these triploids was followed up to advanced stages of organogenesis (Bomsel-Helmreich 1965). Because of their high toxicity and adverse effects on the mitotic spindle of ovulated eggs, both colchicine and colcemid are not in use now for the experimental production of triploidy.

One of the most useful tools for inducing polyploidy in mouse embryos is the cytochalasin B (CB) technique, which is also superior to all other methods employed so far in the experimental production of triploidy (Niemierko 1975). Recently ovulated eggs are released from the oviducts and cultured in drops of medium with CB (5 μg ml^{-1}). After 5–7 h in culture the eggs are checked under an inverted microscope for the presence of three pronuclei. Due to CB block of second polar body emission, 50 per cent of treated eggs revealed triploidy. For more details of the technique the reader is referred to Niemierko and Opas (1978). The usefulness of CB-technique for experimental production of triploidy in other mammalian species is probable but remains to be proved.

A different technique, recommended for the experimental production of triploidy in mice, is based on strong osmotic shock (Opas 1977). When fertilized eggs at the pronuclear stage are subjected to distilled water treatment for 2–6 min, incorporation of the second polar body into the egg ooplasm occurs in 30–60 per cent of the treated eggs, giving rise to digynic triploidy. The frequency of triploidy and other types of ploidy variations due to osmotic shock varies significantly, depending on duration of treatment, strain of mice, pronuclear progression, etc. (Niemierko and Opas 1978).

The efficiency of other experimental approaches to the artificial induction of triploidy is less reliable. Delayed fertilization, i.e. the ageing of gametes before fertilization, results in 40 per cent of triploidy according to some reports (Thibault 1959), but not according to others (Piko 1958; Vickers 1969).

Contradictory results on the incidence of triploidy are also obtained with hormonal treatments used for the induction of superovulation. According to numerous data, eggs obtained after artificial superovulation procedures cannot be distinguished in any way from those of spontaneously ovulated mice if an appropriate hormonal dosage is used (Edwards and Gates 1959). However, a very high incidence (up to 20 per cent) of digynic triploids in mice of A/He strain was obtained after the administration of pregnant mare's serum and human chorionic gonadotropin (Takagi 1971; Takagi and Sasaki 1975). However, superovulation has no significant effect on the incidence of triploidy in rabbits (Beatty and Coulter 1978).

Cytochalasin B treatment and to some extent osmotic shock may be considered as the most suitable procedures for the experimental production of triploidy, at least in laboratory mice.

2.4 Embryonic development in triploids

Most human triploids are eliminated by week 5–6 of gestation (Boué, Boué, Phillippe, and Guegen 1972), or sometimes a little later (Kulazenko 1976). No more than 1 per cent of all human triploids are recovered during the third trimester. In some rare cases, triploid embryos are born alive but they all die within a short time of birth (Niebuhr 1974). Unusually long survival of some triploids (up to two and even to five months of post-natal age) has also been recorded (Cassidy, Whitworth, Sanders, and Lorber 1977; Fryns, van de Kerckhove, Goddens, and van der Berghe 1978).

To our knowledge not a single case of full triploidy (i.e. non-mosaic) was ever reported for newborns of laboratory animals. They are usually eliminated during the major organogenesis stages of embryogenesis.

Let us consider the developmental effects of a triplicated genome in more detail.

2.4.1 Pre-implantation stages

No data are yet available on the developmental effects of triploidy in human pre-implantation embryos. Early manifestations of triploidy have been studied in mice, rats, and rabbits. Most of the experimental data obtained so far are based on a limited number of embryos and so should be considered as preliminary.

Since the classical works of Beatty and Fischberg (1951a,b), triploidy has been known to be compatible with egg cleavage and blastulation. However, four- to five-day triploid blastocysts in mice have an almost 30 per cent reduced cell count compared to normal diploids.

Table 10. Number of diploid, triploid, and tetraploid mouse embryos with different blastomere numbers, on the fourth day of gestation (Baranov 1976b)

Karyotype	Number of embryos studied							
	Total	Blastomere number						
		<14	15–19	20–24	25–32	33–42	43–58	>58
Diploidy $2n = 40$	178	4	12	14	67	48	16	17
Triploidy $3n = 60$	27	2	3	4	7	8	1	2
Tetraploidy $4n = 80$	4	—	4	—	—	—	—	—

Conspicuous reduction of cleavage rate in mouse triploids has also been reported by Edwards (1954a). These results, however, should be considered with some caution as they were obtained in experiments with colchicine treatment, which is known to induce chromosomal aberrations and is very toxic to ovulated eggs (see Section 2.3). Some reduction of cell count was also reported in triploids induced by cytochalasin B (Niemierko 1975). The slower rate of development of these triploids probably resulted from the stress due to *in vitro* culture, rather than from the unfavourable effect of cytochalasin B.

A slight retardation of pre-implantation development was also evident in triploid mouse embryos of spontaneous origin (Baranov 1976a,b, 1983a). Mean cell count in 28 spontaneous triploids recovered on the fourth day of development was only 24.5 ± 3.5 compared to 33.5 ± 1.3 in normal embryos with diploid karyotype. Some of these triploids were composed of almost the normal number of blastomeres while others (12) contained less than 21 cells (Table 10). Thus it looks as if some triploids have a quite normal pattern of pre-implantation development while others appear already to be retarded during cleavage and blastulation (Fig. 5).

Very similar results were obtained for spontaneous triploids of laboratory rats (Dyban *et al.* 1971). Some rat triploids were quite comparable to normal diploid embryos in their cleavage rate, while others had a slower rate of development. According to other data rat triploids do not develop beyond two to three blastomeres (Mikamo and Hamaguchi 1975). But the reduced survival rate of these triploids was probably caused by intrafollicular over-ripeness of the eggs rather than by the addition of an extra chromosome set.

Triploid rabbit blastocysts are smaller than diploid ones of comparable age and have reduced mitotic activity (Bomsel-Helmreich 1965, 1967; Fujimoto, Passantino, and Koenczoel 1975). These changes are at least partly reversible

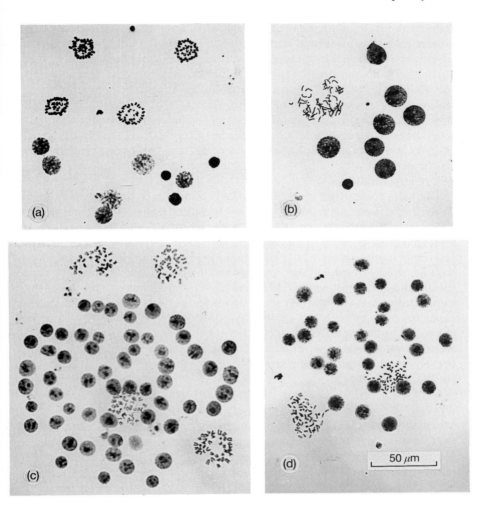

Fig. 5. Total preparations of mouse embryos on the third (a,b) and on the fourth (c,d) days of gestation. Giemsa stain; (a,c) diploid karyotype ($2n = 40$); (b,d) triploids ($3n = 60$).

since early post-implantation rabbit embryos with triploidy closely resemble normal diploid controls.

Contrary to other mammals, pre-implantation development of triploid embryos in Chinese hamsters was found to be quite normal and their viability seemed not to be impaired, at least initially. These results were obtained on so-called giant diploid oocytes which are rather frequent in this species (Funaki and Mikamo 1980; Funaki 1981). In spite of unusually large size, these triploids demonstrate high mitotic activity and their total cell count is greater than that of normal diploids of the same age (Funaki 1981). It was suggested that the nuclear–cytoplasmic ratio in these giant triploid eggs was particularly favourable for pre-implantation development (Funaki and Mikamo 1981). No information on post-implantation development of these triploids is yet available. Nonetheless, the cases of unusually prolonged survival of triploid embryos both in human beings and in laboratory animals might be most probably attributed to the fertilization of giant diploid oocytes (Funaki and Mikamo 1980), and an altered nuclear–cytoplasmic ratio might be suspected as a cause of the abnormal development of triploid embryos before implantation.

There are some direct indications in cell cultures as to the reduced mitotic activity and unusually long S phase of triploid cells (Goldefeder 1965; Lennartz, Schunilfeder, and Maurer 1966; Mittwoch and Delhanty 1972). Some temporal changes of mitotic activity might be expected to take place in pre-implantation stages of triploid embryos in mice and rats. The nuclei of triploid embryos spread and stained on air-dried preparations look very similar and are almost equal in size (Fig. 5), yet diploid mouse embryos of the same age contain two distinct kinds of nuclei, varying in sizes and staining capacity.

Special studies are needed to clarify the cell cycle peculiarities of triploid embryos before implantation. The invention of reliable and simple methods for experimental production of triploidy in laboratory mice makes these studies feasible.

2.4.2 Post-implantation stages

Much information concerning pathomorphological changes of implanted triploid embryos has been accumulated for different animal species as well as for human beings. More than 100 triploid mouse embryos have been studied post-implantation (Fischberg and Beatty 1951; Wroblewska 1971; Vickers 1969; Takagi 1971; Niemierko 1975; Baranov 1976a,b). Some detailed experimental work has dealt with post-implantation triploids in rabbits (Bomsel-Helmreich 1965, 1967), rats (Dyban, Udalova, Akimova, and Chetobar 1971; Piko and Bomsel-Helmreich 1960), and pigs (Thibault 1959; Bomsel-Helmreich 1967). About 450 spontaneous abortuses with triploidy have been described so far in humans (Niebuhr 1974; Beatty 1978; Lauritsen 1982). In spite of such abundance of experimental and clinical data, many questions concerning the developmental effects of triploidy in mammals are still lacking definite answers, for example the so-called 'triploidy syndrome' and the causes of death in triploids.

2.4.2.1 Specific and non-specific phenotypic effects of triploidy. General retardation of growth seems to be the main non-specific developmental

manifestation of triploidy in post-implantation mammalian embryos. In both rabbits (Bomsel-Helmreich 1965, 1967) and pigs (Bomsel-Helmreich 1967, 1976) general growth delay of all embryonic rudiments is known to be the single phenotypic effect of triploidy, and there is no clear-cut malformation that accounts for the death of these triploids.

Pathomorphological changes vary significantly in human triploids. Some of them are even carried to term with multiple minor abnormalities and usually low birth weight. Other human triploids are represented by empty chorionic sacs with or without nodular rests or cylindrical amorphous remnants of embryo inside (Niebuhr 1974; Kulazenko, Kuliev, and Grinberg 1975). The absence of any specific pathomorphological changes was postulated for human triploids in some earlier works (Carr 1969, 1971b). However, a detailed pathomorphological inspection of numerous human triploids seems to substantiate the existence of a 'triploidy syndrome' most clear-cut during fetal development. It includes hydatidiform degeneration of the chorionic villi and syndactyly (hands: digits III–IV, feet: toes I–II or III–IV), combined with rather specific congenital malformations of the nervous, genital, and urinary systems (Kuliev 1974; Kulazenko *et al.* 1975).

A specific pattern of pathomorphological changes has also been described in laboratory mice. Initially it was discovered in CBA strain mice (Wroblewska 1971; Baranov 1976a) and later confirmed in A strain mice (Niemierko 1981). Development in mouse triploids is usually normal up to the seventh day and most form typical early egg cylinders. Characteristic disturbances in embryogenesis of triploids are already evident from the eighth to ninth days of development. At this stage mouse embryos with normal diploid karyotype are composed of three primary germ layers (see Fig. 6), with well-developed cranial folds protruding deep inside the amniotic cavity and large paired heart primordia. The mesoderm is already differentiating into the axial part of the future somites, and into somato- and splanchnopleura which form the two layered amniotic and visceral yolk sac walls, respectively. Triploid embryos of the same age are composed of all three primary germ layers, which, however, are much less differentiated and very scanty. They are formed by a few layers of round cells with large pale nuclei and light basophilic cytoplasm. The borders of the primary germ layers are not sharply outlined. Only the cranial part of these embryos, in the form of paired headfold primordia, is visible by external inspection. No sign of typical chorda or heart primordia can be detected in these triploids. The yolk sac of tenth day mouse triploids is only half normal size, and is pale, lacking the typical net of omphalic vessels. The embryonic part of the yolk sac is significantly shortened, while a disproportionately long allantoic rudiment grows upward towards the base of the chorionic plate. Inside the yolk sac there is a very small amniotic vesicle with a very thick, non-transparent wall

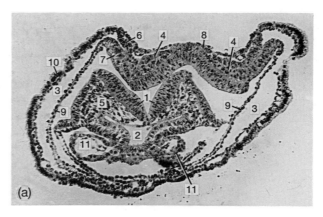

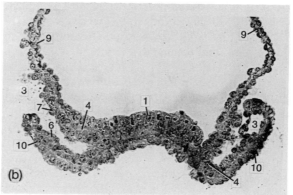

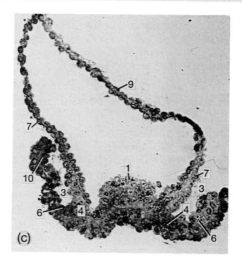

(Fig. 7). The headfolds are flattened and often look asymmetrical. The rear part of the embryo looks disorganized without any traces of the axial elements or somites (Fig. 7).

Thus triploidy in laboratory mice of some strains may induce rather characteristic disturbances of embryogenesis, mainly due to specific delay in the growth of the embryonic part of the egg cylinder. They include a substantial reduction of the neural plate, flattening of the headfolds, disproportionate growth of the allantoic stalk, absence of axial elements, and reduction the of amnion.

The triploid syndrome depends on genetic background and is not typical for all strains of mice. A mouse triploid with advanced organogenesis is shown in Fig. 8. Though slightly retarded in growth, this particular embryo possesses well formed head parts, heart primordia, neural tube, and a few somite pairs. Among spontaneous triploids derived from 16 genetic combinations the most advanced organogenesis and highest viability was achieved by triploids of 129/Sv strain and its crosses (Wroblewska 1978).

Developmental effects of triploidy in rats are found also to depend on genetic background. According to some studies rat triploids eliminated during major organogenesis do not reveal any pathomorphological changes of a specific nature (Piko and Bomsel-Helmreich 1960). On the other hand other studies favour the view of specific growth retardation in the embryonic part of the egg cylinder, with resistance of fetal membranes to this genome anomaly (Dyban et al. 1971).

Thus, both specific and non-specific pathomorphological changes are encountered in mammalian embryos with triploid karyotype. Developmental capacities and malformation patterns of triploid embryos depend on their genetic background.

It was demonstrated in fibroblasts cultured in vitro that triploidy is not a true cell-lethal mutation (Kuliev, Grinberg, Vasileiski, Stepanova, and Urovskaya 1972; Kucharenko, Kuliev, Grinberg, and Teskich 1975). However, there is good evidence that triploid cells require a longer time to double their DNA content than diploid cells in humans (Mittwoch and Delhanty 1972; Cure, Boué, and Boué 1973; Paton, Silver, and Allison 1974) and in rabbit embryos (Bomsel-Helmreich 1976). The survival interval of cultured triploid cells is also significantly diminished (Bomsel-Helmreich 1976). It was suggested therefore that the disorganized growth and limited lifespan of triploid embryos reflects abnormalities in their cell properties

---

Fig. 6. Transverse histological sections of nine-day mouse embryos stained in Mayer's haemalum and eosin. (a) Diploid karyotype; (b,c) triploids. 1 = nerve plate, 2 = foregut, 3 = coelom, 4 = mesoderm, 5 = mesenchyme, 6 = splanchnopleure, 7 = somatopleure, 8 = notochord, 9 = amnion, 10 = yolk-sac endoderm, 11 = heart.

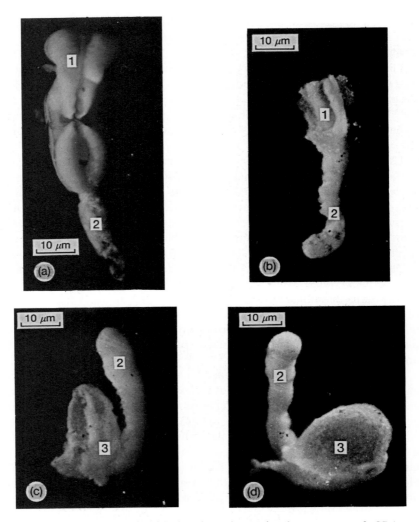

Fig. 7. Control (a) and triploid (b,c,d) embryos in the progeny of CBA♀ × CBA-T6/CBA-T6 ♂ mice. Ninth day of pregnancy. 1 = neural folds, 2 = allantois, 3 = amnion.

(e.g. changes in their mitotic cycle) which affect cellular proliferation, differentiation, and interactions (Bomsel-Helmreich 1976).

Another explanation of fetal death due to triploidy is based on possible impairment of maternal–fetal relationships. The survival rate of a homogeneous group of triploid embryos transferred into the uterus of a pseudopregnant rabbit is significantly less than that of triploid embryos transplanted

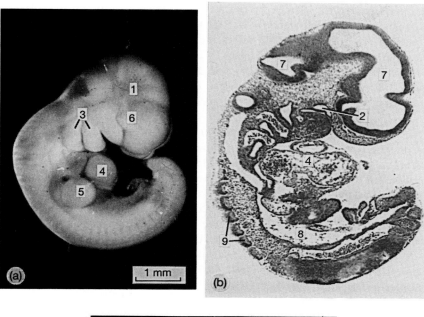

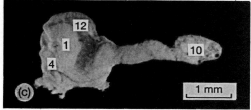

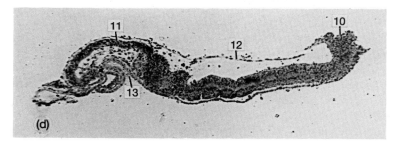

Fig. 8. Triploidy in the progeny of mice CBA/Lac ♀ × ♂ Rb1Iem/Rb1Iem. Ninth day of pregnancy. Bouin's fluid fixation. (a) Control, $2n = 40$; (c) triploid, $3n$.

Sagittal sections of control embryo (b), and triploid (d). 1 = head, 2 = oral cavity, 3 = mandibular and hyoid arches, 4 = heart, 5 = forelimbs, 6 = optic vesicle, 7 = myelocoel, 8 = vitelline vein, 9 = somites, 10 = allantois, 11 = nerve plate, 12 = amnion, 13 = foregut.

simultaneously with normal diploid fetuses (Bomsel-Helmreich 1965, 1967). This suggests a higher vulnerability of triploid embryos to the changes of hormonal level (Dyban 1972) and some deficiency in the endocrinological feto-maternal relationships (Bomsel-Helmreich 1976).

Thus the mechanisms of embryonic death and of congenital malformations induced by triploidy are rather complex and involve pathological changes at both the cellular and supracellular levels.

2.4.2.2 Causes of death of triploid embryos. The death of triploid embryos occurs at any stage of development from implantation up to term. The actual causes of mortality are still obscure. Many suggestions have been made to explain this most puzzling topic of developmental cytogenetics, though none of them is adequately substantiated as yet.

According to some general considerations, the death of triploids is mainly due to impaired balance of regulatory and structural genes (Prokofieva-Belgovskaya, Grinberg, Revazov, Mikelsaar, and Kuliev 1973).

As already mentioned above the survival of triploid embryos correlates in some unknown way with the genetic background of the parents, being better in triploids from interstrain crosses than from intrastrain ones (Fischberg and Beatty 1951; Wroblewska 1971, 1978; Baranov 1976a).

One other plausible explanation relates survival diversity of triploids to the altered ratio of gene products of autosomes and gonosomes (Edwards *et al.* 1967). This hypothesis postulates that severity of pathological changes is correlated with the gonosome set of triploids which might be XXX, XXY, or XYY. There are several lines of circumstantial evidence indicating exceptionally low viability of triploids, with only one active X chromosome (XYY) compared to XXX or XXY groups in humans (Mikamo 1971; Niebuhr 1974; Jacobs, Angell, Buchman, Hassold, Matsuyama, and Manuel 1978; Beatty 1978). XYY triploids in humans are usually represented by empty embryonic sacs with no traces of embryonic remnants (Kuliev 1974; Kulazenko *et al.* 1975). In man a maximum likelihood analysis indicates that XXX and XXY triploids have viability of comparable magnitude (Beatty 1978). The same correlation is probably true for triploidy in laboratory mice (Wroblewska 1971; Baranov 1976a). However, cumulative data on survival capacities of rabbit triploids suggest a higher viability of XXY triploids (Beatty and Coulter 1978).

Because of the rarity of diandry, almost nothing is known on the viability of XYY triploids in mice or rats. Nonetheless, it seems likely that the gonosome–autosome ratio, as well as the type of X-chromosome inactivation, have a significant influence on the embryonic development and survival capacities of XXX and XXY triploids. It has recently been found in digynic mouse triploids of XXX- and XXY-gonosome constitution that cells from the embryonic region tend to have only one active X chromosome, whereas those from extra-embryonic membranes have two active X chromosomes (Endo, Takagi, and Sasaki 1982). The presence of only one active X chromosome in the embryonic region correlates well with the severe growth retardation in the embryonic part of the egg cylinder.

Other facts in favour of Edward's hypothesis come from biochemical studies of triploid cells growing in culture (Bomsel-Helmreich 1976). Total protein content determined by the Lowry method was found to be very similar in $3n$ and $2n$ populations of rabbit cells. Five enzymes controlled by autosomal genes showed no difference between triploid and diploid cells, while one autosomal enzyme (galacturidyl transferase) had a lower activity in triploids. However the relative activity of X-linked enzymes was almost twice that of those determined by autosomal genes. These results suggest abnormal X-chromosome inactivation in triploid cells of rabbits. Observations on nuclear sex show that XXY triploids usually lack a Barr body, while XXX triploids have one and rarely two Barr bodies. Thus the data on X-chromosome inactivation in rabbit embryos are significantly different from those in mice. Nonetheless, all rabbit triploids are eliminated by the sixteenth day of development.

Disturbances of the autosome–gonosome ratio, enhanced by peculiarities of the X-chromosome inactivation mechanism, are important but probably not the only cause of early embryonic developmental failure induced by triploidy.

2.5 Conclusions

Spontaneous triploidy of digynic or diandric type is a common chromosomal aberration in all mammalian species studied cytogenetically. Various techniques for experimental production of triploid embryos in laboratory animals are suggested. Triploidy is fully compatible with pre-implantation development and may not reveal adverse effects during cleavage or blastulation, but the rate of cleavage may be markedly reduced.

Triploid embryos may successfully implant and proceed through gastrulation, neurulation, and organogenesis. Triploidy by itself may not be classified as a true cell lethal since it does not block cyto- and histodifferentiation either *in vivo* or *in vitro*. Nonetheless, triploid cells are distinguished by reduced mitotic activity and low levels of some enzymes, primarily those encoded by X chromosome genes.

At post-implantation stages triploid embryos show signs of general growth retardation as well as a rather specific pattern of malformation—the triploidy syndrome of humans and mice. This may depend on the genetic background of the affected embryo.

The mortality of triploid embryos is extended throughout the post-implantation stages of development, depending on genetic background, gonosome content, and species.

3

Tetraploidy

3.1 Mechanisms of origin

All theoretically possible mechanisms of spontaneous or induced tetraploidy have been covered elsewhere (Austin 1969), yet the role played by each feasible route depicted in Fig. 9 in the production of tetraploidy remains mostly unknown.

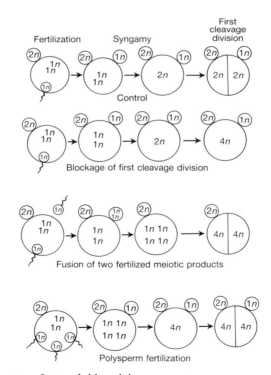

Fig. 9. Basic routes of tetraploidy origin.

3.1.1 Anomalies of gamete maturation and fertilization

A tetraploid zygote may result from a digynic ovum fertilized by two normal haploid sperms or by one sperm with a non-reduced diploid chromosome set. The former mechanism seems to be more plausible as the probability of two abnormal gametes meeting at fertilization must be extremely low. One other route for the origin of tetraploidy is concomitant fertilization of both meiotic products, that is the ovum and its second polar body, with consequent fusion of the two resulting cells. This route is favoured by some genetic data on spontaneous mouse chimaeras, the genetic constitution of which might be attributed to double fertilization of both the ovum and its second polar body (Russell and Woodiel 1966). Fusion of an already extruded polar body with an egg has been recorded (Austin 1969). Polyspermic fertilization of an ovulated ovum by three different haploid spermatozoa should be considered as a third possible route, though it has never been demonstrated.

3.1.2 Disturbances of cleavage

The initially haploid paternal and maternal pronuclei in mammals have to double their DNA content, and the zygote containing a tetraploid amount of DNA then undergoes the first cleavage division, with two diploid blastomeres formed. The complete block of the first cleavage division results in a tetraploid zygote which, if it proceeds to the next cycle, gives rise to two tetraploid blastomeres. This mechanism is responsible for the experimental tetraploidy induced by cytochalasin B (CB). Tetraploidy might also result from the fusion of already separated blastomeres at the two-cell stage, resulting in a synkaryon with a tetraploid number of chromosomes. Blastomere fusion may be achieved using inactivated Sendai virus (Graham 1971a), or by the treatment of blastomeres with special fusogenic substances such as polyethylene glycol (Eglitis 1980). Irrespective of the mechanism, the resulting tetraploid zygote gives rise to a tetraploid embryo if its mitotic spindle is undamaged, and provides a regular chromosomal distribution between the daughter cells. Tetraploid zygotes undergoing multipolar mitoses usually give rise to blastomeres with varying unstable chromosome numbers (haploidy, diploidy, aneuploidy, etc.; Schwarzacher 1976).

3.2 Frequency of spontaneous tetraploidy

Spontaneous tetraploidy may be a rather rare event in mammalian embryogenesis. Only 57 (6.2 per cent) tetraploids were recorded among 921 human abortuses with abnormal karyotype (Boué and Boué 1974; Boué et al. 1975, 1976). Spontaneous tetraploidy is even more exotic in other mammals studied cytogenetically at much earlier stages of embryogenesis than in humans; namely, before implantation. According to Beatty and Fischberg (1951a, 1952), a tetraploid karyotype was recorded in five of 3671 (0.1 per cent) mouse embryos studied before implantation. A very similar incidence of spontaneous tetraploidy was found in our experiments (Baranov 1976a,b, 1983a). Of 2168 eggs studied cytogenetically as late morulae or early blastocysts, seven were tetraploid (0.3 per cent). A low level of spontaneous tetraploidy was observed in both laboratory rats and rabbits (Tables 3 and 4). A single tetraploid embryo was recorded among 259 rat blastocysts karyotyped on the fifth day of gestation (0.4 per cent; Dyban *et al*. 1971) and two tetraploids were found among 611 (0.3 per cent) 6-day rabbit blastocysts (Tables 2 and 3).

An unusually high frequency of tetraploidy (3.4 per cent) was recorded in cytogenetic studies of 88 pig blastocysts (McFeely 1967), but not a single case of tetraploidy was revealed among 169 ten-day-old embryos collected from prepubertal gilts with hormonally induced ovulation (Dolch and Chrisman 1981). The causes of these discrepancies remain obscure.

Spontaneous tetraploidy seems to be more often encountered in humans than in other mammals.

3.3 Experimental production of tetraploidy

Interference with the first cleavage division is the principal method of experimental production of tetraploidy in laboratory mammals. Various experimental procedures may be used, including heat shock of fertilized eggs near the time of the first cleavage division (Pincus and Waddington 1939; Beatty and Fischberg 1952) and colchicine treatment of zygotes (Pincus and Waddington 1939; Edwards 1954a). Both techniques are hard to reproduce and colchicine treatment is very toxic for mammalian eggs and impairs their viability (McGaughey and Chang 1969).

Two main routes for producing tetraploidy in laboratory animals are now routinely in use. These are:

(a) cytochalasin B treatment of fertilized eggs,

(b) experimental fusion of blastomeres.

The former route is generally accepted as more reliable and less complicated, and is recommended for the experimental production of tetraploidy in mice (Niemierko and Opas 1978).

3.3.1 Cytochalasin B treatment

This efficient technique is applicable for both spontaneously ovulated eggs and for eggs from induced ovulation (Snow 1973; Tarkowski, Witkowska, and Opas 1977). It yields between 40 and 75 per cent tetraploid eggs, and about half of them develop into blastocysts. Mouse eggs obtained 38–40 h after fertilization (two-cell stage) or 47–49 h post-HCG treatment are cultured in cytochalasin B supplemented with Whitten's medium (10 μg ml^{-1}). Afterwards they are washed in fresh medium, cultured up to the blastocyst stage, and then surgically transferred to the uteri of pseudopregnant females. For more technical details of the method the reader is referred to the review by Niemierko and Opas (1978).

3.3.2 Fusion of blastomeres

The method was originally invented by Graham (1971a) for two-cell mouse embryos. Initially the zona pellucida is removed and the pairs of naked blastomeres are treated with inactivated Sendai virus suspension in Whitten's medium. The treatment provokes cytoplasmic and subsequent nuclear fusion and as a result giant tetraploid eggs are formed, with a few of them developing into blastocysts if cultured *in vitro*. The tetraploidy of these embryos was proved on air-dried preparations as well as by means of integrating microdensitometer analysis of Feulgen-stained nuclei (see Graham 1971 for more technical details).

Because Sendai virus easily damages the blastomeres, this technique has found limited application so far. In more recent experiments, the fusion of blastomeres is achieved with polyethylene glycol (PEG) (Eglitis 1980). Briefly, the pairs of previously disaggregated four-cell stage embryos are initially aggregated in pairs with phytohaemagglutinin (PHA) and afterwards treated for 2 min with 45 per cent w/v PEG for inducing cell fusion (Eglitis and Wiley 1981).

3.3.3 Microsurgical technique

The method is based on microsurgical transplantation of normal diploid nuclei into fertilized mouse eggs (Modlinski 1978, 1981). The nuclei from 12–16-cell mouse embryos are microsurgically injected into fertilized eggs, so that the nuclear material of the donor and recipient cells are combined to form a single tetraploid genome, which in some cases allows the development of these chimeric embryos into morphologically normal blastocysts. The method, however, is more applicable for studying the developmental capacities of nuclei isolated from different parts of blastocysts—the inner cell mass (ICM) or trophectoderm (TE)—than for routine experimental production of tetraploidy.

3.4 Embryonic development in tetraploids

Tetraploid products of conception in humans are eliminated during the initial stages of embryogenesis. The aborted tetraploids are usually represented by empty chorionic sacs without any traces of embryonic rudiments inside (Carr 1969, 1971b, 1972). Tetraploid embryos die much earlier than human conceptuses with other types of chromosomal abnormality.

Experimental data on the developmental effects of tetraploidy are very limited and are almost exclusively confined to the laboratory mouse.

Relevant information for other mammalian species is either absent or very meagre.

3.4.1 Pre-implantation stages

The cleavage rate of all early mouse tetraploids studied so far is substantially reduced. On the fourth day of gestation they have only half the blastomere number of their diploid littermates (Beatty and Fischberg 1951b). The same correlation of blastomere counts in diploid and tetraploid embryos was also found in our experiments (Baranov 1976b, 1983a). On the fourth day of gestation tetraploids had blastomere numbers below 16 compared to 33.0 ± 1.3 blastomeres in diploid embryos of the same age (Table 10). The nuclei of tetraploid cells were somewhat enlarged and stained lightly.

Conspicuous cleavage delay was also found in tetraploid embryos of rats (Dyban *et al.* 1971), golden hamsters (Yamamoto and Ingalls 1972), and rabbits (Shaver and Carr 1969).

Thus, tetraploidy affects the cleavage rate though which cell cycle stage is sensitive to the excessive genetic load is not known at present.

Development of experimentally induced tetraploids varies substantially, depending on the methods applied. Cytochalasin B (CB)-induced tetraploids on the fifth day of *in vitro* growth had a mean cell count of only 20.4, whereas in control diploid embryos of the same age the count was 54. Sixty-seven per cent of all tetraploids failed to produce normal blastocysts and formed so-called 'empty trophoblastic vesicles', lacking the ICM typical for the blastocyst stage. Nonetheless, about 40 per cent of two-cell tetraploids produced blastocysts with the characteristic knot of ICM inside (Snow 1973, 1975). Trophoblastic vesicles are eliminated if transferred to the uteri of pseudopregnant foster mothers, but advanced tetraploid blastocysts usually invoked normal decidual reactions, successfully implanted, and were capable of further development (Snow 1975, 1976).

The normal transition from morula to blastocyst stage is known to correlate with cell number (Tarkowski 1971; Gardner 1972, 1974, 1976; Herbert and Graham 1974; Gardner and Papaioannou 1975). Snow (1975, 1976) suggested that only tetraploids attaining 22–25 cells before implantation are able to form an ICM and successfully transform into normal blastocysts. The reduction of cell number below the critical count of 22–25 impairs cavitation and differentiation of the ICM, and thus predisposes these tetraploids to form empty trophoblastic vesicles, destined to die during implantation or soon after.

Strains of mice are known to differ markedly in their cleavage rate and this trait is genetically determined (Molinaro, Siracusa, and Monesi 1969; Bowman and McLaren 1970b; Molls, Zamboglon, and Streffer 1983). According to Snow's hypothesis the tetraploid embryos with enhanced cleavage rate have more chance to produce viable blastocysts with well-developed ICMs than mouse embryos with genetically lower cleavage rates. Strain differences in cleavage rate might therefore be responsible for variations in tetraploid survival in mice. The early pre-implantation death of all spontaneous tetraploids might be due to the lower cleavage rate of CBA/lac mice used in our experiments compared to mice of Q strain used by Snow.

One more reason for the developmental failure demonstrated by tetraploid mouse embryos stems from the recent experiments of Eglitis and Wiley (1981). According to their data the activity of some enzymes as well as RNA synthesis in PEG-induced tetraploid embryos did not increase in parallel with the doubling of the genome. This partial gene dosage compensation was most strongly pronounced in tetraploid embryos with a cell number equivalent to intact diploid embryos, while the tetraploids with a highly reduced cell count increased their enzyme activity in parallel with the increase in gene dosage.

Thus, the intriguing possibility remains that viability and survival capacities of mouse tetraploids correlate with their ability to compensate for at least part of their excess genetic material.

The causes for discrepancies of gene dosage compensation in different tetraploids are completely unknown. It should be recalled, however, that a special mechanism of hidden breaks is known to operate in phylogenetically tetraploid fish for dosage compensation of ribosomal genes. Thus regulation of gene activity at a post-translational level is plausible for tetraploid cells of eukaryotes (Leipoldt and Engel 1983).

3.4.2 Post-implantation stages

A small proportion (about 17 per cent) of experimentally produced tetraploid blastocysts in mice are capable of post-implantation development. Thus, a major part of all tetraploids are eliminated before and during implantation.

Post-implantation development of mouse tetraploids has been predominantly studied after cytochalasin B treatment (Snow 1975, 1976). About 60 per cent of tetraploid blastocysts initiated implantation, but only a few were able to develop further. About 6 per cent of all tetraploids developed advanced embryos. The latter resembled normal embryos if inspected before the fourteenth day of gestation but revealed substantial pathomorphological changes thereafter in the embryo itself but not in its membranes or placenta. The mean weights of tetraploid embryos on the fifteenth and sixteenth days were 192 and 407 mg, respectively and the weights of their allantoic placentae were 145 and 210 mg, respectively. The corresponding figures for normal diploid embryos were 491 and 805 mg for embryos and 180–198 mg for placentae. Thus tetraploid embryos weighed significantly less than diploid embryos of a corresponding stage.

The other manifestation of tetraploidy in post-implantation mice involved large blood islands and local necroses in many internal organs. The blood cells were unusually large, and this seemed to be responsible for the breakage of capillary walls and severe haemorrhages. Other histological changes included the reduction of germ cell number in gonads, severe reduction or absence of optic nerves, and abnormalities of brain and kidneys.

Taking into account the low body weight of tetraploids and the increased volume of individual tetraploid cells, tetraploid embryos should contain only about one-quarter as many cells as normal diploids of the same stage (Snow 1975). Some tetraploid fetuses were born alive but all died (Snow 1975).

Out of more than 3000 early post-implantation embryos studied cytogenetically in our experiments, just a single case of tetraploidy was registered (Baranov 1976c). This unique tetraploid was recovered on the ninth day in a CBA mouse: it consisted of a small white trophoblastic vesicle containing no traces of embryonic tissue.

It seems that we cannot distinguish any definite pattern of abnormality due to genome tetraploidization. Pathomorphological changes of tetraploid embryos are not specific, as they are encountered in embryos with other types of chromosomal abnormality. These non-specific changes, found on histological inspection in almost all embryonic rudiments, were probably due to reduced cell number on the one hand and abnormally large cell volume on the other.

Inefficiency of the tetraploid genome in supporting normal development of embryonic cells was recently shown in experimental chimaeras between diploid and tetraploid mouse embryos (Lu and Markert 1980). Two aggregated diploid–tetraploid chimaeras were found to be alive after birth, with one carrying about 10 per cent tetraploid cells and the other about 50 per cent. Both mice looked abnormal. The former was stunted while the latter, besides general retardation, revealed severe neurological abnormalities.

Thus, tetraploid cells cannot participate fully in embryonic development even if they are surrounded by normal diploid cells. These data substantiate some subcellular lesions in tetraploid embryos that were actually proved by direct biochemical studies (see Section 3.4.1).

3.5 Conclusions

Tetraploidy is one of the rare spontaneous chromosomal aberrations in mammalian species. Tetraploid embryos in mice may be successfully produced experimentally by cytochalasin B treatment of recently fertilized eggs or by means of blastomere fusion induced by inactivated Sendai virus or by polyethylene glycol (PEG). The basic information on developmental effects of tetraploidy was obtained in mouse embryos after cytochalasin B treatment.

The initial manifestation of tetraploidy during pre-implantation development is a reduced cleavage rate. Tetraploidy is not a cell lethal and is compatible with pre-implantation development. Formation of the blastocyst is the critical event in the development of early tetraploid embryos. Tetraploid embryos which achieve a sufficient number of blastomeres during cleavage are able to form normal blastocysts and successfully implant. Tetraploid embryos with a cell count reduced below the critical level become empty trophoblastic vesicles and are eliminated. Partial gene dosage compensation mechanism is operative in advanced tetraploid embryos at the blastocyst stage.

Post-implantation development of at least some tetraploids may proceed up to term but they reveal severe developmental abnormalities, which may be the result of reduced cell number, low mitotic activity, and abnormally large cell volume.

Partial tetraploidy in chimaeric mice is compatible in some cases at least with post-natal development but may cause abnormalities of growth and neurogenesis.

Further studies are needed in mice of different strains as well as on other mammalian species to clarify the reasons of developmental failure induced by tetraploidy.

4

Mechanisms of origin of numerical and structural aberrations of individual chromosomes in mammalian embryos

Chromosome aberrations in mammalian embryos may arise through non-disjunction of meiotic chromosomes during spermatogenesis or oogenesis and through chromosome loss or structural damage induced during gametogenesis or initial development.

4.1 Non-disjunction of meiotic chromosomes

Non-segregations of meiotic chromosomes are usually classified by their origin as primary, secondary, or tertiary: that is, occurring in individuals with otherwise normal karyotype, ones with already inherited trisomy, or in chromosome translocation carriers, respectively.

In the former case, i.e. primary non-disjunction, the regular segregation of meiotic chromosomes is disturbed, giving rise to abnormal gametes with disomy or nullisomy for a particular chromosome.

Secondary non-disjunction is due to the presence of an extra chromosome forming a univalent during Metaphase I (M-I). Half of the haploid meiotic products carry the extra chromosome in their genome.

Tertiary non-disjunction results from the abnormal segregation of the chromosomes participating in a complex meiotic configuration (quadrivalent or trivalent) formed at pachytene by the rearranged (translocated) chromosomes and their non-translocated homologues. Translocation heterozygotes usually give rise to both balanced and unbalanced gametes, carrying duplications and deficiencies for chromosomal segments involved in rearrangements in reciprocal translocation heterozygotes, and imbalance for the whole chromosomes involved in centric fusions in Robertsonian translocation heterozygotes.

The pattern of chromosome segregation in Robertsonian translocation heterozygotes is schematically depicted in Fig. 10. Of three feasible ways of chromosome segregation from a trivalent configuration in anaphase II, the

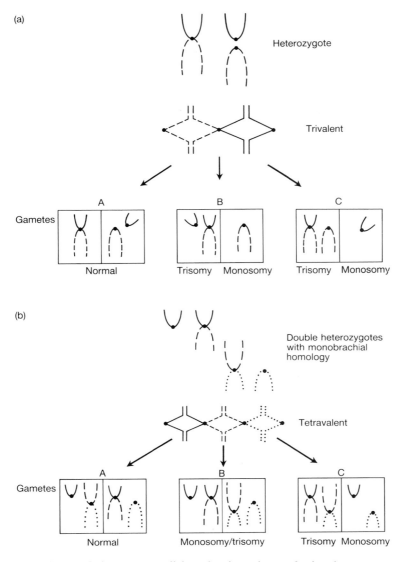

Fig. 10. Scheme of chromosome disjunction in meioses of mice, heterozygous for one Robertsonian translocation (a) and for two Robertsonian translocations with monobrachial homology (b). Modes of disjunction: A = alternate; B = type I; C = type II.

alternate type usually predominates. It yields an almost equal proportion of genetically balanced gametes, with about half of them carrying a centric fusion product and the rest without rearranged chromosomes. Adjacent-1 (homologous centromeres disjoining in M-I) or adjacent-2 segregations (homologous centromeres proceeding to the same pole in M-I) generate gametes with a deficiency or excess of the acrocentric chromosomes involved in the centric fusion (Prokofieva-Belgovskaya 1969; Nusbacher and Hirschhorn 1968).

In the doubly Robertsonian heterozygotes having one chromosome arm in common in both centric fusions, a complex multivalent configuration is formed (Fig. 10 (b)). It substantially potentiates the rate of non-disjunction, with the chromosome in common being most often involved in the malsegregation. This double metacentric technique originally invented in mice heterozygous for Rb(5.19)1Wh and Rb(9.19)163H (White, Tjio, Water, and Crandell 1974a,b—chromosome 19 is common to both) was found to be very convenient for the deliberate production of aneuploidy in mouse embryos (Gropp, Kolbus, and Giers 1975; Gropp, Putz, and Zimmerman 1976; White et al. 1974a,b; Epstein and Trans 1979; Baranov 1983a).

The segregation pattern of rearranged chromosomes in reciprocal translocation heterozygotes is schematically depicted in Fig. 11. The unequal reciprocal translocation T(14;15)6Ca is taken as an example. The alternate disjunction yields two main types of genetically balanced gametes, while both adjacent-1 and adjacent-2 disjunctions may result in an additional series of six different meiotic products with imbalance of genetic segments involved in the rearrangement. Duplications and deficiencies for the proximal segments of translocated chromosomes can occur as a consequence of adjacent-2 disjunction, which is much rarer than normal (alternate) disjunction in mice with reciprocal translocations (Searle, Ford, and Beechey 1971b). The chain multivalent found in T6-heterozygotes often breaks into a trivalent and a tiny univalent formed by the very small T6-marker chromosome (Baranov and Dyban 1968; Ford 1969). The latter can proceed to either meiotic pole, giving rise to aneuploid gametes (Fig. 11, types 7,8). If these gametes take part in fertilization, they can transmit partial tertiary trisomy or monosomy for the genetic material carried in the T6-fragment.

4.1.1 Characteristics of non-disjunction of sex chromosomes

Abnormal behaviour at meiosis is very similar for both autosomes and gonosomes, though the latter more often participate in complex malsegregations (consequent, double, etc.) at meiotic divisions, giving rise to zygotes with an exceptionally high imbalance of gonosomes (Davidenkova, Verlinskaya, and Tysiachnuk 1973).

Another peculiarity of gonosome non-disjunction concerns non-random segregation of the X-chromosome in oocytes. It has been repeatedly found

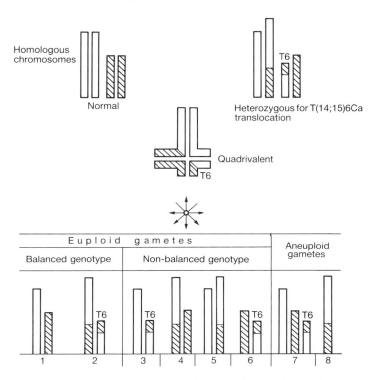

Fig. 11. Scheme of chromosome disjunction in meioses of mice, heterozygous for unequal reciprocal translocation T(14;15)6Ca.

1, Normal chromosomes. 2, Translocated chromosomes. 3–6, Deletions and duplications (see Chapter 9 for details). 7, Tertiary trisomy for T6. 8, Tertiary monosomy.

that chromosome sets without an X-chromosome or with an extra X-chromosome are more likely than normal sets to be eliminated in the first or second polar body (Cattanach 1962; Kaufman 1972; Luthardt 1976). Whether such a mechanism for prezygotic selection is operative in other mammalian species remains obscure.

4.1.2 Frequency of spontaneous anomalies of meiosis

Numerous chromosome aberrations in humans and other mammals take their origin from meiotic non-disjunction and to a somewhat lesser extent from mitotic faults during early cleavage. This abundant and heterogeneous group should be considered as an outcome of primary non-disjunction (see Section 5.1).

In women with supernumerary X-chromosomes (triplo-X syndrome) or with Down's syndrome (extra chromosome 21), meiosis can be successfully completed and in some cases these patients give birth to children with an abnormal karyotype (Grinberg 1969; Jagiello 1981), that is with a secondary trisomy. Only 27 pregnancies of women with Down's syndrome have been documented so far (Jagiello 1981). Thus secondary meiotic non-disjunction is also relevant to human beings, though its input to the total incidence of chromosomal aneuploidy in humans is very low.

The secondary type of aneuploidy (more precisely partial trisomy or monosomy) is not so rare in laboratory mice (Lyon and Meredith 1966; de Boer 1973; Gregorova, Baranov, and Forejt 1981; Beechey, Kirk, and Searle 1980). No proven cases of secondary aneuploidy have been recorded for other laboratory mammals as yet.

Tertiary non-disjunction of meiotic chromosomes is a widespread phenomenon in the animal kingdom, and has been most thoroughly studied in the laboratory mouse. Hundreds of Robertsonian and non-Robertsonian (reciprocal) translocations are known and maintained in special stocks of mice, which are widely used for detailed analysis of abnormal segregation of rearranged chromosomes in translocation heterozygotes (Gropp and Winking 1981; Baranov 1980; de Boer 1974). Tertiary non-disjunction is an obligatory consequence of translocation heterozygosity in all other mammals, including pigs (Popescu and Boscher 1982; Bruere et al. 1981), sheep (Bruere, Scott, and Henderson 1981), cattle (King, Linares, and Gastarsson 1981), and humans (Carr 1969; Hamerton 1968, 1970; Benirschke 1974).

The relative frequency of meiotic abnormalities can be assessed by chromosomal studies of meiotic plates during the first or second maturation divisions, as well as by karyotype analysis of cleaving embryos. Only a few mammalian species have so far been studied from this point of view (Nijhoff and de Boer 1981). The basic information on spontaneous meiotic non-disjunction has been obtained in laboratory mice. The incidence of non-disjunction at meiotic stages was somewhat higher in female than in male gametes. Not a single aneuploid spermatocyte I or II was recovered in non-inbred mice (Ohno, Kaplan, and Kinosita 1969) and this data was confirmed in later studies (Szemere and Chandley 1975). Out of 207 spermatocytes II in NMRI strain only one hyperhaploid cell has been recorded (Tettenborn and Gropp 1970). The mean incidence of spontaneous non-disjunction estimated from 5200 M-II spermatocytes counts in mice of 13 strains was only 0.38 per cent, with a probability value for each of the 20 chromosome pairs of about 0.019 per cent (Beatty et al. 1975). A very low level of non-disjunction (one hyperhaploid M-II spermatocyte in 391 analysed) was found in male mice of CFLP strain (Polani and Jagiello 1976), but chromosome counts on 1142 M-II cells of CS-I strain revealed hyperhaploidy in 10 cells. Such an unusually high level of aneuploidy was probably due to abnormal pairing of the sex chromosomes in male mice of the CS-I strain (Polani and Jagiello 1976).

The data on spontaneous non-disjunction in M-II oocytes recently summarized by Nijhoff and de Boer (1981) indicate a substantial fluctuation of non-disjunction values in mice of different strains. In 305 M-II oocytes of CBA mice, we found 266 with a balanced haploid count ($n = 20$) while 39 were aneuploid, and only three were hyperhaploid ($n > 20$; Chebotar 1976a). The significant prevalence of hypoploids over hyperploids could be due at least in part to an artificial loss of chromosomes during cytological preparation. On the other hand, the hyperhaploidy cannot result from an artefact. Taking into account that the frequency of hypohaploids could be expected to be approximately equal to that of hyperhaploids, the true incidence of non-disjunction in this CBA strain was calculated to be about 1.9 per cent. Very similar levels of non-disjunction (1.2 per cent) were obtained in 1293 M-II oocytes matured *in vitro* from hybrid mice C3H × ICR/Swiss (Uchida and Freeman 1977), but very low non-disjunction rates (0.4–0.5 per cent) or even complete absence of aneuploidy were reported for mice of NMRI, CFLP, CS-1, and some other strains, for M-II oocytes matured *in vivo* or *in vitro* (Reichert, Hansmann, and Röhrborn 1975; Jagiello, Ducayen, and Lin 1972; Jagiello and Ducayen 1973; Jagiello and Lin 1973; Uchida and Lee 1972; Polani and Jagiello 1976) An unusually high incidence of non-disjunction (about 4.8 per cent) was recorded in M II oocytes of C3H strain and F_1 (C3H × 101) (Röhrborn 1972; Hansmann 1974), but extensive recent data (more than 3000 ovulated oocytes studied) showed no significant difference in non-disjunction rate between mice of NMRI, C3H and F_1 (C3H × 101) strains—0.47, 0.62, and 0.81 per cent, respectively (Hansmann and El-Nahass 1979).

Thus, strain differences in meiotic chromosome behaviour should not be ignored, but the actual incidence of meiotic non-disjunction is very low (about 1 per cent in female mice and less in male mice).

The overall incidence of aneuploidy contributed by male and female gametes can be reliably assessed by direct chromosome counting in pronuclei of fertilized ova at the metaphase of the first cleavage division. Due to different states of contraction the maternally and paternally derived chromosome sets can be easily distinguished and cytologically analysed. Unfortunately, this stage has not yet been widely studied. By using *in vitro* fertilization techniques for mice of different strains (CBA, CBA-T6, C57BL, DBA, F_1 (CBA × C57BL) and outbred TO mice, the incidence of aneuploidy ($2 \times (n + 1)$) was found to be almost equal in male and female pronuclei (0.86 and 0.94 per cent, respectively) while *in vivo* studies demonstrated some preponderance of aneuploidy in the female compared to male complements (0.86 and 0.54 per cent, respectively) (Fraser and Maudlin 1979). These data, though obtained on 2743 zygotes with scorable mitotic plates, need further verification in mice of other strains. An estimate of 0.72 per cent non-disjunction has been obtained by doubling the number of hyperhaploid paternal pronuclei in the fertilized oocytes from F_1 hybrid strain C3H $\times$ 101H, while in the female pronuclei the incidence of aneuploidy varied from 0.38 per cent in young mice to almost 3 per cent in old mice (Tease 1982). An unusually high rate of spontaneous meiotic non-disjunction has been reported recently for Swiss albino ICR mice (Martin-Deleon and Boice 1983). Chromosome studies of 321 zygotes showed that five out of eight hyperdiploids arose in the female genome. This gives a female non-disjunction frequency of 3 per cent (1.5×2), a frequency which is the sum of anaphases I and II non-disjunction events, and a male non-disjunction frequency of about 1.8 per cent.

Thus in laboratory mice the frequency of aneuploid zygotes after fertilization *in vivo*, based on the summed frequencies of aneuploid male and female pronuclei, varies widely (from 1.4 to 4.8 per cent), depending on genetic background and maternal age (see Sections 4.3.1 and 4.3.4 for more details).

Data on spontaneous meiotic non-disjunction in other mammals are scarce and fragmentary. To our knowledge the incidence of spontaneous meiotic non-disjunction has not yet been estimated in male rats and hamsters, though some relevant information has been obtained on oogenesis. Of 861 M-II oocytes of the laboratory white rat, 797 (92.7 per cent) had a balanced haploid number of chromosomes and 61 oocytes were aneuploid, with four of them being hyperhaploid ($n > 21$) and the rest (57) hypohaploid ($n < 21$). Taking into account the ease of chromosome loss during cytological preparation the frequency of aneuploidy in female rats was estimated as being about 0.8 per cent (Dyban and Chebotar 1975). This value is much lower than the non-disjunction frequency, estimated in one-cell rat embryos (Chebotar 1978). Out of 174 rat zygotes karyotyped, hyperdiploidy was found in six, giving an overall frequency of non-disjunction of about 6.8 per cent (3.4 per cent $\times$ 2). The chromosomes of the male and female pronuclei could not be distinguished with certainty in metaphase plates of the first cleavage division, so the reason for the almost tenfold increase in non-disjunction rate between M-II oocytes and one-cell rat embryos remains unknown. It could be due either to anaphase II non-disjunction in oogenesis or to an exceptionally high incidence of non-disjunction in male meiosis.

A single hyperhaploid M-II oocyte was found in a group of 307 Syrian hamster cells, while no extra chromosome was evident in 334 Chinese hamster oocytes (Hansmann and Probeck 1979). No hyperploid M-II oocytes were found among 148 Chinese hamster ovulated eggs (Basler 1978). This extremely low level of non-disjunction has not been confirmed by later studies. Chebotar (1979) found that four of 259 karyotyped oocytes from Syrian hamster (*Mesocricetus auratus*) had a hyperhaploid number of chromosomes ($n > 22$), giving a total value for female meiotic non-disjunction of about 3 per cent. This figure is in good agreement with the frequency of chromosome abnormalities analysed in 917 female pronuclei in Syrian hamster eggs fertilized by human sperm. The minimal estimate of aneuploidy was 1.7–2.2 per cent of total chromosome abnormalities (Martin, Balkan, and Burns 1983).

A direct estimate of the incidence of spontaneous meiotic non-disjunction in humans cannot yet be given (Nijhoff and de Boer 1981). All 37 M-II oocytes analysed cytologically had the normal haploid chromosome set (Uebele-Kallhardt 1978). Few cases of abnormal behaviour of meiotic chromosomes have been recorded in the secondary spermatocytes of humans (Pearson, Ellis, and Evans 1970). Determination of human meiotic chromosome non-disjunction by counting the number of Y chromosomes or specific autosomal markers in mature sperm heads after quinacrine or quinacrine mustard staining did not give any conclusive results (see Nijhoff and deBoer 1981). In 1978 a method for direct visualization of the chromosome constitution of mature human spermatozoa was devised (Rudak, Jacobs, and Yanagimachi 1978). The method is based on fertilization of zona-free hamster eggs with freshly ejaculated spermatozoa. Chromosome counts on 1000 human spermatozoa by this method have revealed chromosome abnormalities in 8.5 per cent of sperm, with 2.4 per cent hypermodal (hyperhaploid), 2.7 per cent hypohaploid, and 3.3 per cent with structural aberrations (Martin *et al.* 1983) These data could be interpreted as strong evidence in favour of a very high frequency of meiotic non-disjunction in humans compared to laboratory animals.

Some rare cases of aneuploidy were recorded for M-II oocytes of sheep, cows, and monkeys (Jagiello, Karrecki, and Ryan 1968; Jagiello *et al.* 1973; Jagiello, Miller, Ducayen, and Lin 1974; Edwards and Fowler 1970; Church, Shea, and Ferrit 1974). These fragmentary data do not permit any proper estimate of non-disjunction rates in these species.

4.2 Non-disjunction and loss of mitotic chromosomes

Non-disjunction and the loss of mitotic chromosomes at initial stages of embryogenesis may result in trisomy, monosomy, or chromosomal mosaicism. In normal cleavage, each of the two blastomeres formed after the first division receives the normal diploid number of chromosomes, and this remains constant throughout the whole subsequent development. Non-disjunction at the first cleavage results in aneuploidy of both initial blastomeres. More complex segregational distortion of mitotic chromosomes may give rise to other types of chromosomal abnormality. Viability of both aneuploid cell clones generated from the chromosomally abnormal blastomeres may result in chromosomal mosaicism. Depending on the initial number of cells comprising the abnormal cell clone, and their capacities for proliferation and migration, the proportion of abnormal cells may vary among different embryonic tissues and rudiments. Embryonic development stemming from only one of the two cell populations produced by initial non-disjunction may result in either complete trisomy or monosomy. There is some direct as well as indirect evidence that the mitotic chromosome loss mechanism operates in the production of XO embryos, with preferential loss of the paternal X chromosome during early cleavage (Russell 1964, 1965; Fraser 1963; Race and Sanger 1969). (See Chapter 8.)

It is convenient to classify numerical mosaics according to the number of cell lines and the minimum number of mitotic non-disjunction events that must be assumed to account for them. The existence of only two cell clones with minimal deviations in chromosome number (XO/XXY) is indicative of a single non-disjunction event in early development. Three or more cell clones in a mosaic may be taken as an indication of mitotic non-disjunction at later stages of development or as evidence of two successive non-disjunctions (Ford 1969). According to Ford, participation of abnormal cell clones in the development of any given tissue or organ is influenced by:

(a) development of progenitor cells between the future trophoblast and embryoblast, that is Inner Cell Mass (ICM);
(b) variations in the ICM contribution to the formation of the fetal membranes;
(c) unequal distribution of cells between sites during organogenesis;
(d) the possible influence of cell migration and differential proliferation.

Thus, the ratio of cell clones may vary substantially at different stages of embryonic development as well as in adults. Aneuploid cell clones deficient in whole autosomes or in parts of autosomes are usually eliminated early in development, while the proportion of normal diploid cells and cells with trisomy may be shifted in favour of diploid cells to a varying extent in different organs and tissues (Ford 1969). The presence of an XY germ cell clone in the testes of XYY men is an example of a normal cell clone produced by a non-disjunctional error in a non-mosaic subject, and its subsequent selective proliferation (Thompson, Melnyk, and Hecht 1967; Hulten and Pearson 1971).

These factors account for the uneven distribution of chromosomally abnormal clones among cell populations in different tissues and organs. However, evidence of mosaicism in adult subjects is usually provided by chromosomal analysis of a single-cell type – PGA stimulated lymphocytes. More thorough cytological analysis of other cell types is needed to estimate the level and possible origin of chromosomal mosaicism. This would be easier in an embryo with all its organs accessible for chromosomal analysis. More information on chromosomal mosaicism might be obtained in cytogenetic studies of cleaving embryos, but technical limitations at these early stages include the difficulty of doing chromosome counts on all blastomeres, and the high frequency of chromosome loss and breakage during preparation.

4.3 Factors influencing the frequency of numerical chromosome aberrations in gametogenesis and embryogenesis

4.3.1 Genotype

All stages of meiotic maturation are known to be under the control of particular genes, which if mutated (so-called mei-mutations in plants) cause meiotic blocks or substantially potentiate meiotic non-disjunction (Golubovskaya 1979). Data on mei-mutations have been obtained in maize and for *Drosophila*. No true mei-mutation has yet been discovered in mammals, though a number of mutants are known among laboratory mice in which impairments of gametogenesis, predominantly spermatogenesis, are part of a wider pleiotropism (Searle 1982). The similarity of meiotic processes in the whole animal kingdom leaves little doubt of the relevance of mei-mutations to mammalian species.

As already mentioned (see Section 4.1.2) the incidence of primary non-disjunction in mice is dependent on genetic background. The same is true for mice with chromosomal translocations. In our early experiments this was demonstrated in mice heterozygous for Robertsonian translocation Rb(8.17)1Iem and in mice with unequal reciprocal translocations T(14;15)6Ca (Baranov 1976a; Baranov and Dyban 1968). Table 11 gives the results in Rb1Iem/+ mice. Non-disjunction rates in mice of different strains carrying the T6 translocation are even more impressive (Baranov and Dyban 1972c) The frequency of T6 tertiary trisomics in the progeny of female mice C3H-T6/+ is 28.6 per cent, in CBA-T6/+ females 19 per cent, in C57BL-T6/+ females 10 per cent, and in the progeny of double translocation carriers Rb1Iem-T6/+/+ 8.9 per cent. The difference of T6 trisomy frequency was found to be statistically significant between C3H-T6/+ and C57BL-T6/+ females ($p < 0.01$ $\chi^2 = 9.7$) and was probably real ($0.05 < p < 0.10$, $\chi^2 = 3.2$) between the C3H-T6/+ and CBA-T6/+ females.

A positive correlation of non-disjunction rate and incidence of univalents at the first meiotic division has been reported in some early cytogenetic studies (Chyi-Chyang, Tsushida, and Morris 1971; Berry, Beechey, and Searle 1973; Luthardt, Palmer, and Yu 1973). Recently, de Boer and van der Hoeven (1980) reported a parallel relationship between univalents and non-disjunction with respect to a particularly rearranged chromosome of female mice homozygous for a reciprocal translocation. The existence of this relationship has not been confirmed by other studies on mice (Polani and Jagiello 1976; Speed 1977) and Chinese hamsters (Sugawara and Mikamo 1983).

Numerous investigators reported an increase in univalent number to be

Table 11. Incidence of trisomic embryos in the progeny of mice of different strains, heterozygous for Rb(8.17)1Iem translocation (Baranov 1976a). Chromosome analysis on the eighth to tenth days of pregnancy

Crosses			No. of embryos with known karyotype											
Female	Male	Total	Diploid (NF[a] = 40)		Heteroploid		Trisomy		Mosaics isochromosomes		Triploidy			
			No.	% ± m	No.	% ± m	No.	% ± m	No.	% ± m	No.	% ± m		
Rb1Iem/ + – C3HA	CBA	182	158	86.8 ± 2.4	24	13.1 ± 2.4	21	11.5 ± 2.3	3	1.6 ± 1.2	—	—		
Rb1Iem/ + – CBA	CBA	146	132	94.3 ± 2.8	8	5.7 ± 2.6	6	4.3 ± 2.1	1	0.7 ± 0.3	1	0.7 ± 0.3		
Rb1Iem/ + – T/t^{12}	CBA	91	78	85.7 ± 3.1	13	14.3 ± 2.8	12	13.1 ± 2.6	—	—	1	1.2 ± 0.4		

[a] NF = nombre fondamental: basic number of chromosome arms.

concomitant with a decline in chiasma frequency (Luthardt *et al.* 1973; Henderson and Edwards 1968; Polani and Jagiello 1976; Speed 1977). Average chiasma count per bivalent in the M-I spermatocytes of mice is in the range 1.03–2.02, and is somewhat higher in laboratory strains than in wild mice (Searle, Berry, and Beechey 1970). Mutations affecting or completely preventing chiasma formation are well known in mice. The direct relationship between chiasma frequency and recombination index has also been confirmed (Forejt 1972). Statistically significant differences in chiasma frequency have been reported for strains varying in the position and amount of C-heterochromatin blocks. Polymorphism of C-heterochromatin in pericentromeric regions is a well-known phenomenon in mice of different strains (Forejt 1972, 1973; Dev, Miller, and Miller 1973) and sometimes even in mice of the same strain (Dyban and Udalova 1974). The effects of C-heterochromatin blocks on chiasma formation and multivalent configuration, and in consequence on the incidence of univalents and meiotic non-disjunction have been suggested (Dyban 1972; Polani 1972). However, no direct relationship between univalent incidence in M-I and non-disjunction rate in M-II or during the first cleavage division has been found in more recent studies on mice (Polani and Jagiello 1976; Speed 1977), and Chinese hamsters (Sugawara and Mikamo 1983). Thus the mechanism of the genetic effects on meiotic non-disjunction and on the incidence of aneuploidy remains obscure.

4.3.2 Specificity of the type and the number of aberrant chromosomes in gametes

The incidence of trisomy or monosomy in the progeny of mice heterozygous for different Robertsonian translocations may significantly vary both before (Baranov 1983a) and after implantation (Gropp, Giers, and Kolbus 1974; Gropp *et al.* 1975; Baranov 1976b; Baranov and Udalova 1975; see Tables 12 and 13). It can be inferred from our data that some chromosomes involved in centric fusions are much more prone to participate in non-disjunctional events and thus to cause aneuploidy in the progeny than their translocated partners. For instance, in +/+ females crossed to Rb(6.15)1Ald/+ males, all aneuploid embryos were trisomic for chromosome 6 but not for 15. Crosses of male mice heterozygous for Rb(1.3)1Bnr translocation to +/+ females almost exclusively produced trisomy for chromosome 1 but not for chromosome 3 (Baranov and Udalova 1975). However, an almost equal proportion of aneuploid embryos with an excess or deficiency of translocated chromosomes 1 and 3 were recovered before implantation in crosses of heterozygotes Rb1Bnr/+ × Rb1Bnr/+ (Baranov 1983 a). Discrepancies between the pre-and post-implantation data might stem from the survival capacities of aneuploid embryos as well as from some peculiarities of the non-disjunction pattern in male and female meiosis. Nonetheless, it looks very

Table 12. Incidence of embryos with trisomy for different autosomes in the progeny of mice, heterozyogous for different Robertsonian translocations: Rb(8.17)1Iem-Rb1Iem;   Rb(6.15)1Ald-Rb1Ald;   Rb(5.19)1Wh-Rb1Wh; Rb(9.19)163H-Rb163H; Rb(5.15)3Bnr-Rb3Bnr (Baranov 1976a). Chromosome analysis on the ninth to tenth days of pregnancy

Crosses		No. of embryos studied	No. of embryos with trisomy of autosomes							
Males	Females		Total	%	5	8	9	15	17	19
Rb1Iem, Rb1Ald/ + +	CBA	128	6	4.9				6		
CBA	Rb1Iem, Rb1Ald/ + +	116	11	9.8			4	1	6	
Rb1Iem, Rb1Ald, Rb163H/ + + +	CBA	112	9	8.0				2	4	3
Rb1Wh/ +	Rb1Wh/ +	94	3	3.1	2					1
Rb163H/ +	Rb163H/ +	100	3	3.0						3
Rb3Bnr/ +	CBA	109	10	9.9	7			3		

Table 13. Incidence of embryos with trisomy for different autosomes in the progeny of mice heterozygous for different Robertsonian translocations: Rb(1.3)1Bnr-Rb1Bnr; Rb(9.14)6Bnr-Rb6Bnr; Rb(16.17)7Bnr-Rb7Bnr (Baranov 1976a). Chromosome analysis on the ninth to tenth days of pregnancy

Crosses		No. of embryos studied	No. of embryos with trisomy of autosomes					
Males	Females		Total	%	1	9	14	16	17
Rb1Bnr/ +	CBA	100	9	9.0	9				
Rb6Bnr/ +	CBA	110	9	8.6		4	5		
Rb7Bnr/ +	CBA	100	4	4.0				2	2

probable now that translocated chromosomes possess different probabilities of participating in non-disjunction events during meiotic maturation, and thus differ in their capacity for production of aneuploidy. Supporting examples are well-known from medical genetics, e.g. D/G translocation carriers in humans more often give rise to fetuses with G-type trisomy than with imbalance of chromosomes in the D group (Grinberg 1969; Walbaum, Dehaene, and Breynaert 1971; Hamerton 1968, 1970).

Examples of non-random segregation of translocated products are abundant in classical genetics of *Drosophila* (Novitsky 1967). The configuration of the quadrivalent and the position and distribution of chiasmata along the chromosomes are probably responsible for the non-random segregation of translocated chromosomes (Russell 1962, 1964).

The presence of other chromosomal abnormalities in the karyotype of translocation carriers may significantly influence the non-disjunction rate, in both humans (Grinberg 1969) and mice. About 50 per cent of all gametes were aneuploid in male mice heterozygous for seven different Robertsonian translocations of Bnr series (Tettenborn and Gropp 1970; Gropp 1973). It has been concluded that a large portion of non-disjunction in tobacco mouse ×
laboratory mouse heterozygotes may result not from structural hetero-zygosity *per se* but rather from genetic differences between the two species (Cattanach and Moseley 1973).

A high incidence of non-disjunction (20 per cent) has been recorded in mice heterozygous for two different Robertsonian translocations, Rb(6.15)1Ald and Rb(5.19)1Wh (White *et al.* 1972a,b). The increase of non-disjunction was not so evident in another double heterozygote Rb1Iem, Rb1Ald/+ + and was almost negligible in the triple heterozygote Rb1Iem, Rb1Ald, Rb163H/+ + + (Baranov 1980; see also Tables 14 and 15).

In double heterozygotes with one translocated chromosome in common (so called 'monobrachial homology') the incidence of aneuploidy increases dramatically and a high proportion of embryos with trisomy or monosomy for the particular chromosome shared by both metacentrics is produced. This very convenient method of breeding for specific aneuploidies (trisomic or monosomic) was independently introduced into the experimental cyto-genetics of mice by White *et al.* (1974a,b) and by Alfred Gropp and his collaborators (Gropp *et al.* 1975). The former group showed that in mice heterozygous for Rb163H and Rb1Wh translocations (chromosome 19 in common), almost 12 per cent of all embryos are trisomic for chromosome 19. Trisomy 17 has been recorded in 15–25 per cent of all post-implantation progeny of female mice heterozygous for translocations Rb(8.17)1Iem and Rb(16.17)7Bnr (Gropp 1975). This double translocation technique has found especially wide application in the fundamental studies of Alfred Gropp's group on the developmental effects of particular types of trisomy in mice (see Chapter 6).

Thus the simultaneous participation of several translocations in the karyo-type may either enhance or reduce the incidence of non-disjunction of meiotic chromosomes.

4.3.3 Sex

The incidence of aneuploidy in the progeny of translocation heterozygotes is sex-dependent. This conclusion can be inferred from both clinical and experimental data. In humans there is a greater chance of a trisomic child being born if the mother rather than the father is heterozygous for a Robertsonian translocation D/G (Hamerton 1968, 1970). The tendency for

Table 14. Incidence of embryos with chromosomal aberrations in the progeny of mice heterozygous for different translocations (Baranov 1976a). Chromosome analysis on the ninth to tenth days of pregnancy

Crosses		No. of embryos with known karyotype						
		Total	Diploid (NF^a = 40)	Heteroploid				
Males	Females			Total No. (%)	Trisomy (%) (NF = 41)	Monosomy (NF = 39)	Mosaics; isochromosomes	Triploidy (2n = 60)
+/+	T6Ca/+	159	128	31(19.4)	26(16.3)			5
+/+	Rb1Iem/+	182	158	24(13.1)	21(11.5)		3	2
Rb1Ald/+	+/+	131	125	6(4.5)	2(1.5)		2	2
Rb163H/+	+/+	101	94	7(6.8)	3(2.9)		1	3
Rb1Wh/+	+/+	94	87	7(7.3)	3(3.1)		1	3
Rb1Bnr/+	+/+	132	111	21(16.0)	18(13.6)	2		1
Rb6Bnr/+	+/+	117	102	15(12.7)	9(7.6)	3	1	2
Rb7Bnr/+	+/+	241	227	14(5.7)	8(3.3)		1	5
+/+	Rb1Iem, Rb1Ald/++	167	152	15(8.9)	15(8.9)			
Rb1Iem, Rb1Ald/++	+/+	128	119	9(7.0)	7(5.4)		1	1
Rb1Iem, Rb1Ald, Rb163/+++	+/+	150	138	12(8.0)	9(6.0)	1	1	1
+/+	Rb1Iem, T6Ca/++	78	70	8(10.1)	4(5.1)	3		1
+/+	Rb1Iem, Rb7Bnr/++	121	98	23(19.7)	17(4.5)			6

a NF = nombre fondamental: basic number of chromosome arms.

Table 15. Incidence of embryos with chromosomal aberrations in the progeny of mice heterozygous for different Robertsonian translocations (Baranov 1976a). Chromosome analysis on the fourth day of pregnancy

Crosses		Aberrant autosomes	No. of embryos with known karyotype								
Females	Males		Diploid $NF^a = 40$	Heteroploid							Mosaics
				Total (%)	3n	4n	Trisomy		Monosomy		
							$NF = 41$ (%)	$NF = 42$	$NF = 39$ (%)	$NF = 38$	
+/+	+/+	—	203	10(4.7)	8	1	1(0.5)				
+/+	Rb1lem, Rb1Ald, Rb163H/+++	?	93	31(25.0)	4	1	8(6.4)	9	4(3.2)	3	2
Rb1lem, Rb1Ald, Rb163H/+++	+/+	?	181	47(20.6)	7		13(5.7)	4	18(7.8)	3	2
+/+	Rb6Bnr/+	9.14	118	33(21.8)	3		16(10.5)	1	9(5.9)	3	1
Rb6Bnr/+	+/+	9.14	74	38(33.9)	4	1	19(16.7)		11(9.8)	2	1
Rb3Bnr, Rb1Ald/++	+/+	15^b	93	31(25.0)	1	1	15(12.7)		14(11.3)		
Rb7Bnr, Rb1lem/++	+/+	17^b	39	19(32.7)	1	1	17(29.3)				
+/+	Rb3Bnr/+	5.15	55	16(22.5)	1	1	10(14.0)	1	3		
Rb1Wh, Rb163H/++	+/+	19^b	48	15(23.8)			8(12.7)		6		1

a NF = basic number of chromosome arms.
b Chromosome in common for both centric fusions.

female heterozygotes to produce more chromosomally unbalanced progeny than do heterozygous males was shown in our early studies in mice with Rb(8.17)1Iem translocation (Baranov and Dyban 1971a,b; Baranov and Udalova 1974, 1975). The female carriers of Rb1Iem mated to +/+ males produced 14.3 per cent trisomic embryos if studied on the eighth day of gestation, whereas heterozygous males sired only chromosomally balanced progeny (see Section 4.3.5 for more details, Baranov and Dyban 1971b). Very similar results have been obtained in mice with reciprocal translocations. Females heterozygous for the translocation T(14;15)6Ca have produced a high frequency of the tertiary trisomic 14^{15} (T6), while males heterozygous for the same translocation have produced no translocation trisomic offspring (Eicher 1973). These data were substantially extended by Oshimura and Takagi (1975), who have demonstrated that in heterozygous carriers for the T6 translocation the frequency of non-disjunction in M II involving the T6 marker chromosome (14^{15}) was 22.2 per cent in females compared to only 4.4 per cent in heterozygous males. The proportion of aneuploid gametes was almost six times greater in female than in male heterozygous carriers of another reciprocal translocation, T(1;13)70H (Oshimura and Sandberg 1978). Thus the higher level of tertiary trisomics in the progeny of female heterozygotes probably reflects the higher incidence of non-disjunction in primary oocytes compared to spermatocytes. We found that the frequency of aneuploidy in T6 heterozygotes correlates well with the rate of univalent formation in M-I (Baranov and Dyban 1968).

The tendency for female heterozygotes to give a higher yield of aneuploidy in the progeny than do heterozygous males has also been reported for Robertsonian translocations—Rb6Bnr, Rb4Bnr, and Rb7Bnr (Table 14; see also Gropp 1973a,b)—but not for Rb1Iem, and Rb1Ald, Rb163/+ + + (Tables 14 and 15; Baranov 1980).

In contrast, White *et al.* (1974a) found more trisomic embryos in the progeny of male rather than female double heterozygotes for the Rb1Wh and Rb163H translocations, with chromosome 19 in common.

Thus the frequency of non-disjunction is sex-dependent but each structural rearrangement is unique, so different translocations may behave differently in male and female meiosis.

4.3.4 Age

4.3.4.1 Male meiosis. No conclusive data on the age-dependent increase of aneuploidy in spermatocytes or mature gametes of men are yet available (Hamerton 1971), but an increased frequency of Down's syndrome with advanced paternal age has been reported (Stene, Stene, and Stengel-Rutkowski 1981; Matsunaga, Tonomura, Oishi, and Kikushi 1978) and confirmed by amniocentesis results. Fathers above the age of 39 appear to have an elevated risk of having trisomy 21 offspring, with a strong effect from the age of 41 (Stene *et al.* 1981). The recently invented method for chromosome analysis of human spermatozoa by penetration of zona-free hamster eggs will probably clarify this puzzle soon (Rudak *et al.* 1978; Martin *et al.* 1983).

The available experimental data on this topic are still rather contradictory. The incidence of chromosomal abnormalities varied significantly between C57BL males, but did not change much between 30 and 750 days of age (Leonard and Leonard 1975). By contrast, in mice of NMRI strain no abnormal M-I was found in young males, but the number of spontaneous translocations in spermatocytes I of aged males (600 days) approached 0.75 per cent (Schleirmacher 1972). According to Muramatsu (1974) the incidence of meiotic abnormalities in three-year-old male mice may reach 1.12 per cent. These early observations to some extent conflict with more detailed recent observations. Studies of male meiosis in mice of Cs1 and CFLP strains revealed a tendency for autosomal chiasma counts to rise slightly with age, while the number of univalents significantly dropped with age (Polani and Jagiello 1976). Very few hypermodal spermatocytes II have been found in these mice, either in the young (one month) or in the old (18 months) males (Polani and Jagiello 1976). These observations have been extended and confirmed in mice of Q, CBA, and C57BL strains (Speed 1977). Males showed a slight but non-significant rise in chiasma frequency and a non-significant increase in univalents between the ages of 2 and 15 months. The same was true for Swiss-Cann strain mice (Jagiello and Fang 1979). Thus, most of the experimental data available at present indicate that there are no age effects on meiotic chromosome disjunction in males.

However, some recent data demonstrate a high mutagenic effect of cell aging on mature spermatozoa in mammals. For mature spermatozoa in the male genital tract, a delay of 6 days before natural mating had no effect on chromosome anomalies detectable in one-cell mouse zygotes. But when the delay extended beyond 6 days there was a highly significant increase in the number of all chromosome anomalies including trisomy (9.7 per cent) and monosomy (7.4 per cent), with the male genome involved significantly more often than the female one in the origin of aneuploidy. A selective mechanism operating between chromosomally balanced and unbalanced sperm during their sojourn in the male reproductive tract has been suggested (Martin-Deleon and Boice 1982).

4.3.4.2 Female meiosis. Relationships between maternal age and the incidence of human trisomies are well established. It is widely believed that advanced maternal age is correlated with non-disjunction of chromosomes in oocytes and is the major cause of Down's syndrome (trisomy for chromosome 21: Carr 1965, 1967a,b; Boué *et al.* 1975, 1976), although this is not universally accepted (Penrose 1966; Hamerton 1971). We have found that in young CBA mice the frequency of non-disjunction in secondary oocytes is about 2.14 per cent, while in M-II oocytes of aged mice (10–12 months) it reaches 8.5 per cent (Chebotar 1980). In contrast, no hypermodal ($n + 1$) oocyte was observed in the young (2 months) or old (11 months) group of CBA mice, but 5.2 per cent of hypermodal oocytes occurred in mice of an intermediate age group (Martin *et al*. 1976).

The most clear-cut effects of ageing on female gametes were shown in experiments with artificial insemination (Maudlin and Fraser 1978). There was a significantly higher incidence of aneuploidy in oocytes from the aged group (8–10 months) of TO mice than from young (2–4-month-old) virgin females (7.5 per cent versus 3.3 per cent). A substantial increase of aneuploidy was recorded in 10.5-day-old mouse embryos from aged females (Yamamoto, Endo, and Watanabe 1973), but no significant increase in the non-disjunction rate with age was found in oocytes from mice carrying the Robertsonian translocation Rb(9.19)163H (Donahue and Karp 1973).

The detrimental effect of ageing on meiotic chromosomes of oocytes is a well-established phenomenon for other laboratory species also. The incidence of aneuploidy in M-II oocytes from young rats was estimated to be about 0.8 per cent, almost five times less (4.14 per cent) than in oocytes from aged rats (10–12 months; Dyban and Chebotar 1975; Chebotar 1976b). The same correlation was found in recent studies with Chinese hamsters (Sugawara and Mikamo 1983). The frequency of both M-I oocytes with univalents and M-II oocytes with aneuploidy due to meiotic non-disjunction was higher in the 'aged' oocytes than in the 'young' ones.

Thus with few exceptions to date, the tendency for meiotic chromosome non-disjunction to rise substantially with advanced maternal age might be considered as a general phenomenon in mammals.

4.3.4.3 The mechanism of aneuploidy in the aged gametes. The reasons for the increase of meiotic chromosome non-disjunction with advanced maternal age remains mostly unknown.

Henderson and Edwards (1968) put forward the so-called 'production line hypothesis', according to which oogonia committed to meiosis earlier in pre-natal life generate oocytes with more chiasmata, and ovulate earlier in adult life than those entering meiotic prophase later. The lower chiasmata frequency in oocytes formed later would lead to more univalents at the first meiotic division and this would result in more frequent chromosomally unbalanced gametes and more aneuploid zygotes from the older females. The findings of Henderson and Edwards on chiasma formation and univalent incidence in oocytes from 'young' and 'aged' females have been repeatedly confirmed in mice (Luthardt *et al*. 1973; Polani and Jagiello 1976; Speed 1977) and in Chinese hamsters (Sugawara and Mikamo 1983). However, these studies have failed to prove the second part of the Henderson–Edwards hypothesis, that is a positive correlation between the univalents in M-I and non-disjunctional frequency in M-II or in zygotes. Indeed, it was demonstrated that the univalents in M-II of mice or Chinese hamsters do not necessarily undergo non-disjunction, and thus there was no parallelism between the frequency of univalents and chromosome aneuploids that could be identified at M-II or during early development (Polani and Jagiello 1976; Speed 1977; Sugawara and Mikamo 1983).

According to Evans (1967), non-disjunction is mostly due to the persistence of the nucleolus during meiotic arrest in aged oocytes. The persistence of the nucleolus causes aberrant disjunction of the nucleolar-associated chromosomes at the first meiotic division. Proved association of human trisomies for the D and G chromosomes known to carry nucleolar organizing regions (NOR) with advanced maternal age seems to confirm this suggestion. Moreover, it has been shown that the nuclear envelope of meiotic cells contains some still unknown colcemid-sensitive mechanism, which regulates the movements of the chromosomes during meiotic pairing and disjunction (Salonen *et al.* 1982).

Another reason for abnormal chromosome disjunction in the oocytes may be a hormonal disbalance in the aged females. This mechanism might be especially relevant to the mutagenic effects of so-called 'pre-ovulatory over-ripeness' (Dyban 1972, 1973a,b, 1974). The hypothesis suggests inadequate reactions of follicular cells and the oocyte itself to hormonal stimuli and thus lack of synchronization between the development of the pre-ovulatory follicle as a whole and the oocyte it contains. More advanced meiotic development of not yet ovulated oocytes would predispose to pre-ovulatory over-ripeness, and hence to non-disjunction and aneuploidy during meiotic divisions (Dyban 1970a,b).

It has been demonstrated that meiotic non-disjunction may be associated with a defect of spindle microtubules in oocytes that have undergone intrafollicular pre-ovulatory over-ripeness (Mikamo 1968; Mikamo and Hamaguchi 1975). The result of the spindle defect is detachment of chromosomes, which subsequently form micronuclei and are then lost. Meiotic chromosome non-disjunction in the oocytes of aged females might therefore be due to inadequate tubulin polymerization, which causes incomplete formation of the spindle microtubules (Sugawara and Mikamo 1983). Cell ageing (over-ripeness) of oocytes has long been known in many species and has been well established as a teratogenic factor (Mikamo and Hamaguchi 1975).

It should be remembered that the meiotic oocytes of mammals do not contain typical centriole bodies (Szollosi 1975); these are replaced during the first meiotic division by a ring of electron-dense material. Because of the less firm spindle the microtubules anchoring the chromosomes could easily disperse in over-ripe pre-ovulatory oocytes.

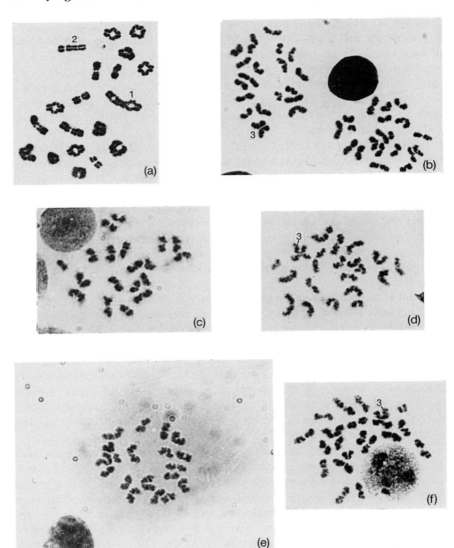

Fig. 12. Meioses in mice heterozygous for Robertsonian translocation. Lacto-aceto-orcein.

(a) Metaphase I; (b–f) Metaphase II; (a,b) Rb1Ald/+ (euploid); (c,d) Rb1Iem/+ ; (e,f) Rb1Ald/+ (aneuploid); (c,e) n = 19; (d,f) NF = 21.

1 = trivalent, 2 = sex-bivalent, 3 = marker centric fusion—translocated chromosomes.

Thus the exact mechanism responsible for the increased rate of meiotic chromosome non-disjunction in aged mammalian oocytes still remains unknown.

4.3.5 Elimination of gametes (prezygotic selection)

The frequency of chromosome aberrations detected during embryogenesis depends on the efficiency of the mechanisms that select against chromosomally abnormal gametes during maturation divisions or at fertilization (so-called 'prezygotic selection'). However, the idea of an efficient mechanism of prezygotic selection against genetically unbalanced gametes is not widely accepted. The results of many experiments suggest that gametes take part in fertilization regardless of the genome abnormalities that they bear (Ford 1972; Ford and Evans 1973a,b; Ford 1975). The data in mice heterozygous for reciprocal or non-reciprocal (Robertsonian) translocations are consistent with the assumption that spermatozoa take part in fertilization in numbers proportional to those produced regardless of whether they have a normal haploid genome or show a severe chromosome imbalance (Oshimura and Takagi 1975; Ford and Evans 1973a,b). The close coincidence between the aneuploidy rates in the M-II genome of maturing gametes and those in pre-implantation embryos is strong evidence in favour of this assumption.

However, some of our experimental data seem to conflict with this view (Baranov and Dyban 1976). In male mice heterozygous for the Rb1Iem translocation the frequency of aneuploid spermatocytes at M-II stage was almost three times higher than in Rb1Ald/+ males at the same stage of maturation. One would expect therefore to get a similar proportion of aneuploidy in the progeny of female mice crossed to the appropriate male translocation carriers (Fig. 12), but this proved not to be so (see Tables 16 and 17). The frequency of trisomic embryos in the progeny of Rb1Ald/+ carriers was 1.5 per cent (two out of 131), corresponding well to meiotic studies in

Table 16. Incidence of euploid and aneuploid metaphase II(M-II) in spermatocytes of mice, heterozygous for Robertsonian translocations Rb(8.17)1Iem and Rb(6.15)1Ald (Baranov and Dyban 1976)

M-II with basic number of chromosome arms (NF)	No. of metaphase II scored			
	Rb1Iem/+		Rb1Ald/+	
	Total	$\% \pm m$	Total	$\% \pm m$
18	6	1.8 ± 0.7	4	1.1 ± 0.5
19	45	13.8 ± 1.9	18	5.1 ± 1.2
20	240	73.8 ± 2.4	319	90.3 ± 2.6
21	31	9.6 ± 1.6	9	2.6 ± 0.8
22	3	0.9 ± 0.3	3	0.8 ± 0.2

Table 17. Incidence of autosomal trisomy in the progeny of CBA and C3HA females crossed to males heterozygous for the Robertsonian translocations Rb1Iem and Rb1Ald (Baranov and Dyban 1976). Chromosome analysis at 8–12 days of pregnancy

Crosses		No.of embryos with known karyotype			
Males	Females	Total	$2n = 39$, NF = 40 (%)	$2n = 40$, NF = 41 (trisomy) (%)	$2n = 40$, NF = 40 $2n$ (%)
Rb1Iem/ +	+ / +	142	81(56.7)	—	61(43.2)
Rb1Ald/ +	+ / +	131	65(49.6)	2(1.6)	64(48.8)
+ / +	Rb1Iem/ +	182	93(51.1)	21(11.5)	68(37.4)
+ / +	Rb1Ald/ +	73	35(47.0)	—	38(53.0)

Rb1Ald/ + males. By contrast, not a single 10-day-old C3HA female embryo crossed to a Rb1Iem/ + male had trisomy, though at least 11 per cent of aneuploid embryos were recorded in the progeny of female heterozygotes crossed to + / + males. The karyotype of aneuploid embryos as shown by chromosome banding had almost equal proportions of trisomies 8 and 17, both involved in the Rb1Iem centric fusion.

Thus in male mice heterozygous for the Rb1Iem translocation, the aneuploid gametes arising through non-disjunction of translocation products do not appear to participate in fertilization, while aneuploid gametes in female heterozygotes complete the second meiotic division and give rise to embryos with trisomy or monosomy (Baranov, Vaisman, and Udalova 1982; Baranov 1983a). A substantial difference in trisomy rate has been recorded between the progeny of male and female mice heterozygous for the Rb(16.17)7Bnr translocation (Gropp 1973).

Both of these translocations involve chromosome 17 in the centric fusion. Chromosome 17 in the mouse is famous for its complex *H-2* and *t*-loci (Klein 1971; Committee on Standardized Genetic Nomenclature for Mice 1972). The numerous *t*-haplotypes affect morphological and physiological properties of the mature sperm, providing evidence for post-meiotic gene expression during mammalian spermatogenesis (Braden and Gluecksohn-Waelsch 1958; Beatty 1970; Olds 1971; Braden 1972; Yanagisawa, Bennett, Boyse, Dunn, and Dimeo 1974a; Yanagisawa, Pollard, Bennett, Dunn, and Boyse 1974b). The possibility for selective elimination of aneuploid gametes due to some peculiarities of gene function in the *t*-region of chromosome 17 involved in the Robertsonian translocations Rb1Iem and Rb7Bnr remains to be clarified.

Although there is strong evidence for post-meiotic RNA synthesis during spermatogenesis in mice (Kierszenbaum and Tres 1975; Erickson 1978), definitive proof of haploid genome selection before fertilization is still lacking.

However, other examples of prezygotic selection operating in mammals are known. Rams heterozygous for the Robertsonian translocation of Massey I type regularly produce secondary spermatocytes with aneuploidy, but none of the karyotyped blastocysts in the progeny of these rams have an excess of the autosome. It has therefore been inferred that male aneuploid gametes in the rams with M-I Robertsonian translocation do not take part in fertilization (Bruere 1975; Long 1977).

Thus, in at least some cases prezygotic selection seems to be quite an efficient mechanism for reducing aneuploidy rate in mammalian embryos. Whether prezygotic selection of aneuploid gametes should be considered as a rule or as unique exceptions relevant to particular autosomes of particular species still remains uncertain.

4.4 Conclusions

Numerical and structural aberrations of autosomes and gonosomes in mammalian embryos may arise through primary, secondary, or tertiary non-disjunction of meiotic chromosomes in males and females. Similar chromosome aberrations could be due to mitotic chromosome loss during early cleavage divisions.

The frequency of spontaneous numerical aberrations of meiotic chromosomes depends on some still unknown peculiarities of genetic background, specificity of the type and the number of aberrant chromosomes in the gametes, and the sex of the translocation carrier. The age of the father has no adverse effect on non-disjunction of meiotic chromosomes in spermatogenesis, but the ageing of the already mature spermatozoa before fertilization dramatically impairs the male haploid genome. The ageing of female gametes both before and after ovulation predisposes to abnormal disjunction of the meiotic chromosomes of the secondary oocytes. Possible reasons for non-disjunction include reduced chiasma formation, persistence of nucleolar membranes, the absence of true centrioles in meiotic oocytes, inadequate tubulin polymerization in meiotic spindles, and defective hormonal mechanisms regulating follicular growth and oocyte maturation in aged females.

Some rare cases of prezygotic selection known from cytogenetic studies in mice and rams suggest elimination of some male gametes with chromosomal aberrations at post-meiotic stages of gametogenesis, that is, before fertilization. Whether prezygotic selection of aneuploid gametes is relevant to other chromosomes and to other mammalian species remains to be established.

5

Experimental production of embryos with numerical and structural aberrations of individual chromosomes

Embryos with trisomy or monosomy for a particular chromosome are regularly produced in the progeny of mice heterozygous for Robertsonian translocations, due to non-disjunction of chromosomes involved in the centric fusion. Embryos with structural chromosomal aberrations may result from the direct mutagenic effect applied during gametogenesis and early embryogenesis or they may arise in the progeny of mice heterozygous for a reciprocal translocation.

5.1 Monosomy and trisomy of individual autosomes in studies on mice with Robertsonian translocations

Robertsonian translocations result from the centric fusion of two non-homologous acrocentric chromosomes into a single biarmed metacentric or submetacentric chromosome. As a result of centric fusion the total chromosome number in the karyotype is reduced but the number of chromosomal arms, the so-called 'fundamental number' or 'nombre fondamental' (Matthey 1965) remains constant. Thus the Robertsonian type of translocation is not accompanied by the gain or loss of genetic material of rearranged chromosomes.

Robertsonian translocations are the most common spontaneous chromosomal aberrations in the animal kingdom, including mammals and humans. It is generally accepted that Robertsonian translocations are of major importance for karyotype evolution and speciation (Matthey 1965; Vorontzov 1966; White 1969, 1973, 1978). In mammals Robertsonian translocations are most common in rodents (Lyapunova, Voronzov, and Korobitsyna 1980), being particularly widespread in *Mus musculus domesticus* of laboratory and feral origin. Centric fusions (whole arm translocations) were first reported for laboratory stocks of mice almost simultaneously in Great Britain (Evans, Lyon, and Daglish 1967) and in Belgium (Leonard and Decknudt 1967a). Their number has gradually increased since that time in different laboratory stocks throughout the world (White and Tjio 1968; Baranov and Dyban 1971a; Cattanach and Savage 1976; Jameela and Murthy 1979; Yadov and Yadov 1981; Baranov 1981).

The major output of new Robertsonian translocations into mouse cytogenetics has been obtained from wild populations. Free-living house mice with centric fusions of chromosomes were first described by Gropp and co-workers (Gropp, Tettenborn, and Leonard 1970a; Gropp, Tettenborn, and von Lehmann 1970b). These were the so-called 'tobacco mice'— *Mus poschiavinus*, the karyotype of which had only 26 chromosomes including 14 unusually large metacentrics. Meiotic studies of these mice has revealed the homozygosity of the initial wild population for seven different Robertsonian translocations, called Rb1–7Bnr (Bonn-Rhein) (Gropp *et al.* 1970a,b). It has also been demonstrated that all acrocentrics involved in the corresponding central fusions fully retain their genetic constitution and thus remain homologous with the corresponding original acrocentric chromosomes in spite of cytological rearrangements (Gropp and Zech 1973), which have been assessed by meiotic studies of F_1-hybrids (*Mus musculus* × *Mus poschiavinus*) as well as by direct analysis of mitotic chromosomes after banding (Zech *et al.* 1972; Gropp and Zech 1973). Since then several such wild populations, distinguished by varying numbers (up to nine pairs) of Rb-metacentrics, have been discovered in different parts of Europe, including Italy, Southern Germany, Sicily, Spain, Scotland, and Greece (Capanna, Gropp, Winking, Noack, and Citivelli 1976; Adolph and Klein 1981; Gropp, Winking, Redi, Capanna, Britton-Davidian, and Noack 1982; Nash, Brooker, and Davis 1983). According to our most recent information, the total number of Robertsonian translocations known in mice is over 100 (*Mouse News Letter* 1983). Every autosome of the whole acrocentric karyo-type of the 'normal' mouse was found to be involved in several different centric fusions (Baranov 1980). Some chromosomes (numbers 1,2,4,6,8,13, and 15) take part in seven to eight different Robertsonian translocations. Chromosome 19 is rather rarely involved in rearrangements in mice of wild populations, though it participates in at least two different centric fusions in laboratory stocks—Rb(9.19)163H and Rb(5.19)1Wh.

Unlike the autosomes, the gonosomes (X or Y) are rarely found in centric fusions, probably because of the total sterility of mice with such translocations. Up to now only a single case of a spontaneous Rb-translocation between the X chromosome and an autosome has been reported in the mouse (Nombela and Murcia 1978). Unfortunately, the fertility of this unique mouse was not studied.

By crossing mice of wild populations with different sets of Rb-translocations to 'all-acrocentric' mice of laboratory stocks the isolation of individual Rb-translocations on a predominantly *Mus musculus* genetic background has been reached (Cattanach, Williams, and Bailey 1972; Klein 1971).

By successive crosses of F_1-hybrids (*Mus musculus* × *Mus poschiavinus*, kindly given to one of us (A.P.D.) by Professor A.Gropp) with CBA/Lac mice with an all-acrocentric karyotype, followed by individual metacentric identification and backcrosses of heterozygous carriers, mice homozygous for each of four different Rb-translocations of the Bnr series have been produced (Baranov 1976a).

The abundance of Robertsonian translocations in the house mouse on the one hand and the high non-disjunctional rate of rearranged chromosomes in translocation heterozygotes on the other were two basic prerequisites for the deliberate generation of zygotes bearing trisomy or monosomy for any particular chromosome in the mouse karyotype.

The use of Robertsonian translocations in experimental studies of developmental effects induced by trisomy or monosomy for any specific chromosome has been applied independently and almost simultaneously in several different laboratories (Gropp *et al.* 1970a,b; Gropp 1973; White *et al.* 1972a, 1974; Baranov and Dyban 1971a,b, 1972b). However, the basic contribution to the systematic approach to studying the developmental effects of aneuploidy, and especially of trisomy, in the mouse has been made by the German group in Lübeck (FRG) under the leadership of the recently deceased Professor Alfred Gropp, whose tremendous personal impact in Robertsonian translocation studies and in developmental cytogenetics of the mouse in particular will always be remembered and highly appreciated by anyone touching on these fields.

The efficiency of this approach based on the use of Robertsonian translocations may be illustrated by our cumulative results both before and after implantation. Out of 1157 embryos karyotyped on the eighth to ninth days of gestation—that is, soon after implantation—125 (10.3 per cent) had numerical chromosomal aberrations. Thus the level of aneuploidy in mice with Rb-translocations, studied soon after implantation (see Table 14), was on average three times higher than in mice with an 'all-acrocentric' karyotype (see Fig. 13). The predominant type of aberration at this particular stage was trisomy of autosomes (7.7 per cent). Other chromosomal aberrations (triploidy, chromosomal mosaicism, etc.) did not exceed 2.5 per cent and thus remained within control limits. The incidence of aneuploidy before implantation was about twice that after implantation, and varied between 11 and 36 per cent (see Table 15). Of 849 embryos with reliable chromosome counts on the third to fourth days of gestation, 179 were aneuploids, including 87 trisomic (NF = 41) and 90 monosomic for one or occasionally two different autosomes (Baranov 1983a).

The number of aneuploid embryos generated in the progeny of hetero-zygotes can vary substantially depending on the type of Robertsonian trans-

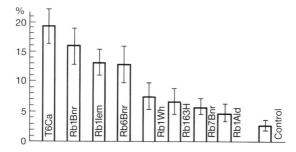

Fig. 13. Incidence of chromosomal aberrations (primarily trisomies) at post-implantation stages (9–10 days) of embryogenesis in mice, heterozygous for different translocations. Ordinate—incidence of heteroploid embryos in per cent.

location used. Those translocations generating either a high (Rb(1.3)1Bnr; Rb(8.17)1Iem; Rb(9.14)6Bnr) or a low level of aneuploidy (Rb(9.19)163H; Rb(5.19)1Wh, Rb(16.17)7Bnr; Rb(6.15)1Ald) have been determined. Ford and Evans (1973a) obtained values of 14 and 5 per cent from trisomic fetuses in mice heterozygous for Rb(11.13)4Bnr and Rb(16.17)7Bnr, respectively. The very low level of non-disjunction in Rb7Bnr/+ male mice has been confirmed by Gropp (1973). However, the non-disjunction rate estimated in different laboratories for the same Robertsonian translocation may vary significantly (Searle 1982). For instance, in Rb(9.14)6Bnr heterozygotes the non-disjunction frequency estimated from M-II counts was found by Cattanach and Moseley (1973) to be 8 per cent, whereas Gropp *et al.* (1975) found a value of 10 per cent, and Ford (1975) 23 per cent. The cause of these variations still remains unknown, though probably they depend on the specific genetic backgrounds of the translocation used in the studies.

Several designs of breeding animals heterozygous for metacentric chromosomes are commonly used for inducing aneuploidy in the progeny:

(a) backcross of single metacentric heterozygotes with 'all-acrocentric' laboratory mice;

(b) use of double metacentric heterozygotes with monobrachial homology (White *et al.* 1974a,b; Gropp *et al.* 1975). Both of these breeding designs are depicted schematically in Fig. 10 (see also Section 4.1). Mice heterozygous for two different Robertsonian translocations sharing a common chromosome arm have proved particularly useful, since these allow aneuploids for a specific chromosome (usually common to both translocations) to be produced with high frequency, and also make them easier to identify cytologically in metaphase plates. This opens the possibility of obtaining nullisomics. Unfortunately, male mice carrying two Robertsonians with a common arm are frequently sterile (Evans 1976), though differences in the background genotype may substantially influence spermatogenesis in double heterozygotes (Searle and Beechey 1982). By contrast, female mice heterozygous for two different translocations with monobrachial homology are usually fertile and generate an especially high proportion of aneuploid gametes. Recently, a much more efficient stock has been suggested by combining three Robertsonians such as Rb(4.6)2Bnr, Rb(4.15)4Rma, and Rb(6.15)1Ald into a single compound with tribrachial homology (Searle and Beechey 1982). If such a tester compound is viable and fertile it would generate only aneuploid gametes, disomic or nullisomic for chromosomes 4,6, or 15, and thus would be especially useful for obtaining nullisomics.

Unlike these so-called 'complex heterozygotes', no single Robertsonian translocation in homozygous or heterozygous form much impairs fertility in males or females (Cattanach and Moseley 1973; Baranov 1980). The frequency of aneuploidy in the progeny of these mice may be very high, which makes them very useful tools for studying the developmental effects of aneuploidy (Baranov and Dyban 1971a, 1973; Gropp *et al.* 1974; Baranov and Udalova 1975).

Thus, both breeding designs (single metacentric heterozygotes and double metacentric heterozygotes with monobrachial homology) have their advantages and disadvantages and can be successfully used to produce aneuploidy for any whole chromosome in mouse embryos.

5.2 Experimental production of embryos with structural chromosome aberrations

Mammalian embryos with different structural aberrations of chromosomes can be obtained by applying a mutagenic effect to gametes or zygotes, as well as by experiments with reciprocal translocations.

5.2.1 Effects of mutagens

Different chromosomal aberrations may be readily induced in gametes or zygotes by means of different chemical or physical mutagens. If they take part in fertilization, these gametes will transmit their abnormal genetic content to the progeny. Most of these mutagens affect chromosomal structure but they may also induce pathological changes incompatible with normal fertilization or early development, so many chromosomal aberrations induced at pre-meiotic stages of gametogenesis are never transmitted to the progeny but are eliminated during meiotic maturation divisions (Ford 1969; Searle 1974). The same may also be true for mammalian zygotes whose genetic load may be readily influenced by numerous environmental chemicals (Watanabe and Endo 1979). Some of these substances, however, are very toxic for fertilized eggs and thus cause early embryonic death (Baranov and Chebotar 1980). Moreover, non-specific consequences of induced chromosome damage make it impossible to predict the type of aberration or chromosome involved in rearrangements, which makes this approach hardly applicable for developmental cytogenetics.

5.2.2 Heterozygous carriers of reciprocal translocations

Since the classical works pioneered by Snell and Hertwig some 40 years ago, it has been known that mice heterozygous for reciprocal translocations (symmetrical exchanges of chromosomal material between non-homologous chromosomes) are usually 'semi-sterile' or partially sterile due to the production in meiosis of a great number of genetically unbalanced gametes with structural or occasionally numerical chromosomal aberrations (Ford and Clegg 1969; Searle, Beechey, Evans, Ford, and Papworth 1971a; Searle *et al.* 1971b). Mice heterozygous for reciprocal translocations should therefore be considered as very convenient biological tools suitable for studying the function of particular chromosome regions in embryogenesis. Furthermore, available methods for chromosome banding by differential staining substantially facilitate the identification of abnormal chromosomal regions, while an abundant collection of reciprocal translocations exists, involving each of the 19 autosome pairs as well as the X chromosome (Searle 1982). Each autosome in the laboratory mouse is subdivided into several pieces by known reciprocal translocation breaks, thus providing an opportunity to compare the developmental effects of genes in different parts of the same chromosome. Unfortunately, up to now such an experimental approach has been applied to only a few reciprocal translocations. These are T(14;15)6Ca (Baranov and Dyban 1968, 1970; Oshimura and Takagi 1975), T(7;14)2Iem (Baranov 1981), T(1;13)70H (de Boer 1973), T(16;17)43H (Forejt, Gregorova, and Goetz 1981), T(9;17)138Ca(Baranov 1983b).

5.3 Conclusions

Embryos with aneuploidy (trisomy, monosomy, nullisomy) may be deliberately obtained in mice with Robertsonian translocations. Numerous translocations of Robertsonian type have been isolated from laboratory and wild stocks of mice and are maintained as inbred or outbred colonies of homozygotes. Each of the 19 autosome pairs in the mouse is involved in several different types of centric fusion, thus giving ample opportunity for studying the developmental effects of any particular type of trisomy or monosomy. Any specific type of aneuploidy may be produced either by breeding the appropriate Robertsonian translocation heterozygote to 'all acrocentric' mice or by the use of a double translocation breeding design. The latter is generally more efficient but is often hampered by the sterility of male double heterozygotes with monobrachial homology.

Embryos with nullisomy may be produced in crosses of mice with double translocations sharing one translocated chromosome, or by a special breeding design using mice with tribrachial homologous Robertsonian translocations.

Embryos with structural aberrations of a particular chromosome, especially with duplications or deficiencies for a specific chromosomal segment, may be obtained by crossing mice heterozygous for an appropriate reciprocal translocation of that particular chromosome.

6

Autosomal trisomy

6.1 Frequency of spontaneous trisomy

Autosomal trisomies make up most human aneuploids. About 50–60 per cent of all spontaneous abortions in the first trimester of pregnancy are chromosomally abnormal. Autosomal trisomy is the most common aberration, encountered in about 50–60 per cent of all abnormalities (Prokofieva-Belgovskaya 1969; Carr 1971a,b; Kuliev 1976). The relative frequency of each type of trisomy may, however, differ significantly. According to some early data the incidence of trisomy for the chromosomes of E group is over 36 per cent, that is 1.5–2 times more than that of trisomies for the chromosomes D, G, and C (22.7 per cent, 19.1 per cent, and 17.9 per cent, respectively; Boué and Boué 1970, 1974). Application of banding techniques has significantly extended our knowledge on the frequency of each type of trisomy in human development. Cumulative data on autosomal trisomies in induced and spontaneous abortions as well as before and after birth have been repeatedly illustrated in numerous reviews (Chandley 1980, 1981, 1984; Sankaranarayanan 1979; Ayme 1980; Lauritsen 1982) and summarized in a recent book devoted to aneuploidy (Bond and Chandley 1983). Useful information on the frequency of trisomy in laboratory mammals is given in a comprehensive review by Hansmann (1983).

Cumulative data on 1926 spontaneous abortuses with chromosomal aberrations unequivocally demonstrated the presence of autosomal trisomy in 981 (50.9 per cent) (Chandley 1984), confirming that autosomal trisomy is the most widespread chromosomal aberration in human embryos. Trisomies for all members of the human karyotype but chromosome 1 have been repeatedly reported.

The relative involvement of each autosome in triplication may vary substantially. Analysis of cumulative data (Chandley 1982, 1984) is consistent with frequencies of trisomies 5, 6, 11, and 19 being in the range 0.1–0.3 per cent, compared to 4.5–7.3 per cent for trisomies 7, 13, 14, and 15. Trisomy 16 accounts for about one-third (32.4 per cent) of all trisomic abortuses, and so is undoubtedly the most frequent chromosomal abnormality in aborted embryonic sacs. The second and third places in this range are shared by trisomy 22 (10.1 per cent) and trisomy 21 (8.3 per cent), respectively.

Uneven participation of different autosomes in spontaneously aborted human embryos could be attributed to two different mechanisms. First, it might be due to the unequal involvement of different autosomes in meiotic non-disjunction events, giving an excess of some particular types of aneuploid gametes.

On the other hand, it might be produced by the selective elimination of some types of autosomal trisomies during the earliest stages of pregnancy, long before they would be used for routine cytogenetic studies in clinical practice; that is, before the second month of gestation.

Direct evidence in favour of the latter suggestion was recently obtained by chromosomal analysis of mature human spermatozoa. This was achieved by interspecies *in vitro* fertilization of denuded (deprived of zona pellucida) golden hamster eggs by mature human sperm. This impressive technique allows visualization of the sperm chromosome complement during metaphase of the first cleavage division. Combined with G-banding it provides an ample opportunity for the precise identification of each element in the human karyotype (Rudak *et al.* 1978). The method has already been applied to the evaluation of the non-disjunction rate in karyologically normal men as well as in men with balanced chromosomal translocations (Martin, Lin, Balkan, and Burns 1982, 1983b; Martin 1984). Detailed chromosomal analysis undertaken by this group on 1000 human sperm revealed chromosomal aberrations in 8.2 per cent, with 5.2 per cent carrying numerical aberrations (2.4 per cent hyperhaploidy; 2.7 per cent hypohaploidy; 0.1 per cent double aneuploidy). It might be inferred from these data that shortage or excess of one chromosome does not seem to interfere with the formation of phenotypically normal sperm, which on taking part in fertilization may give rise to trisomy or monosomy in almost equal proportions of human conceptuses. Moreover, according to these studies, any element of the human karyotype including chromosome 1 may be in excess. The latter type of trisomy, however, has never yet been detected in spontaneous abortuses. These data could be considered as strong evidence favouring unequal postzygotic selection of abnormal genomes rather than differences in meiotic non-disjunction as being chiefly responsible for the unequal participation of different autosomes in karyologically abnormal abortuses (Martin 1984).

Although quite plausible, this view still needs further experimental evidence. The most conclusive results may be expected to come from direct cytogenetical studies of ovulated human eggs or zygotes, as the major part of human aneuploidy seems to arise from meiotic non-disjunction of maturing oocytes (Chandley 1981, 1984). According to recent findings, 72 per cent Ts21 in newborns (Mikkelsen, Poulsen, Grinsted, and Lange 1980) and 92 per cent Ts 13, 14, 15, 16, 21, 22 in spontaneous abortuses (Jacobs *et al.* 1980) are induced by chromosome non-disjunction in female gametes, predominantly during the first meiotic division. The reason for the substantial predominance of maternally derived trisomy remains completely unknown. Whether it reflects the higher frequency of chromosomal non-disjunction in oogenesis compared to spermatogenesis or is due to poor survival of all trisomies of paternal origin still awaits resolution.

Experimental data in mice with Robertsonian translocations unequivocally indicate a higher level of non-disjunction in oogenesis than in spermatogenesis (Baranov and Dyban 1971b; Gropp 1973b; Gropp and Winking 1981), but exceptions to this general rule are sometimes encountered. Heterozygous carriers of translocations involving autosomes 11 and 13 or 8 and 12 generate almost equal proportion of aneuploid gametes in female and male meiosis (Gropp and Winking 1981).

Nonetheless it is tempting to speculate that the actual impact of meiotic faults in human oogenesis also may be substantially higher than in spermatogenesis: but interspecies differences in reproduction as well as in chromosome number and shape make this generalization premature and warn against accepting experimental data without great caution.

Programmes of infertility treatment by *in vitro* fertilization of mature human ova, with subsequent transplantation into the uterus, are now launched in many countries. Besides their clinical value, these studies make a great contribution to our knowledge of the earliest stages of human development, including the type and frequency of chromosomal aberrations in oocytes and cleaving eggs. Only in this way may one expect to get clear-cut information on the actual frequency of aneuploidy in human oocytes and zygotes, on the actual impact of chromosome non-disjunction in male and female meiosis on human aneuploidy, and on the causes of the quite specific pattern of autosomal trisomies in spontaneous abortuses.

Unbalanced karyotypes and autosomal trisomy in particular are much less common in laboratory mammals including mice than in humans (Ford 1971) (see Chapter 11). Cytogenetic studies of normal and abnormal (degenerating) post-implantation as well as pre-implantation embryos of murine rodents (mice, rats) have been carried out in our laboratory. Five-day rat embryos are comparable to human specimens 5 to 7 days old, while 11-day rat embryos are roughly equivalent to 3-week human conceptuses. Abnormal post-implantation embryos in both mice and rats represented by empty extra-embryonic membranes are comparable to human abortuses or 'blighted ova' lacking proper embryonic tissues inside the empty chorionic sacs. These are known to show an exceptionally high rate (up to 70–80 per cent) of chromosomal aberration, including trisomy (Boué and Boué 1970, 1974), so cytogenetic analysis of abnormal embryonic specimens on the 11th day of gestation in rats (Dyban *et al.* 1971) and on the eighth to ninth days of gestation in mice (Baranov and Dyban 1972a) is of special interest.

Of 229 pre-implantation rat embryos with known karyotype, 22 (8.4 per cent) were classified as heteroploids, with 4.7 per cent triploids, 0.4 per cent tetraploids, 1.5 per cent trisomics, 1.2 per cent monosomics, 0.4 per cent chromosomal mosaics, and 0.4 per cent with structural chromosomal aberrations. The pattern of chromosomal abnormalities detected in day-11 rat embryos was quite different. Of 778 embryos examined under the stereomicroscope, 50 malformed or severely stunted specimens and 46 externally normal littermates were chosen for cytogenetic analysis. All embryos of the control group and most of the malformed embryos (44) were found to be diploid. Triploid was revealed in five and tetraploidy in one (2 per cent) morphologically abnormal embryonic specimens. Thus the actual incidence of heteroploidy in day-11 rat embryos related to the total number of embryos studied under the stereomicroscope (778) was only about 0.8 per cent, and did not exceed 12 per cent if related to the total number of abnormal embryonic sacs. Not a single case of spontaneous autosomal trisomy has been detected in post-implantation rat embryos.

Thus, autosomal trisomy is an exceptionally rare event in laboratory rats from the earliest stages of embryogenesis. A very low incidence of spontaneous trisomy is typical also for laboratory mouse embryos (Baranov and Dyban 1972). Of 181 mouse embryos examined under the stereomicroscope, 156 were normal neurula stage embryos whereas 25 (13.8 per cent) were degenerating or malformed. All phenotypically normal embryos and most of their malformed littermates (20) possessed a balanced diploid karyotype; three stunted embryos (12 per cent) revealed triploidy, and two (8 per cent) had isochromosomes and chromosomal mosaicism.

Thus, heteroploidy has been detected in only 2.7 per cent of all embryos with known karyotype and in 20 per cent of all abnormal looking specimens. Spontaneous heteroploidy in mice of early post-implantation stage is represented predominantly by triploidy, while, as in laboratory rats, not a single well-proven case of autosomal trisomy has been detected in these studies so far.

Cumulative data of different authors summarized in Table 1 establish the absence of spontaneous trisomy in post-implantation mouse embryos (Vickers 1969; Baranov and Dyban 1972a; Ford 1970), but a few cases of autosomal trisomy have been recorded among 10–14-day mouse embryos from old females (Yamamoto et al. 1973; Fabricant and Schneider 1978).

The data on cytogenetic studies of pre-implantation mouse embryos are summed up in Table 2. They suggest that the average frequency of trisomy in embryos of different strains is rather low (about 0.3 per cent). However, these results are based on chromosomal analysis of mouse embryos at the morula or blastocyst stage, so do not reflect the true incidence of aneuploid zygotes in mice.

A recently published comprehensive review by Hansmann (1983) covers most of the known publications on the frequency of aneuploidy in ovulated eggs as well as in zygotes at the first cleavage division and in morulae and blastocysts (see also Table 2 and Nijhoff and de Boer 1981). According to these cumulative data the frequency of hyperhaploid oocytes ($n = 21$) in mice may vary significantly (from 0.2 per cent to 2.6 per cent). Maximal levels of hyperhaploidy (4 to 6 per cent) are reached in old females of particular strains. These data corroborate well with our results on cytogenetic studies in ovulated oocytes of murine rodents of different ages. Only 1.9 per cent ovulated eggs were found to have aneuploidy in 2–3-month-old mice of CBA strain (Chebotar 1976a), while the incidence of aneuploidy in CBA females of 10–12 months reached 8.5 per cent (Chebotar 1978). A similar correlation is true for rats: 0.8 per cent of aneuploid oocytes have been found in 2–3 month-old rats (Dyban and Chebotar 1975), but the rate increased almost fivefold (up to 4 per cent) in old females (10–12 months old) (Chebotar 1976b).

An essentially new approach to the assessment of the non-disjunction rate in ovulated mouse oocytes has recently been elaborated in our department, based on a precise chromosome count in previously diploidized one-cell parthenogenetic embryos (Dyban and Noniashvili 1985a,b).

This approach allows visualization of the chromosomal sets of both the ovulated oocyte and its second polar body, and thus assures reliable estimation of hyper- and hypohaploids. Chromosomal analysis of 557 parthenogenetically activated eggs from young hybrid female mice of F_1 (CBA × C57BL) stock revealed aneuploidy in 1.7 per cent, with 0.56 per cent of them hyperhaploid and 1.14 per cent hypohaploid. Thus, meiotic non-disjunction in these young female mice gives rise to aneuploidy in about 2 per cent of all ovulated eggs. Moreover these results indicate appreciably higher levels of hypohaploidy than hyperhaploidy. The predominance of hyperhaploidy cannot be attributed to technical artefacts, as both chromosome sets of the diploidized oocytes had identical chromosomal counts. Predisposition of ovulated oocytes to chromosome loss during meiotic maturation divisions has recently been demonstrated in Syrian hamster eggs fertilized with human sperm (Martin 1984). Out of a total of 3.1 per cent aneuploid eggs penetrated by human sperm, 2.2 per cent were found to be hypohaploid and only 0.9 per cent hyperhaploid. It seems likely that the frequency of chromosomal loss due to incorporation of an extra chromosome into the second polar body occurs more often than the opposite event, persistence of an extra chromosome in the oocyte genome. If this suggestion is confirmed by further studies, existing methods used for the estimation of chromosome non-disjunction rate in mammalian eggs may need to be revised, since they are based on the assumption of equal proportions of hypo- and hyperhaploidy in ovulated eggs. The total level of aneuploidy is usually calculated by doubling the number of hyperhaploid oocytes, in order to reduce the error due to artefactual chromosomal loss. This approach will underestimate the actual level of aneuploidy if there is a real difference between the frequency of hyperhaploid and hypohaploid complements in mammalian eggs.

Spontaneous aneuploidy of early mammalian embryos may also stem from non-disjunction during cleavage as well as from incorporation of an aneuploid male genome at fertilization. The total frequency of chromosomal aberrations in pre-implantation mammalian embryos may therefore be somewhat higher than in ovulated eggs. A few cytogenetic studies of rat and mouse zygotes at the first cleavage division provide direct support for this view. The frequency of trisomy in zygotes from young TO strain females was found to be 1.2 per cent, compared to 4.8 per cent in zygotes from old females (Maudlin and Fraser 1978). In F_1 mice (C3H × 101) these values were 0.6 per cent and 2 per cent, respectively (Tease 1982; see Chapter 4 for more details). An attempt to elucidate both the incidence of aneuploidy and the type of trisomy has been undertaken in zygotes of ICR mice (Martin-Deleon and Boice 1983). Of 321 one-cell embryos of known karyotype, 7.5 per cent were aneuploid (3.6 per cent monosomy, 2.5 per cent trisomy). The incidence of Ts19 was found to be higher than those of Ts12, 13, 14, 15, and 17, while Ts1, 2, 4, 6, 8, and 11 were not detected at all. Of a total of 15 trisomies five possessed an extra chromosome of maternal origin, three of paternal origin and the origin of the extra chromosome in the remaining seven embryos was unknown. The number of trisomic enbryos described is rather small, but nevertheless these data may be taken as the first indication of unequal participation of different chromosomes in aneuploid mouse gametes.

In rats also, the rate of aneuploidy is significantly higher in one-cell zygotes than in ovulated eggs. The overall frequency of chromosomal aberrations in cleaving zygotes was 15 per cent, with 6.5 per cent trisomic and monosomic embryos, 5.5 per cent triploids, and the remaining 2.5 per cent carriers of structural chromosomal aberrations (Chebotar 1976a). Hence, the incidence of aneuploidy during the first cleavage division was almost three times higher than in ovulated rat eggs from old females, and 15–20 times higher than in oocytes of young rats. Thus the non-disjunction rate at fertilization and during spermatogenesis seems to be substantially higher in rats than in mice. A high level of spontaneous aneuploidy during the one-cell stage has been shown for Golden hamsters (Martin *et al.* 1983a).

Thus, spontaneous autosomal trisomy is not a rare event during the first cleavage division in mice, rats, and hamsters, especially in the progeny of old females. However, cytogenetic studies by many authors (see Table 2, Chapter 1) have revealed an exceptionally low frequency of trisomy in mouse embryos at the morula and blastocyst stages. One possible reason for this contradiction lies in the efficiency of chromosomal studies. Because of the known timetable of the first cleavage division, chromosomal analysis of one-cell embryos is almost 100 per cent successful. However, routine chromosomal analysis at the morula and blastocyst stages is usually much less efficient mostly because of the absence or poor quality of mitotic figures. Hence, if one accepts that there will be slightly reduced mitotic activity in trisomic embryos, the chances of picking them up will be rather low. All studies on mice with Robertsonian translocations give clear evidence that no autosomal trisomy affects pre-implantation development (see below), but this does not imply that the survival capacities of trisomic embryos remain unchanged. Trisomic embryos could be more vulnerable to the action of various noxious agents (chemical, biological, physical, etc.) or to unfavourable environmental conditions than their chromosomally normal littermates. Hence, they might be prone to early selection from the very onset of embryogenesis.

According to a few publications devoted to the chromosomal analysis of cleaving zygotes, spontaneous trisomy in mice varies significantly depending on the mother's age as well as on genetic background (see Chapter 4), from 0.6–2.5 per cent in young to 2–5 per cent in old females. In laboratory rats the relevant figures are 1.5–6 per cent and in Golden hamsters 1–2.5 per cent. No information on spontaneous trisomy is available so far for any other mammalian species. Preliminary and rather fragmentary data exist for spontaneous trisomy in pre-implantation embryos of rabbits and pigs (see Tables 2 and 3). A few cases of trisomy have been recorded in cattle and monkeys during post-natal life (McClure, Beiden, and Pieper 1960; Herzog and Hohn 1971).

Thus, autosomal trisomy of spontaneous origin occurs in all mammalian species that have been studied cytogenetically. The impact of trisomy on embryonic wastage may vary substantially between different mammalian species. The frequency of spontaneous trisomy is especially high in human beings, and seems to be low in other mammals, including laboratory rodents.

6.2 Influence of trisomy on development

6.2.1 Pre- and post-natal human development

The characteristic phenotypes of various trisomies in humans are described in a number of papers, reviews, monographs, and textbooks (Prokofieva-Belgovskaya 1969; Prokofieva-Belgovskaya et al. 1973; Bochkov 1971; Davidenkova et al. 1973; Kuliev 1972, 1976; Boué et al. 1976; Boué and Boué 1976; Jacobs 1982), but data on the developmental effects of extra autosomes before implantation are still lacking. It is generally accepted that autosomal trisomies affect morphogenesis more strongly and block embryonic development earlier than other chromosomal abnormalities except autosomal monosomies (see Chapter 7). The progressive increase in the relative frequency of trisomy during the early stages of gestation confirms the latter statement, and suggests that some chromosomal trisomies may be aborted before the pregnancy is recognized.

Aborted trisomies are usually represented by membranous or chorionic sacs with or without embryonic remnants inside. Most types of trisomy in humans are incompatible with advanced embryogenesis and provoke early embryonic death. Macroscopic and microscopic studies of spontaneous abortuses with trisomy B chromosomes revealed severe hypoplasia of amniotic cavities, abnormalities of internal organs, and underdevelopment of amniotic stalk and chorionic villi (Kulazenko 1976). Identical or very similar morphological findings are common for all other types of trisomy, with very variable embryonic damage (Mikamo 1970; Carr 1972).

However, in spite of abundant clinical data, no clear-cut correlation between any type of trisomy and any embryonic phenotype has yet been established (Carr 1972; Jacobs 1982). An extra copy of the same autosome may either devastate the developing embryo and make it totally disorganized with no recognizable pattern, or may be compatible with its almost normal development. Thus current genetic data on genotype–phenotype inter-relationships in human trisomies are very disappointing (Carr 1969; Boué et al. 1976; Boué and Boué 1976). However, half of all embryos with D trisomy have gross facial anomalies (Roux 1970) and detailed morphological studies of fetal membranes in abortuses with certain types of trisomy have revealed both specific features in some and overlapping features between different trisomies (Phillips and Boué 1970; Carr 1972).

Trisomy during the fetal stage is usually restricted to certain chromosomes and is accompanied as a rule by a definite pathological syndrome. The chromosomes concerned are trisomy 21, giving rise to Down's syndrome, trisomy 13, giving rise to Patau syndrome, trisomy 18, giving rise to Edward's syndrome, trisomy 22, and trisomy 8. Of all these trisomies only trisomy 21 is compatible with survival to adulthood and comprises a substantial part of all congenital chromosomal defects (Jagiello 1981). Detailed clinical studies of these trisomies have revealed both non-specific (overlapping) features, and highly specific features characteristic for each type of trisomy. The non-specific features include general stunting, anomalies of heart, brain, face, etc. The specific features are rarely numerous but are of special clinical value for preliminary diagnosis of chromosomal imbalance. The developmental effects of autosomal trisomy in humans should always be considered with at least two basic problems in mind:

1. Early human embryos are only rarely available for morphological or cytogenetic studies and thus we are still almost completely ignorant of the developmental effects of trisomy or any other chromosomal abnormality during the initial stages of human embryogenesis.

2. Clinical geneticists are usually dealing with the final product in a long chain of pathological processes, and most often with only the remains of completely disorganized embryos. Extra-embryonic membranes, especially chorion, display high resistance to any intrinsic or extrinsic injury and often

may retain highly abnormal or already dead embryos for long periods. These macerated embryos usually acquire cylindrical or nodular forms and undergo progressive degeneration before being expelled from the uterus (Dyban 1959). Thus, even the most thorough morphological studies on aborted specimens may provide little reliable information on the pathogenesis of any chromosomal disorder and will fail to detect which embryonic rudiments are most vulnerable to any particular type of trisomy. This natural limitation of clinical cytogenetics should always be kept in mind if the developmental effects of trisomy or of any other chromosomal anomaly are considered. The rare occurrence of some rather specific changes in early human trisomies should therefore be taken with great caution. The development of model systems to study chromosomal aneuploidy is therefore gaining credence.

6.2.2 Development of other mammals

Our knowledge of the developmental effects of autosomal trisomy in mammals other than laboratory mice is very scarce, and is based on just a few cases. In laboratory rats trisomy for chromosomes 3–10 does not affect pre-implantation development and is compatible with the completion of major organogenesis (Dyban *et al.* 1971). The few trisomic rat embryos recovered before implantation had a reduced cell count and low mitotic index compared with diploid littermates on the sixth day of gestation (Fujimoto *et al.* 1975).

In some rare cases autosomal trisomy has also been recorded in post-natal life. One case of an apparently trisomic juvenile water vole (*Arvicola terrestris*) has been reported (Fredga 1968). This unique specimen had no visible abnormalities except a relatively short tail and minor dental anomalies.

Two morphologically normal Chinese hamsters with complete trisomy for chromosome 10 have been obtained in the progeny of an irradiated male (Sonta 1980).

Contrary to the findings for laboratory and non-laboratory rodents, the morphological peculiarities of a post-natal trisomic chimpanzee much resembled those of Down's syndrome (McClure *et al.* 1960). Fluorescent staining of the trisomic chimpanzee chromosomes revealed an extra small acrocentric chromosome similar in its banding pattern to human chromosome 21 (Benirschke, Bogart, McClure, and Nelson-Rees 1974).

6.2.3 Influence of trisomy on development in mice

Since there are 40 chromosomes in the metaphase plates of laboratory mice and 20 pairs of homologous chromosomes in the karyotype (19 pairs of autosomes and one pair of sex chromosomes), a total of 19 different autosomal trisomies can be expected.

The developmental profiles of all these trisomies has already been studied, though to a variable extent. The greatest contribution to our knowledge on phenotypes of mouse trisomies has been contributed by Gropp[a] and his collaborators in Lubeck (FRG) (Gropp *et al.* 1974, 1975, 1976; Gropp 1975, 1981a,b, 1982; Gropp and Winking 1981; Gropp, Winking, and Herbst 1983). Valuable information on the phenotype of trisomy 19 embryos has been obtained by White, Tjio, and co-workers (White *et al.* 1972, 1974a,b; Bersu, Crandall, and White 1982, 1983). Pre- and post-natal development of mouse embryos with trisomy for chromosomes 1, 2, 3, 5, 8, 9, 14, 15, 16, 17, and 19 has also been studied in our group (Baranov and Dyban 1971b, 1972b, 1973; Baranov and Udalova 1975; Baranov 1974, 1976a, 1980, 1983a).

6.2.3.1 Pre-implantation stages. Some preliminary data indicated the absence of pre-implantation lethality in mouse embryos with autosomal trisomy (Ford 1971; Gropp 1973a). This has since been confirmed for autosomes 1, 12, 17, and 19 (Epstein and Travis 1979). Data on pre-implantation development of trisomics have also been obtained by Baranov (1980, 1983a), who studied 175 embryos with different types of autosomal trisomies on the third and fourth days of gestation. One hundred and sixty-one embryos had one and 15 had two different extra autosomes. The main types of cross and trisomy are represented in Tables 15 and 21 (see Chapters 4 and 7).

The preliminary studies were made in mice heterozygous for three different Robertsonian translocations—Rb(8.17)1Iem, Rb(6.15)1Ald, and Rb(9.19)163H, which could be expected to generate gametes with disomy or nullisomy for any of the six chromosomes involved in the centric fusions, 6, 8, 9, 15, 17, and 19.

On the fourth day of the pre-implantation development 22 embryos were found with one extra chromosome, and 13 with two extra chromosomes. Double trisomics predominated in the progeny of heterozygous females. Examination under a stereo- or phase-contrast microscope could not distinguish between trisomic embryos and their normal diploid littermates. Mean cell counts in trisomics and in diploids scored on air-dried preparations were 30.2 ± 1.5 and 33.4 ± 1.0, respectively.

Effects of particular trisomies have been examined in mice heterozygous for one or two different Robertsonian translocations with monobrachial homology (Tables 15 and 21). This breeding design was chosen for trisomies

[a]The great personal contribution made by Alfred Gropp to experimental cytogenetics goes far beyond the discovery of jumping (tobacco) mice with seven pairs of metacentric chromosomes in their karyotype in a remote mountain valley in Switzerland. He was one of the first to adapt these wild mice for the deliberate production of laboratory stocks of mice carrying individual Robertsonian translocations in their genome, and independently with other authors (B. White and J.H. Tjio) he introduced the double translocation technique (see below), suitable for the production of a large number of embryos with deficiency or excess of some specific chromosome.

Table 18. Mean cell count in pre-implantation (fourth day) embryos with different types of trisomy and in their diploid littermates (Baranov 1983a)

Chromosome studied	Number of trisomics	Mean cell count ± m	
		Trisomics	Diploids
1	6	39.6 ± 2.3	38.5 ± 3.9
2	4	47.2 ± 4.1	52.6 ± 2.3
3	8	38.5 ± 3.9	39.6 ± 2.3
5	15	32.5 ± 3.6	29.1 ± 2.4
6	6	50.1 ± 5.3	52.3 ± 4.5
9 or 14	35	31.2 ± 1.3	32.5 ± 0.8
15	19	30.2 ± 1.8	35.4 ± 1.3
16	4	31.3 ± 3.6	30.8 ± 1.6
17	21	38.1 ± 2.3	39.6 ± 1.5
19	14	29.8 ± 3.7	35.6 ± 3.7

5, 15, 17, and 19. The number of embryos with any particular type of trisomy recovered on the third or fourth day of gestation as well as the average cell count from air-dried preparations are presented in Table 15. Embryos trisomic for autosomes 1, 2, 3, 5, 6, 9, 14, 15, 16, 17, or 19 appear to proceed quite normally through pre-implantation development, with no adverse effects on cleavage, compaction or blastulation. The mean cell count in each type of trisomy was always in good agreement with the corresponding figure for normal diploid littermates (Table 18). Some minor reduction of the cell count was seen in trisomy 15 embryos (Ts30.2; Diploid 35.4) and some increase of the cell count was observed in trisomy 17 embryos (Ts40.0, Diploid 38.0). However, neither of these deviations was statistically significant.

Thus, in our studies trisomy for any of 11 autosomes is quite compatible with normal cleavage and blastulation. Taking into account some earlier studies of pre-implantation trisomies (Ford 1971; Gropp 1973b; Epstein and Travis 1979) it seems justifiable to suggest that an extra copy of any autosome in mice has no harmful effect on pre-implantation development.

6.2.3.2 Post-implantation stages. There is good evidence against selection of trisomies before implantation in laboratory mice (Gropp 1973; Ford 1972; Baranov 1976a). In agreement with this, we found that the proportion of trisomic embryos recovered at pre-implantation stages in mice heterozygous for three different Robertsonian translocations—Rb1Iem, Rb1Ald, and Rb163H—was almost identical to that at post-implantation stages.

All 19 types of single trisomy have been found at the stage of major organogenesis in mice, but double trisomies (NF = 42), though recorded before implantation, were rather rarely encountered later, either in our experiments or in those of Gropp. It therefore seems that at least some double trisomies cannot support post-implantation development in mice.

The developmental profile for each type of trisomy is rather specific and probably depends on the genetic content of each particular chromosome and on the expression of its genes during embryogenesis. Let us now consider the characteristics of post-implantation embryogenesis in mouse embryos trisomic for each of the 19 pairs of autosomes.

Chromosome 1. Over 80 embryos with trisomy 1 have been studied on the tenth to fifteenth days of gestation. Most (about 60) have been collected by Gropp's group in the progeny of double translocation heterozygotes with monobrachial homology Rb(1.3)1/Rb (1.10)10 Bnr (Gropp *et al.* 1975). The rest have been obtained in the progeny of single heterozygotes Rb(1.3)1 Bnr by our group. All the data show that the development of Ts1 embryos proceeds normally during major organogenesis but is progressively retarded thereafter. After the twelfth day, general hypoplasia with variable degrees of retardation becomes the common feature of all Ts1 embryos irrespective of their genetic background. The average body weight of Ts1 embryos in our studies did not exceed 10 mg and the crown–rump length was only 6–8 mm, compared with 16 mg and 11–12 mm for diploid littermates. Gross inspection of these trisomics under a stereomicroscope, in transverse sections made by a razor blade, or in routine histological preparations, failed to show any specific abnormality in any of our specimens (Fig. 14). By contrast, most trisomics in the progeny of NMRI females crossed to double heterozygotes in the experiments of Gropp and his collaborators displayed heavy cranial dysmorphy. The heads of such embryos were small, flattened, and oddly shaped because of disproportionate delay in the development of the mandibles and tongue. The picture of facial dysplasia with microgenia was most frequent with advancing developmental age of the trisomic embryos. Some of the most severely affected Ts1 embryos had disproportional retardation of their brain, especially of the telencephalon.

The nature of this developmental anomaly in Ts1 embryos became evident from histological investigations of the head region. Impairment of prosencephalon development, defective ocular primordia with typical holoprosencephaly, and cyclopia were the most common features of malformed embryos. Very rarely, Ts1 embryos developed a typical hypotelorism with holoprosencephaly and missing olfactory nerves (Domarcus 1983). In some other Ts1 embryos no malformations other than hypotelorism were found. The most severe manifestation of this trisomy included holoprosencephaly combined with aprosopia or cyclopia.

One more trait of Ts1 embryos common to specimens of different genetic background was the developmental retardation and small size of the placenta. According to our data the labyrinth portion of placentae was

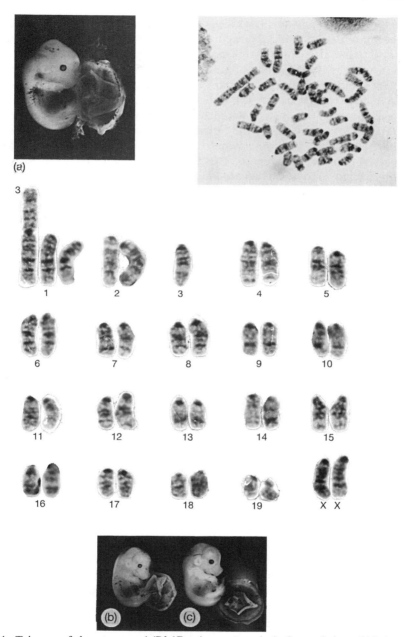

Fig. 14. Trisomy of chromosome 1 (Rb1Bnr/♂ × ♀ + / +). General view of 15-day embryos fixed in Bouin's fluid, trypsin–Giemsa staining of chromosomes; (a,b) NF = 41; (c) control specimen, $2n = 40$.

disproportionately small, with undifferentiated cytotrophoblast occupying the major part of the fetal placenta. The failure of placental growth, differentiation, and efficiency was probably one of the principal causes of trisomic death by the sixteenth day of gestation.

In addition to morphological observations, biochemical dosage studies have been carried out in Ts1 embryos on the twelfth or thirteenth days of development. The specific activity of the enzyme isocitrate dehydrogenase (IDH-1) controlled by a gene carried in chromosome 1 was 1.53 times more than in normal embryos, while the activity of other enzymes (glucose-6-phosphate dehydrogenase, G6PD; 6-phosphogluconate dehydrogenase, 6PGD) known to be carried on other chromosomes was not altered (Epstein, Tucker, Travis, and Gropp 1977).

Thus an extra copy of chromosome 1, the largest in the karyotype of the laboratory mouse, is compatible, with post-implantation development and advanced organogenesis, although it is inevitably lethal thereafter. The phenotype of Ts1 embryos shows a remarkable range of variation, which probably stems from genetic background effects or from the number of Robertsonian translocations in their karyotype.

Chromosome 2. Data on the manifestation of Ts2 are still very meagre. Twenty-one Ts2 embryos have been recorded in the progeny of double heterozygotes Rb(2.18)6Rma/Rb(2.17)11Rma (Gropp *et al.* 1983) and only three in the progeny of single heterozygotes Rb(2.6)4Iem/+ in our studies. Twenty-one trisomics were collected on the ninth–tenth days of gestation and four were found on the twelfth day. Severe developmental retardation and hypoplasia were predominant features for all Ts2 embryos irrespective of their developmental stage. Smallness of the cranial region and failure of neural plate closure were common to all Ts2 embryos in our studies.

Thus Ts2 may be considered as a rather early embryonic lethal in laboratory mice. In spite of genetic background differences it does not permit survival beyond the tenth day of gestation, with most embryos being eliminated during major organogenesis.

Chromosome 3. The triplication of this autosome belongs to the group of still rather poorly studied chromosomal abnormalities distinguished by their early adverse effects on post-implantation development. About 14 Ts3 embryos have been obtained in the progeny of double heterozygous females Rb(1.3)1Bnr/Rb(3.8)2Rma crossed to NMRI males (Gropp *et al.* 1983). Severe general retardation with a disorganized phenotype was the most common manifestation of Ts3 during major organogenesis and none of the trisomic embryos survived beyond day 13 of gestation.

Chromosome 4. Information on developmental effects of Ts4 is still very scarce. According to Gropp and his collaborators (Gropp *et al.* 1983), 50 per cent of 11-day-old embryos displayed trisomy 4 in the progeny of NMRI females crossed to the doubly heterozygous males—Rb(4.17)13Lub/Rb

(4.15)4Rma. All resembled their normal littermates with diploid karyotype. However, all 13 trisomics at more advanced stages (days 12 and 13) were moderately retarded and hypoplastic. The recovery rate of Ts4 embryos on the twelfth day was less than half that on day 10 (20 per cent and 49.1 per cent, respectively) and on the thirteenth day only a few were found.

Whether this type of trisomy is compatible with more advanced embryonic development, and if so to what extent, is still unknown.

Chromosome 5. The only data collected so far on this type of trisomy come from the 10.5-day embryos obtained by our group in the progeny of mice heterozygous for translocations Rb(5.19)1Wh or Rb(5.15)3Bnr. A single Ts5 embryo has occasionally been found in the progeny of three Robertsonian translocation heterozygotes (Rb1Ald, Rb1Iem, Rb163H/ + + +), none of which involve chromosome 5 in the centric fusion. All ten trisomics looked alike and displayed severe hypoplasia and retardation similar to that shown in Fig. 15. The average weight was about 0.5–1.0 mg compared to 12–13 mg for their diploid littermates. The trisomic embryos were still not completely separated from the yolk sac wall, hence their rotation was delayed and they still remained U-shaped. The neural tube was open throughout its length, with no signs of otic or eye placodes in the cranial region. The absence of heartbeat, maldevelopment of allantoic stalk, and very poor vascularization of the yolk sac indicated that the developmental block of Ts5 embryos occurred at the onset of major organogenesis, and by the eleventh day of gestation they were dead.

Chromosome 6. The breeding system for the production of Ts6 embryos is based on double metacentric heterozygotes—Rb(4.6)2Bnr/Rb(6.15)1Ald (Gropp *et al.* 1975), but we know little of the developmental capacities and phenotype of the affected embryos. Some preliminary data indicate that Ts6 embryos may develop up to organogenesis (Gropp 1982). All Ts6 embryos display a syndrome of slight to moderate, sometimes even severe retardation and hypoplasia. Individual phenotype may range from near normality to severe hypoplasia (Gropp *et al.* 1976). Thus, Ts6 may be regarded as a rather late embryonic lethal, causing elimination of all affected embryos by 15–16 days of gestation.

Chromosome 7. Developmental profiles of Ts7 embryos were studied in almost 100 specimens obtained from 'all acrocentric' females crossed to the doubly heterozygous males Rb(6.7)13Rma/Rb(7.8)28Lub (Jude, Winking, and Gropp 1980; Gropp *et al.* 1983). The number of trisomics progressively decreased from 32 per cent on the ninth day to 2.9 per cent on the thirteenth day of development. Although we unfortunately still lack detailed phenotypic description of the embryos, Ts7 may be considered as a rather early embryonic lethal.

Chromosome 8. Twenty-one embryos with triplication of this chromosome have been studied on the tenth to twelfth days of gestation. Eleven were

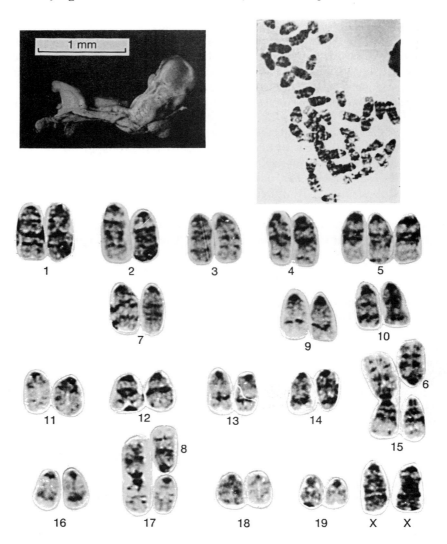

Fig. 15. Trisomy of chromosome 5 (Rb1Iem, Rb1A1d, Rb163H/+ + +)♂ × +/+ ♀. General view of 10.5-day embryo, fixed in Bouin's fluid. Severe retardation, craniorachischisis, reduction of head region; optic vesicles, hyoid and mandibular arches are absent. Trypsin–Giemsa staining of chromosomes.

found in the progeny of double heterozygotes—Rb(8.12)5Bnr/Rb(8.17)1Iem (Gropp *et al.* 1975)—and 10 more in the progeny of single heterozygotes—Rb(8.17)1Iem (Baranov and Udalova 1975). The results of both groups coincide well and favour general growth retardation as a basic developmental

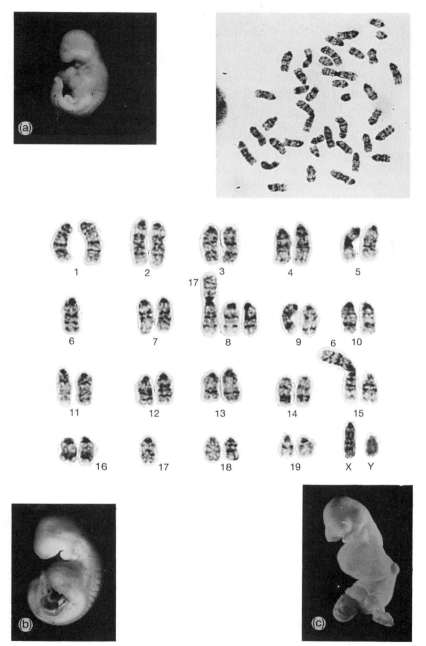

Fig. 16. Trisomy of chromosome 8(Rb1Iem, Rb1Ald/ + +) ♀ × ♂ + / + . General
view of tenth (a,b) and eleventh (c) day embryos fixed in Bouin's fluid. General
retardation, reduction of hyoid mandibular arches, expanded heart chambers, blunt
allantoic stalk(c). Trypsin–Giemsa staining of chromosomes.

manifestation of Ts8. According to our observations, Ts8 embryos of the tenth day were completely separated from their yolk sacs. Their crown–rump length was about 3–4 mm and the somite number approached 20. Inspection under a stereomicroscope revealed a disproportionately reduced cranial region in all trisomics and non-closure of the middle part of the neural tube in two (Fig. 16). Some Ts8 embryos displayed a failure of allantoic growth. Most of the 12-day trisomics showed empty and much dilated heart chambers, with an extremely thin heart wall as well as vast haemorrhages into the trunk and leg rudiments. Histological findings included necrotic changes in the outer layer of the neural tube, in the eye and otic placode, as well as in the maxillary processes. Undifferentiated brain vesicles and other morphological features of Ts8 embryos indicated that their development was blocked by the tenth day of gestation. A significant range of morphological variation occurred in most of the trisomics studied histologically. A notochord anlage was found in two day-12 trisomics but was absent in two others. Maldevelopment of liver rudiments and allantoic vessels gave evidence in favour of inefficiency of the allantoic circulation as the most probable cause of the early embryonic death. Whether this defect was due to delay in allantoic growth or to some major defects of the chorio-allantoic placenta and liver rudiment remains to be found out.

Thus, Ts8 should be considered as a rather early embryonic lethal. The failure of chorio-allantoic placenta function is most likely to be responsible for the early elimination of these trisomics.

Chromosome 9. Most of these trisomics (22) have been found in 11–12-day embryos of NMRI mice mated to double heterozygous males— Rb(9.14)6Bnr/Rb(9.16)9Rma (Gropp *et al.* 1983). Only four Ts9 embryos have been recovered in the progeny of single heterozygotes—Rb(9.14)6Bnr/ + and Rb(9.19)163H/ + (Baranov and Udalova 1975). Developmental features of day-10 trisomics in our studies included non-specific general retardation, delay in allantoic growth, marked hypoplasia of the head with non-closure of the neural tube segments of the middle and hind brains (Fig. 17), reduction of maxillary processes, mandibular and hyoid arches. Examination of histological sections revealed prominent dilatation of heart chambers with concomitant thinning of the heart walls, and absence of liver anlage. Eye and ear placodes remained non-invaginated or were missing. Initial stages of exencephaly were even more prominent on the eleventh to twelfth days of

Fig. 17. Trisomy of chromosome 9 Rb1Iem, Rb163H/ + + $\circ$ × $\circ$ + / + . General view of 10.5-day embryo fixed in Bouin's fluid. General retardation, cranioschisis, expanded heart primordium, failure of allantois growth, reduction of hyoid and mandibular arches. (a,b) Transverse histological sections; (a) Ts9, (b) control, NF = 40.

1 = diencephalon, 2 = myelencephalon, 3 = optic vesicles, 4 = anterior cardinal vein, 5 = internal carotid artery, 6 = maxillary processes, 7 = Rathke's pocket, 8 = oral cavity. Haematoxylin–eosin.

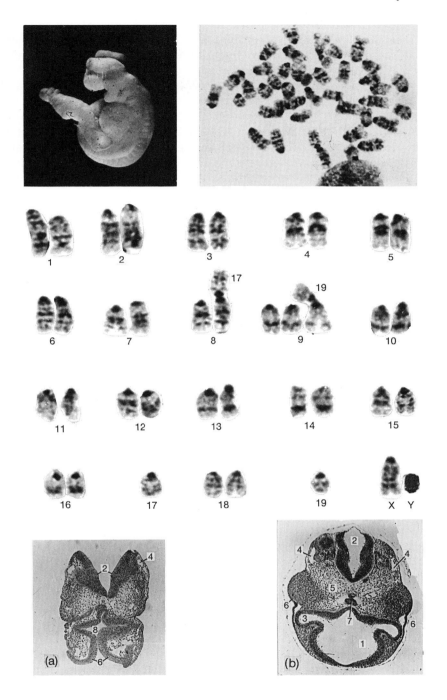

development. No Ts9 embryos were recovered on the thirteenth day, indicating their inability to survive beyond major organogenesis. The failure of allantoic growth with subsequent delay of proper chorio-allantoic placentation may play a major role in the elimination of Ts9 embryos.

Chromosome 10. The triplication of this autosome has been recorded both in the progeny of single heterozygotes such as Rb(1.10)10Bnr and in double metacentric heterozygotes—Rb(10.11)8Bnr/Rb(1.10)10Bnr (Gropp *et al.* 1974, 1975). The number of Ts10 embryos is small and we are still almost totally ignorant of the developmental capacities and phenotypic characteristics of this trisomy. Ts10 embryos display developmental retardation and are smaller than their normal littermates. Most show slight to moderate hypoplasia. Some survived to the fifteenth day of gestation. No gross external malformations were observed, but 62 per cent had ventricular septal defects.

Thus, chromosome 10 seems to belong to that small group for which an extra copy is compatible with fairly lengthy survival. The affected embryos are eliminated not long before birth. More experimental studies are needed to extend and support these preliminary conclusions.

Chromosome 11. The developmental profile of Ts11 embryos has been studied from the ninth to the twelfth day of gestation in the progeny of NMRI female mice mated to the double translocation males—Rb(11.13)4Bnr/Rb(10.11)8Bnr (Gropp *et al.* 1975, 1983). The data on morphological features of Ts11 embryos are still rather meagre. Tenth day trisomics displayed severe retardation and hypoplasia. In extreme cases the trisomics were represented by disorganized degenerative masses. The number of trisomics gradually decreased from the ninth to the eleventh day of gestation (13.8 and 10.8 per cent respectively) but none has been found alive on the twelfth day. The causes of this massive degeneration remain completely unknown.

Thus the available data suggest a very early lethal effect of Ts11, like that of Ts4 and Ts8.

Chromosome 12. The trisomy for this autosome is undoubtedly one of the most thoroughly studied aberrations in laboratory mice so far. Over 120 Ts12 embryos have been studied under the stereomicroscope, in histological sections, and by scanning electron microscopy. The developmental age of the trisomics varied from the eighth day to almost term. Most Ts12 embryos were found in the progeny of NMRI or other 'all acrocentric' females crossed to Rb(8.12)5Bnr/Rb(4.12)9Bnr male heterozygotes (Gropp *et al.* 1975, 1976, 1983). In contrast to many other autosomal trisomies, Ts12 embryos possess a very specific phenotype. Nearly all trisomics encountered between days 12 and 17 displayed a malformation syndrome with exencephaly as the most conspicuous feature. They made up 7–17 per cent of all implants in NMRI females (Gropp *et al.* 1976). However, in the progeny of C57BL/6J females crossed to the double heterozygotic males only half of all trisomics displayed exencephaly (Gropp 1982). Thus the expression of Ts12 depends on genetic background. At fetal stages the defect of the cranial region was always complete, with brain tissue protruding beyond the cranium to make a 'cap-like' outgrowth. The eyes were remarkably small, hypoplastic or often completely absent by external inspection. In more severe cases the brain was even more severely affected and the trisomic embryos displayed more or less pronounced anencephaly (Gropp and Kolbus 1974). The trisomic embryos with exencephaly were usually of normal or near normal size while all anencephalic trisomies were always severely retarded and hypoplastic.

The developmental characteristics of exencephaly and accompanying involvements of the skull base, eyes, and inner ears have been thoroughly studied (Putz and Morris-Kay 1981). Detailed morphometric and stereo-logical measurements indicated that the malformations of neural tube closure in the region rostral to the otic pits was primary, and constituted the most specific effect of Ts12. All other malformations, including eye and skull defects, should be considered as secondary effects of brain hypoplasia. Examination by means of scanning electron microscopy revealed hypoplasia and reduced somite number in the early pre-exencephalic Ts12 embryos on the ninth day. External differences from control embryos were already evident at the late pre-somite stage (eighth day). Everted lateral edges of the neural tube in the forebrain segment remained apart and their free margins grew laterally, failing to oppose and fuse. Pre-exencephalic Ts12 embryos were found to be a very useful experimental tool for further study of the various cellular and extracellular components during the early stages of abnormal neurulation (Putz and Morris-Kay 1981).

A special study was undertaken to compare the morphological characteristics of exencephaly in 13- to 15-day-old mouse embryos caused by Ts12 and by excessive doses of vitamin A (Putz, Krause, Garde, and Gropp 1980). Significant similarities as well as distinct differences were disclosed. In Ts12 embryos, malformation of the cranial region affected the neural tube only, with paraxial mesoderm being the primary target; whereas vitamin A excess apart from exencephaly caused generalized mesenchyme abnormalities, especially prominent in the skull base and face.

More recently, other morphological traits have been discovered in at least half of all Ts12 embryos. These include congenital cardiovascular malformations and ventricular septal defects (Pexieder, Miyabara, and Gropp 1981).

The lifespan of Ts12 embryos may vary significantly depending on the genetic background of the parental strains. The Ts12 embryos hardly survived beyond the twelfth day in C3H or certain other strains, while they remained viable almost until term on the C57BL/6J genome (Gropp and Grohe 1981). On NMRI genetic background the death of a number of Ts12 embryos occurs between days 15–17 while at least some found on the nineteenth day were already dead and autolysed.

Underdevelopment of the embryonic part of the chorio-allantoic placenta, with hypoplasia and differentiation arrest of its circulatory system, is the most probable reason for the death of all Ts12 embryos (Gropp 1981a,b), but the developmental defects displayed at earlier stages were probably induced by some other cause. Retarded growth of Ts12 embryos could originate from the impaired proliferation capacity of trisomic cells. However, the *in vitro* growth pattern of isolated fibroblast-like cells appears to show no clear-cut changes of cell cycle kinetics (Gropp *et al.* 1976; Gropp 1981), and normal differentiation and proliferation of Ts12 embryonic cells have been clearly shown by Gropp and his associates. Trisomic haemopoietic stem cells from the fetal liver could be readily rescued if injected into lethally irradiated adult mice. Transplanted Ts12 stem cells are able to restore the host's haemo- and lymphopoiesis successfully and permanently with documented survival of more than 220 days (Herbst, Pluznik, Gropp, and Uthgenannt 1981; Plunznik, Herbst, Lenz, Sellin, Herzog, and Gropp 1981). Moreover, some slight submicroscopic changes found in Ts12 cells disappeared after grafting into radiation chimaeras. The erythropoietic cells, including the organelles of haemoglobin biosynthesis, were morphologically normal at the electron microscopic level in Ts12 radiation chimaeras. These results suggest that the haemopoietic stem cells of Ts12 embryos are able to undergo complete or almost complete maturation (Gropp 1982). However, the immune reactivity of spleen lymphocytes with Ts12 was rather restricted compared to controls (Sellin, Schlizio, and Herbst 1982), and if Ts12 stem cells were transplanted into the competitive (mixed cell) system with normal cells, inferiority of trisomic cells was observed (Gropp 1982).

Thus, not all abnormal characteristics of Ts stem cells are restored to normal after their grafting to normal diploid irradiated recipients.

In summary, Ts12 is compatible with rather prolonged survival of mouse embryos. It may induce specific malformations, with exencephaly being the most prominent trait. Both the survival time and malformation pattern of trisomic embryos depend on genetic background. Trisomic cells may be rescued by grafting to previously irradiated adult recipients. This cellular system, isolated from the trisomic organism, gives ample opportunities for the study of phenotype–genotype relationships in trisomic embryos.

Chromosome 13. Ts13 embryos were obtained from NMRI females mated to male mice with double translocations—Rb(11.13)4Bnr/Rb(6.13)3Rma (Hongell and Gropp 1982). The triplication of chromosome 13 is compatible with prolonged survival, as many Ts13 embryos were found at all stages between days 12 and 19, and some (13.6 per cent) were born alive. Over 80 trisomics were studied morphologically, both with a stereomicroscope and in histological sections. All Ts13 embryos, irrespective of their age, revealed general developmental delay. Both crown–rump length and the weight of trisomics corresponded to those of normal diploid embryos with 1–1.5 day lag.

The most conspicuous external features of developmentally advanced trisomics included marked transient oedema of the neck and back, delayed ossification, and cleft palate. About 86 per cent of Ts fetuses from day 16 onwards had cleft palates.

Microscopic examination of some serially sectioned Ts13 embryos uncovered numerous cardiovascular malformations, including pulmonary stenosis (86.2 per cent) and 'double outlet right ventricle'(10 per cent) (Pexieder *et al.* 1981). Pulmonary stenosis is a rather rare abnormality in other trisomies, so might be considered as a specific manifestation of Ts13.

The lifespan and phenotype of Ts13 embryos correlate with the genetic background of the parental strains. Although Ts13 embryos showed better survival *in utero* than Ts12 embryos, the haemopoietic cells were ineffective in fully or permanently restoring the haemo- and lymphopoietic systems of lethally irradiated recipients (Gropp 1982).

Thus, an extra copy of chromosome 13 is harmful for differentiation and for haemopoietic stem cell function.

Chromosome 14. Developmental effects of Ts14 have been analysed both in mice heterozygous for a single Robertsonian translocation Rb(9.14)6Bnr (Baranov 1974; Baranov and Udalova 1975) and in mice heterozygous for double translocations with monobrachial homology—Rb(9.14)6Bnr/Rb(6.14)16Rma (Putz and Morris-Kay 1981). About 75 Ts14 embryos have been studied altogether by both groups (30 in the progeny of single and 45 in double heterozygotes). Most were collected during pre-somite or early somite stages (52), about 20 during fetal development, and only three in newborns. The average frequency of Ts14 was in the range 7.6–8.6 per cent for single heterozygotes and 15–25 per cent for double, and did not change much during organogenesis and fetal development. Morphological studies included examination with a stereomicroscope, serial histological sections, scanning electron microscopy. In agreement with other trisomies, general retardation and moderate hypoplasia were the two most common manifestations of Ts14 during fetal development (Fig. 18). The average crown–rump length of day 18 trisomics was below 15 mm and their weight was less than 600 mg (Table 19). In a single case trisomy 14 was proved solely on chromosomal preparations, as the length and weight of the trisomic were within normal limits. This unique embryo, however, had a common placenta (fused placenta) with a karyologically diploid littermate. A small proportion of diploid cells was found in this trisomic embryo and vice versa—the otherwise normal littermate contained a few Ts cells in its liver. This suggests some exchange of blood cells through placental anastomosis, so both embryos in a strict sense should be considered as natural spontaneous chimaeras. Whether the normal diploid blood cells, the improved placental function or some other unknown

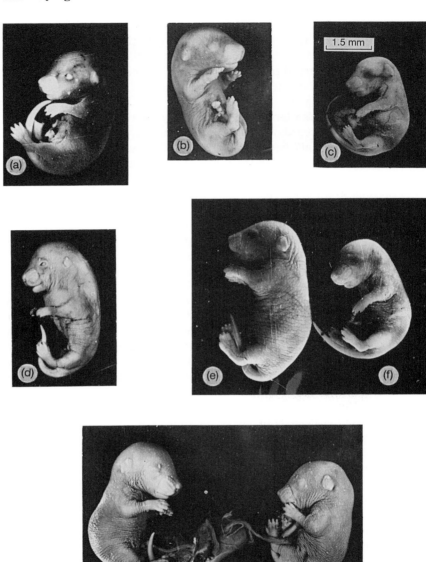

Fig. 18. Trisomy of chromosome 14, Rb6Bnr/ + ♂ × ♀ + / + . Eighteenth (a–f) and nineteenth (g,h) day embryos fixed in neutral 10 per cent formalin (a,b) or in Bouin's fluid (c-h). (a,b,d) General subcutaneous oedema, haemorrhages; (c-f) haemorrhages, stunted in development; (e) control diploid (NF = 40); (g,h) fused placentae (right specimen with predominated Ts14 cells).

Table 19. The number of Ts14 mice embryos with different abnormalities studied on the sixteenth to eighteenth days of gestation (Baranov 1974)

Fetus phenotype	Number
Normal	1 (fused placentae)
Retarded without visible abnormalities	6
Retarded with gross abnormalities	9
Exencephaly	5
Hydrocephaly	1
Cleft palate	1
Oedema and severe haemorrhage	2
Total	16

factor restored the developmental capacities of this Ts14 embryo remains obscure.

The most conspicuous feature of Ts14 embryos was an exencephaly (Fig. 19), which quantitatively as well as qualitatively differed from the similar anomaly in Ts12 embryos. Only half of all Ts14 embryos displayed any pathological outgrowth of neural tissue (Fig. 19). This reduced incidence of exencephaly was found in the progeny of both single and double heterozygotes, mated to mice of different genetic background (Putz and Morris-Kay 1981; Baranov 1974).

Two different types of Ts14 embryo (with and without exencephaly) can already be distinguished at the late pre-somite stage. One type of embryo showed crumpled neural folds at the zero to two-somite stage while the other displayed no gross abnormalities of neural folds at pre-somite or early somite stages (Putz and Morris-Kay 1981), though later development might proceed abnormally (Baranov 1976a). The neural tube defect of Ts14 embryos is somewhat less severe than in Ts12 embryos of equivalent age. Scanning electron microscopy has shown that, unlike Ts12, Ts14 embryos usually display normal closure of the neural tube up to the hindbrain/midbrain junction. Only the forebrain region of the neural tube remained open throughout the somite stage, giving rise to the typical exencephaly of Ts14 fetuses. Thus, exencephaly in Ts14 embryos is morphologically different from the similar brain abnormality in Ts12 embryos (Putz and Morris-Kay 1981).

Gropp has suggested that the term 'semispecificity' of malformations in chromosome disorders may be appropriate for this sort of finding. One more difference is that the eye rudiments are of normal size and appearance

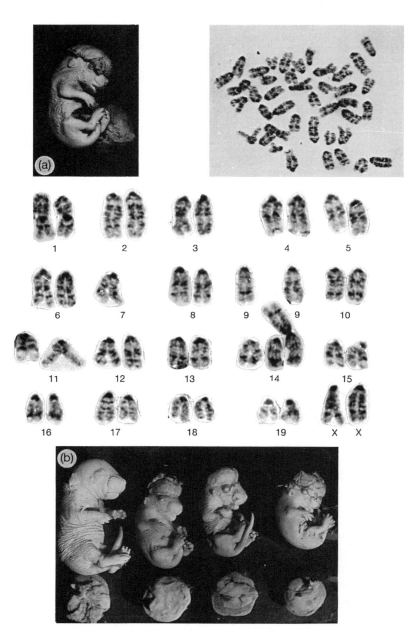

Fig. 19. Trisomy 14, Rb6Bnr/ + ♂ × ♀ + / + . (a) General view of eighteenth day specimen and its trypsin–Giemsa karyotype. (b) Control fetuses (left), eighteenth day, Bouin's fixative, and three Ts14 specimens with exencephaly and their placentae.

in Ts14 embryos, while they are usually severely affected in Ts12 embryos (see above).

One Ts14 embryo on the eighteenth day had a cleft palate and two were distinguished by massive haemorrhages and severe oedema. No visible skeletal changes could be detected on alizarine-stained preparations of 16–18 day trisomics. Three trisomics were still alive immediately after birth and could easily be distinguished by their small size, general hypotrophy, and cyanotic colour. Two had no visible abnormalities while one displayed hydrocephaly and cleft palate. Non-expanded small lungs suggested that inability to breathe was the major cause of death.

Most Ts14 embryos display various types of cardiovascular malformation. About half show ventricular septal defects, 30 per cent double outlet right ventricle, and 10 per cent transposition of great arteries (Pexieder *et al.* 1981). No conspicuous pathological changes could be identified on histological sections of the liver, kidney, and internal secretion glands of trisomic embryos (Fig. 20), but retardation of differentiation and growth of thymus and gonads was always seen. Regions of extensive cell necrosis were especially prominent in the ovaries of some trisomic female fetuses. Whether this degeneration was due to elimination of Ts14 oocytes bearing an unpaired extra chromosome 14, or was delayed physiological death of some oocytes during the pachytene stage of meioses remains unknown.

The developmental delay of placental growth was most prominent in the labyrinthine part. Poor vascularization of the yolk sac with very small hypotrophic villi of the visceral endoderm was also characteristic of all Ts14 embryos (Fig. 21).

Transplanted Ts14 stem cells are able to restore haemo- and lymphopoiesis successfully and permanently in lethally irradiated adult mice. Recipient mice survived for a long time, with no evidence of secondary karyotype changes. (Gropp 1982.)

Thus, Ts14 is compatible with embryonic and fetal development and even with birth, though all newborn trisomics die because of breathing problems. Ts14 may result in specific (semispecific) changes of phenotype, including exencephaly, cardiovascular defects, necrosis in the ovaries, developmental delay of thymus, yolk sac, and placenta. Non-specific changes of Ts14 embryos are represented by the general growth retardation, oedema and severe haemorrhages. Ts14 is not a typical cell lethal as trisomic cells demonstrate normal proliferation and differentiation capacities in embryonic and adult chimaeras.

Chromosome 15. Trisomy 15 is the most frequent chromosome abnormality in spontaneous leukaemia of AKR mice (Herbst *et al.* 1981), as well as in some T- and B-cell leukaemias of experimental origin (Chan, Ball, and Sergovich 1979; Hagemeier, Smit, Govers, and Both 1982). Little is known, however, of its developmental effects. Male double heterozygotes

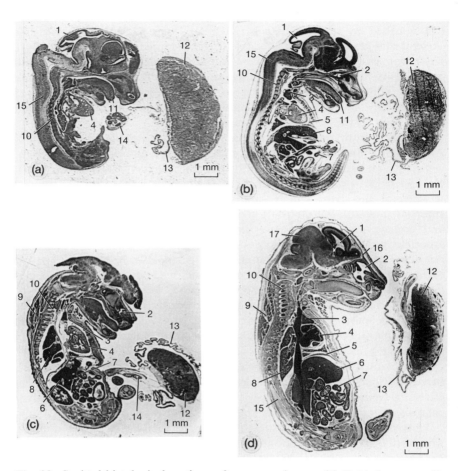

Fig. 20. Sagittal histological sections of mouse embryos with Ts14, haematoxylin and eosin. (a) Sixteenth day specimen—spinal cord hernia; (b) control, NF = 40; (c) Eighteenth day specimen (Ts14), exencephaly; (d) control, NF = 40.

1 = mesencephalon, 2 = nasal chambers, 3 = thymus, 4 = heart, 5 = diaphragm, 6 = liver, 7 = intestine, 8 = lungs, 9 = back muscles, 10 = vertebra, 11 = tongue, 12 = placenta, 13 = yolk sac, 14 = umbilical stalk, 15 = spinal cord, 16 = dura mater, 17 = mesencephalon.

(Rb(9.15)3Bnr/Rb(6.15)1Ald have been mated to 'all acrocentric' females of various genetic backgrounds both by Gropp and his associates (Gropp *et al.* 1975) and by our group (Baranov and Udalova 1975). Over 20 embryos were identified in our experiments. At 10.5 days of gestation, five trisomics were

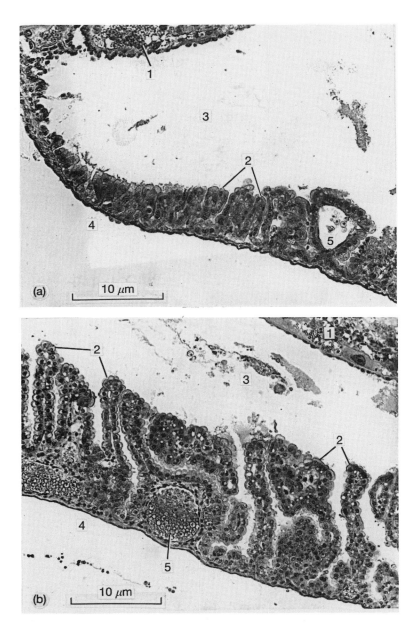

Fig. 21. Transverse sections of the yolk sac of sixth day mouse embryo (a) with Ts14 and (b) with normal diploid karyotype; Mayer's haemalum and eosin. 1 = chorio-allantoic placenta, 2 = yolk sac villi, 3 = yolk sac cavity, 4 = coelom, 5 = omphalic vessels.

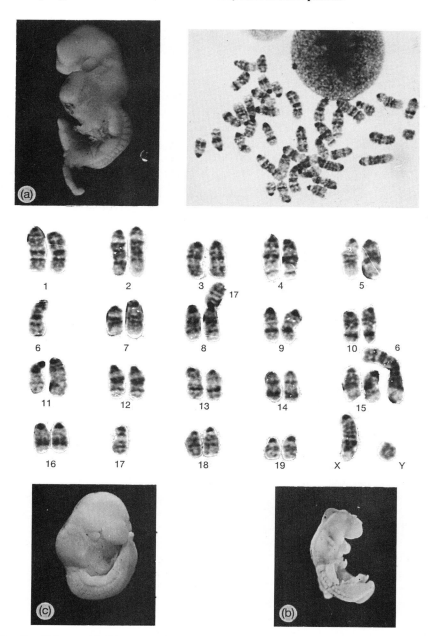

Fig. 22. Trisomy 15; Rb1Iem, Rb1Ald/ + + ♂ × ♀ + / + .; tenth (a,b) and eleventh day (c) embryos fixed in Bouin's fluid. General retardation, pathological bend of spinal cord, cranioschisis; trypsin–Giemsa staining of chromosomes.

represented by small, pale vesicles containing the remains of unorganized embryonic tissue. The other trisomics, though severely retarded and hypoplastic, were nevertheless completely separated from the visceral yolk sac, had accomplished their rotation, and had normal allantoic stalks and closed neural tubes with already three brain vesicles in the cranial region (Fig. 22). Defective neural tube closure in the hind brain region was found in only one severely retarded 10-day embryo. The crown–rump length of the most advanced 10.5-day trisomics was about 3–4 mm, their somite number near 20 and the mean weight 2.5 mg. Eight of the 20 trisomics displayed rather specific malformations of the neural tube, with an abnormal curve in the breast or lumbar regions visible even on external examination (Fig. 23). Microscopic studies of serially sectioned trisomic embryos with pathological neural tube curvature showed unusual outgrowth of unorganized neural tissue, occupying the lumen of the neural tube or making asymmetrical protrusions into the surrounding mesenchyme (Fig. 23). The nature of these pathological protrusions is still completely unknown. Keeping in mind that an extra copy of chromosome 15 is closely related to various forms of neoplasia in mice (see above), these protrusions might represent either initial steps of malignant transformation or some unusual internal properties of neuroectodermal cells acquired with triplication of chromosome 15. Severe retardation and hypoplasia of Ts15 embryos have also been reported by the Lübeck group, but no specific abnormalities of the neural tube were detected in their studies. Whether the disorganized growth of neural tissue depends solely on the genetic background of the parental mice in our studies, or whether they reflect specific properties of neuroectodermal cells with triplicated chromosome 15 is a matter for clarification in future experiments. The absence of liver anlagen and highly dilated heart chambers with very thin heart walls were other non-specific traits of Ts15 embryos.

Thus, Ts15 should be regarded as a rather early embryonic lethal, which may induce a highly specific malformation of the neural tube in at least some affected embryos.

Chromosome 16. Pathological studies of Ts16 mouse embryos deserve special attention since there is strong evidence of genetic similiarity between chromosome 16 of laboratory mice and human chromosome 21 (Pearson, Roderick, Davison, Lalley, and O'Brien 1982). The evidence is predominantly based on the syntenic groups of genes conserved in both species on these chromosomes (Cox, Epstein, and Epstein 1980). Trisomy 16 has been studied in about 45 mouse embryos from the tenth to the twentieth day of gestation, with one trisomic born alive. Almost all were obtained in the progeny of males doubly heterozygous for Robertsonian metacentric chromosomes—Rb(16.17)7Bnr/Rb(9.16)9Rma mated to NMRI or C57BL/6J females (Miyabara, Gropp, and Winking 1982). A few have been detected in

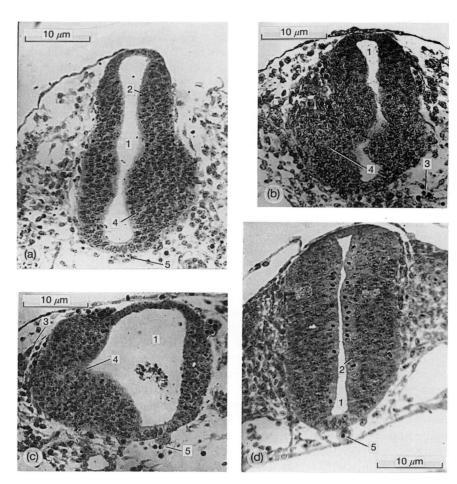

Fig. 23. Transverse sections of spinal cords at the lumbar level. (a,b,c) Abnormal embryos with Ts15; (d) control specimen. (a) Feulgen's staining; (b,c,d) Mayer's haemalum and eosin. 1 = spinal canal, 2 = neuroectoderm, 3 = mesenchyme, 4 = abnormal neural outgrowth, 5 = notochord.

the progeny of CBA females crossed to Rb(16.17)7Bnr/+ males (Baranov 1976a).

The frequency of Ts16 progressively decreased in the progeny of double heterozygotes from more than 20 per cent on day 14 to values between 4 and 7 per cent shortly before them. Some definite strain-dependent variations have been observed in Ts16 embryos of various breeding groups. The Ts16 embryos from doubly heterozygous males mated to females of NMRI strain survived to day 16 only, while they remained viable almost until term in the cross between male and female double heterozygotes (Miyabara *et al.* 1982). In the crosses of the single heterozygote Rb(16.17)7Bnr/ + the viability of Ts16 embryos was drastically reduced. Ts16 should be considered compatible with prolonged survival of the affected embryos, all of which are gradually eliminated during fetal development.

External examination of Ts16 embryos revealed moderate general hypoplasia with slight developmental retardation and occasional cleft palates, but without cleft lips. Microscopic studies of the serially sectioned 14–20 day fetuses showed cardiovascular anomalies in almost all trisomics. Most of the heart malformations belonged to the transposition type, i.e. riding aorta, double outlet right ventricle (DORV) (60 per cent), and ventricular septal defects (60 per cent) (Pexieder *et al.* 1981). Complete transposition of the great arteries with persistent truncus arteriosus was also found (Miyabara *et al.* 1982). The transposition defects accounted for 85 per cent of all observed cardiovascular anomalies. Association of DORV with common atrioventricular channel has been repeatedly reported in patients with Down's syndrome (Vogel and Motulsky 1979). Other abnormalities of Ts16 embryos included generalized transient oedema, 'open eyelids', hydronephrosis, and hydroureter. All of these should probably be attributed at least in part to the effects of general retardation.

Severe cardiovascular malformations in trisomic embryos are usually associated with pronounced hypoplasia of the fetal part of the placenta. Insufficiency of placental functions multiplied by heart malformations were two main reasons for elimination of trisomic embryos.

In spite of the relatively long survival of Ts16 embryos compared to other trisomies, including Ts12, isolated liver cells bearing an extra copy of chromosome 16 were unable to restore the haemo- and lymphopoietic systems of lethally irradiated recipients fully or permanently (Herbst *et al.* 1981; Gropp *et al.* 1983). Transplantation chimaeras with Ts16 stem cells showed barely 40 per cent survivors after two months and almost none after three months. Spleen colony assays with Ts16 fetuses demonstrated considerably smaller and poorer colonies than with controls or Ts12 fetuses. Poor and delayed response of T-lymphocytes from Ts16 embryos have also been reported recently. Distinct morphological disturbances of haemopoietic cells become visible at the electron microscope level (Gropp *et al.* 1983). All these results point to some serious defect of the granulocyte/macrophage and T-lymphocyte compartments of haemo- and lymphopoietic systems in embryos with Ts16.

Further detailed studies of Ts16 embryos and especially the behavioural and biochemical characteristics of Ts16 cells in embryonic and post-natal chimaeras will facilitate the acceptance of murine Ts16 as a model of Down's syndrome (Ts21) in humans.

Thus, Ts16 should be regarded as a chromosomal aberration with rather long survival, at least as long as Ts14 and a little longer even than Ts12. Ts16 is not accompanied by any specific malformations except cardiovascular abnormalities and innate defects of lympho- and haemopoiesis.

Chromosome 17. Developmental effects of Ts17 have been studied both in the doubly heterozygous breeding system (Gropp *et al.* 1974) and in crosses of single heterozygotes (Baranov and Udalova 1975). About 25 Ts embryos were analysed, seven from the progeny of Rb(16.17)7Bnr/Rb(8.17)1Iem females mated to NMRI males and 18 in the progeny of single heterozygotes— Rb(8.17)1Iem/ + ♀ ♀ or Rb(16.17)7Bnr/ + ♂ ♂ crossed to 'all acrocentric' CBA strain mice. The survival profiles of these trisomics depended on the strain background. According to Gropp and his collaborators Ts17 embryos survived to the twelfth or even to the thirteenth day of gestation, but in our studies they inevitably died by the tenth day.

Developmental profiles of all trisomics obtained in the single heterozygote progeny were very similar. At 8–9 days of development, 50 per cent of them had flattened head folds and the allantoic stalk failed to reach the chorionic plate. The implantation chamber surrounding these pathological vesicles was usually full of maternal blood, indicating progressive resorption of these trisomics. A few proceeded to the tenth day and revealed severe disturbances in morphogenesis of the neural tube and heart. The cranial and caudal regions of the neural tube were open and often disproportionately reduced in size (Fig. 24). No optic or otic anlagen could be detected in serial histological section of these embryos. The heart primordia were usually extremely expanded with very thin walls.

Trisomic embryos with relatively long survival (days 12–13 of gestation) often displayed caudal hypoplasia, focal haemorrhages or necrosis in the tail region (Gropp *et al.* 1983). Gross anomalies of the caudal region have been considered as a rather specific trait of Ts17, probably due to the triplication of the t-gene complex located in chromosome 17.

Thus, Ts17 is undoubtedly one of the early lethals causing disorganization and pronounced runting prior to death in embryos with an unfavourable genetic background, and rather specific caudal malformations in more advanced embryos with prolonged survival.

However, the extra copy of chromosome 17 cannot be considered as a cell lethal. This has been clearly demonstrated by the impressive experiments of Epstein's group on Ts17–diploid chimaeric mice (Epstein *et al.* 1982a). Rescue of aneuploid cells has been achieved by the aggregation of trisomic embryos with their diploid litter mates at the 8–16 cell stage. Chimaeric mice with 20–40 per cent of their cells trisomic for chromosome 17 developed quite normally before birth. Some were born alive and reached advanced stages of

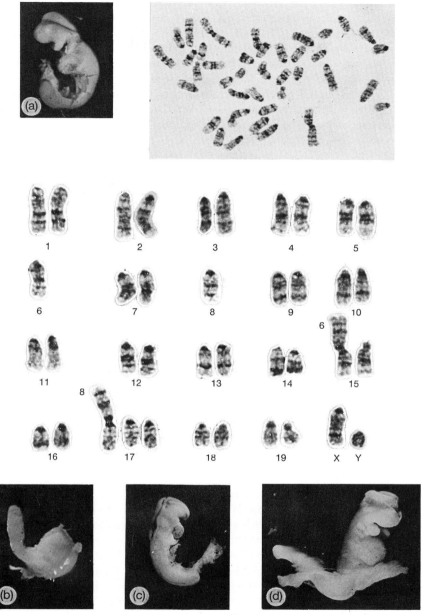

Fig. 24. Trisomy 17; Rb11em, Rb1A1d/ + + ♀ × ♂ + / + . Ninth (b) and tenth (a,c,d) day embryos fixed in Bouin's fluid. Extreme general retardation, craniorachischisis, extreme reduction in the head region.

post-natal life. The organs from these chimaeras, including brain, liver, kidney, lung, muscle, heart, thymus, spleen, bone marrow, blood cells, and skin, contained from 5 to 85 per cent Ts cells, as judged both by enzymatic markers and by karyotype analysis. The growth rates of the living Ts17 $2n$ chimaeras were in the lower half of the normal range. No progeny was obtained from Ts17 germ cells in the two fertile chimaeras. Considerable longevity (more than a year) of the chimaeras and the absence of gross phenotypic abnormalities indicate that the trisomic cells were not only rescued but were induced to undergo normal differentiation and function.

Chromosome 18. This trisomy is still rather poorly studied. The developmental profile of Ts18 embryos appears to resemble that of Ts16 embryos. They may survive to term, and sometimes are even born alive and may survive for several hours after birth (Gropp *et al.* 1981). However, in other studies all Ts18 embryos died by day 12 (Evans, Brown, and Burtenshaw 1980). These considerable variations in survival are probably attributable to genetic background differences. Ts18 embryos with prolonged survival frequently displayed a cleft palate (Gropp, Gropp, and Winking 1981; Gropp *et al.* 1983).

Chromosome 19. This trisomy is undoubtedly the most extensively investigated of all autosomal trisomies in laboratory mice. Over 300 trisomics of this type have been recovered in total for morphological, cytological, and biochemical analysis in different laboratories throughout the world.

Since the pioneering work of White and Tjio's research group in Bethesda (White *et al.* 1972, 1974a,b) Ts19 is routinely generated by means of the doubly heterozygous breeding design. Males heterozygous for two different Robertsonian translocations—Rb(5.19)1Wh/Rb(9.19)163H—are crossed either to 'all acrocentric' females or to doubly heterozygous females.

Data from different laboratories are consistent with exceptionally long survival of the trisomic embryos, the longest among all trisomies in laboratory mice. Though most are eliminated by term, some are live born and in the most favourable cases they have survived for at least 17 days after birth (Gropp *et al.* 1983). Thus, an extra copy of the smallest autosome of the mouse karyotype allows the longest survival of affected embryos, though their capacities for post-natal life are very limited.

Detailed morphological analysis of Ts19 embryos from early post-implantation stages up to term revealed general retardation and hypoplasia as the most consistent features of this trisomy. Ts19 has a recognizable effect on morphogenesis even prior to day 10 (Baranov and Udalova 1975; Bersu 1984). At later stages most trisomics show one day's lag in weight gain and size increase (White *et al.* 1974a,b). Some display extreme growth retardation (Fig. 25), but these probably die during organogenesis. A moderate delay in differentiation was much more common, with trisomics on days 14 and 15 resembling 13–14-day normal fetuses. This relationship remained constant up to term (Bersu 1984; Baranov *et al.* 1982).

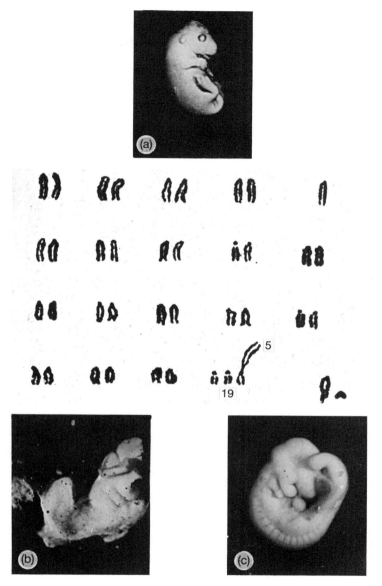

Fig. 25. Trisomy 19; Rb1Wh/ +   ♂ × ♀ Rb1Wh/ + . The embryos of the eighteenth(a) and 10.5th (b,c) days fixed in Bouin's fluid. (a) Microcephaly, brain hernia, micrognathia, cleft palate. (b) General retardation, head reduction, craniorachischisis. (c) Externally normal specimen with Ts19.

Lacto-aceto-orcein staining of chromosomes.

There was no striking change in the number of viable trisomic conceptuses throughout gestation, indicating very moderate selection of Ts19 embryos before birth. Placenta weight was also reduced, and at each developmental stage the ratios of placenta to body weight of normal and trisomic mice were very similar (White *et al.* 1974b).

Inspection of Ts19 embryos by stereomicroscope did not usually reveal any specific malformations. Some of the late Ts19 fetuses had a cleft palate (White *et al.* 1974a) but this was not found by Gropp *et al.* (1983). The persistence of cleft palate in Ts19 newborns has been reported by Bersu (1984). Almost all the cases of cleft palate have been reported in offspring with two chromosome 19 of Rb163H origin and one chromosome 19 from Rb1Wh. Embryos with the reversed combination of translocated chromosomes (2 Rb1Wh + Rb163H) never displayed this malformation (White *et al.* 1974a; Bersu *et al.* 1983). Some unknown genetic factors associated with the fusion of chromosome 19s to chromosome 9 in Rb163H translocation may therefore be responsible for the cleft palate in Ts19 embryos. Some other gross malformations are even less regular, including umbilical hernias, hydrocephaly, oedema, and some cardiovascular abnormalities (double outlet right ventricle, transposition of the great vessels etc.; Bersu 1984).

Comparison of serial histological sections of fetal thymus, kidney, adrenal, and lung from the normals and trisomics showed consistent differences. No specific histopathological changes were found in the kidney, adrenals, and lung, but patches of local necrosis were observed in the ovaries of some Ts19 females (White *et al.* 1974b). Analysis of single-cell suspensions prepared from thymuses of Ts19 embryos at 17, 18, and 19 gestation days has been made by means of a TPS-1 cell sorter (Bersu *et al.* 1982). Consistent increase in the relative numbers of cells in the S and G_2 + M phases of the cell cycle has been shown. The cell cycle alterations may contribute to the reduced cell number, delayed development, and small size of trisomic thymuses. Ts19 blood cells transplanted to lethally irradiated donor mice were able to restore the host's haemo- and lymphopoiesis successfully (Herbst *et al.* 1981). The erythroid, granulocytic, and lymphocytic lines showed histologically full maturation (Pluznick *et al.* 1981). The cell cycles of the transplanted Ts stem cells were not studied, though probably they were normal in the chimaeras. The absence of important differences of growth kinetics between trisomic and euploid cells has been reported for isolated fibroblast-like cells grown *in vitro* (Gropp 1982). More studies are needed to estimate the effects of a particular trisomy on the cell-cycle kinetics of the whole organism, *in situ* and in experimentally produced chimaeras.

Prolonged survival of Ts1 embryos provides a good opportunity for detailed studies of biochemical changes associated with this anomaly. Activity patterns of the enzymes involved in the Ebden–Meyerhof chain and isolated energy pathways and of amino acid metabolism have been measured in Ts19 embryos (Fundele, Bucher, Gropp, and Winking 1981). No significant alterations were observed for any enzyme system studied except glutamate oxaloacetate transaminase (GOT-1) and phosphoglycerate mutase (PGAM). The specific activity of GOT-1 was about 1.5-fold more than that of normal littermates with diploid karyotype. The clear-cut gene dosage effects in all day 14 trisomics (Waisman, Dyban, and Udalova 1981) and in the liver, brain, heart, and skeletal muscles of the term trisomics (Fundele *et al.* 1981) confirmed the assignment of *Got-1* to chromosome 19. The dosage effect for PGAM was assessed in liver, brain, and erythrocytes but not in skeletal muscle, providing the first linkage data for the gene controlling one subunit of this dimer enzyme to chromosome 19 (*Pgam-1*). The cellular activity levels of the enzymes catalyzing reactions of the glycolytic pathway and the citric acid cycle remained unchanged.

Chromosome 19 is known to possess highly repetitive DNA sequences for ribosomal genes. This has been demonstrated by direct *in situ* hybridization with labelled rRNA and by positive NOR-reaction (Henderson, Eicher, Yu, and Atwood 1976; Dev, Tantravahi, Miller, and Miller 1977). The dosage of rDNA genes in Ts19 cells is presumably increased, yet in our data the rate of ^{3}H-uridine incorporation into 28S and 18S rRNA in the Ts embryos was decreased 1.3–1.4-fold on average as compared with control diploids. Taking into account the reduced weight of all trisomics, one may conclude that the concentration of cytoplasmic RNA in Ts19 cells was not much less than that in the control embryos (Baranov *et al.* 1982). Thus, there was no clear-cut dosage effect for the products of ribosomal genes in our studies. Dosage effects of another gene (cell surface determinant—*lyt 1*) assigned to chromosome 19 have also been studied and confirmed in the spleen and in lymphocytes stimulated for growth, differentiation, and antibody formation *in vitro* and *in vivo* (Sellin *et al.* 1982).

Ts19 should be considered as a lethal mutation with a very late lethal effect. Unlike other trisomies in mice, Ts19 is compatible with at least some post-natal survival.

The phenotypic characteristics of Ts19 embryos include a one-day lag in development during fetal stages, and a cleft palate in crosses with two Rb163 H metacentrics. Ts19 on other genetic background may display abnormalities of eyes, brain, and heart. Hence no consistent phenotype can yet be attributed to Ts19. Retardation of growth in Ts19 embryos is probably due to changes of cell cycle. Ts blood stem cells can be successfully rescued by transplantation to lethally irradiated mice and demonstrated full capacity for normal growth, differentiation, proliferation, and function including antibody formation. The GOT-1 and PGAM-1 enzymes encoded by the genes assigned to chromosome 19 demonstrated a clear-cut gene dosage effect in Ts mice. No gene dosage effects have been demonstrated for rDNA genes carried on the triplicated chromosome 19.

6.3 Conclusions

Spontaneous trisomy is one of the most common chromosomal aberrations in the animal kingdom as well as in humans. Over 50–60 per cent of all human spontaneous abortuses in the first trimester of pregnancy bear chromosomal abnormalities. All human trisomies other than for chromosome 1 have now been identified in abortuses, but the relative involvement of the different chromosomes in spontaneous triplication varies.

Spontaneous autosomal trisomy occurs much less frequently in other mammalian species, and we are still almost totally ignorant of the developmental effects of autosomal trisomy in animals other than the laboratory mouse.

In spite of abundant clinical data on autosomal trisomies in humans, no correlation between any particular trisomy and embryonic phenotype has been established so far. The current genetic data regarding phenotype–genotype relationship in human trisomies are disappointing, since from spontaneous or therapeutic abortions it is impossible to obtain the specimens that would be most appropriate for a developmental analysis. This goal must therefore be almost exclusively achieved using experimental animal models.

Mice with Robertsonian translocations provide the best natural model for such studies. With the help of different breeding designs, trisomies for all the 19 autosomal pairs have been obtained, and developmental profiles of all amenable trisomies have already been assessed, though to a variable extent. For many autosomes in mice, it has already been shown that triplication is not harmful for pre-implantation embryos, though after implantation development may proceed differently. Survival capacities and malformation patterns as well as other phenotypic characteristics relevant to each type of trisomy are briefly outlined in Table 20. These data show that a few trisomies in mice survive until term and may even be born alive. These are trisomy 19, in which the longest documented survival is day 17 after birth, and trisomies 16 and 18, in which viability until term is usually observed. Thus, in laboratory mice in contrast to humans, excess of any autosome is not compatible with prolonged post-natal survival. Some of the trisomies—Ts12, 13, and 14—survive up to days 16–18 and rather rarely to term. The elimination of Ts1, 6, and 10 occurs on days 14–15, while most of the other trisomies (Ts3, 4, 7, 8, 9, 11, and 12) are usually dead on days 12–13. Trisomies 2, 5, and 17 as well as trisomies 3 and 11 on a particular genetic background should be considered as the earliest lethals, since affected embryos die during major organogenesis (days 10–11 of gestation). The survival profiles of trisomic embryos may vary substantially depending on the genetic background of the parental strains.

Developmental disorders and survival profiles in murine trisomies show no

Table 20. Survival age and phenotypic characteristics of mouse trisomics

Trisomy	No. of specimens	Age at death (days)	Phenotypic characteristics
1	Over 80	14–15	Moderate hypoplasia with variable retardation from the twelfth day. Heavy cranial dysmorphy, facial dysplasia, hypotelorism with holoprosencephaly; retardation and smallness of placenta; absence of specific malformations on some genetic backgrounds.
2	24	10–11	Severe retardation and hypoplasia; smallness of head; non-closure of neural tube.
3	14	10–11	Small, unorganized, severely retarded.
4	appr. 43	12–13	Moderate retardation and hypoplasia.
5	10	10–11	Severe hypoplasia and retardation; non-closure of neural tube, maldevelopment of allantois, poor vascularization of yolk sac.
6	?	12–15	Moderate to severe retardation and hypoplasia; extreme variation of phenotype.
7	appr. 100	12–13	?
8	21	12–13	Very severe retardation, hypoplasia; delay in allantoic growth disproportionate hypoplasia of head; dilated heart chambers; non-closure of neural tube (middle part); huge haemorrhages; developmental block by day 10.
9	26	11–12	General retardation; marked hypoplasia of head; non-closure of neural tube day 10, exencephaly days 11–12; reduction of maxillary processes, mandibular and hyoid arches; dilated heart.
10	?	?15	Slight retardation, slight to moderate hypoplasia; no gross malformations externally; ventricular septum defect in most.
11	?	10–11	Small, unorganized, severely retarded.

Table 20. *Contd*

Trisomy	No. of specimens	Age at death (days)	Phenotypic characteristics
12	over 120	15–17	Exencephaly, microphthalmia; anophthalmia; anencephalia in defined genetic background, defects of para-axial mesoderm; cardiovascular malformations; severe hypoplasia of chorio-allantoic placentae, full viability of Ts blood stem cells in irradiated chimaeras.
13	over 80	16–18	General retardation, transient oedema of the neck and back, delayed ossification, cleft palate, pulmonary stenosis and double outlet right ventricle; ineffectiveness of Ts blood stem cells for restoration of haemopoiesis.
14	about 75	18– term	Exencephaly due to non-closure of forebrain region of neural tube, cleft palate and oedema in some Ts fetuses; different types of cardiovascular malformations, growth retardation in gonads and thymus; delay of placenta growth and vascularization of the yolk sac; full viability of transplanted blood stem cells in radiation chimaeras.
15	over 20	11–12	Severe retardation, hypoplasia; pathological nerve tube curvature, outgrowth of unorganized nerve tissue, absence of liver; highly dilated heart chambers.
16	over 45	18– term	Moderate general hypoplasia, slight developmental retardation, occasional cleft palate, cardiovascular anomalies; transient oedema, open eyelids; hypoplasia of placenta; inefficiency of Ts blood stem cells in radiation chimaeras.
17	about 25	10.5 12–13	Small, disorganized, retarded; non-closure of neural tube; caudal hypoplasia in survivors to days 12–13; viability of Ts cells in all tissues of chimaeric mice.
18	?	18– term	Slight to moderate retardation, cleft palate

Table 20. *Contd*

Trisomy	No. of specimens	Age at death (days)	Phenotypic characteristics
19	over 300	18– before birth 17– after birth	One day lag in weight gain and size increase; cleft palate on defined genetic background, degenerating oocytes; alterations of cell cycle; full restoration of haemo and lymphopoiesis in radiation chimaeras with Ts19 blood stem cells; gene dosage effect for PGAM-1 and GOT.

correlation with the size of the extra autosome. For example, chromosome 17 is only half the length of chromosome 1, but the development of Ts17 embryos is affected much more severely and their lifespan is significantly shorter than that of Ts1 embryos. Thus, not the absolute size but the specific genetic content of each particular chromosome is responsible for the characteristic developmental manifestations of autosomal trisomy in mice.

The main developmental disorders provoked by trisomies comprise general retardation, hypoplasia, and gross malformations. The former two traits are common for all murine trisomies, though their manifestation may vary substantially. In extremely severe cases (Ts2, 3, 5, 8, 11, 15, and 17), an early developmental block may result in small, unorganized, and severely retarded embryonic remnants. Moderate hypoplasia with or without marked developmental retardation is common for trisomics with longer survival (Ts1, 12, 14, 16, 18, and 19). These, and more rarely 'early' trisomics, display gross malformations. So Ts1 is often associated with abnormal brain development, which may result in holoprosencephaly or aprosopia or cyclopia in more severe cases. Exencephaly is a constant feature of Ts12, often occurs in Ts14, and occasionally in Ts9 or 17. Each type of exencephaly is morphologically distinct and for this sort of finding the term 'semi-specificity' of malformation in chromosome disorders has been suggested by Gropp (1978). Unusual outgrowth of the neural tissue is a specific trait of Ts15, while cleft palate is regularly associated only with Ts13, 18, and frequently with Ts19. Cardiovascular malformations are common in trisomies 10, 12, 13, 14, 16, and 19, though each type of trisomy has its own specific pattern of these anomalies. All developmental disorders studied so far show considerable variation, most probably due to the peculiarities of genetic background.

The mechanisms underlying the developmental effects of trisomies have already been studied at organismal, cellular, and subcellular levels.

Trisomic cells may be rescued either by inclusion in tetraparental chimaeras (Ts ↔ diploid) during pre-implantation development or by transfer of Ts blood stem cells from the fetal liver to lethally irradiated adult hosts. Both methods provide an ample opportunity to follow the fate of the Ts cell clone, i.e. for estimating its capacity to survive in an otherwise normal diploid organism. Proliferation, differentiation, and specific function have already been studied for cell clones with an extra copy of chromosomes 12, 13, 14, 16, 17, and 19.

Biochemical dosage studies have also been carried out in trisomies 1 and 19. The specific activity of all enzymes known to be encoded by the genes of triplicated chromosomes was about 1.5 times higher than that of normal diploids. These data should be considered as additional confirmation of gene mapping and provide valuable information on the genetic regulation of embryonic development.

However, the exact mechanisms of developmental disorders in trisomic specimens still remain completely unknown. There are no clear answers as to why trisomy results in abnormal development and why the excess of genetic information contained in the extra chromosome inevitably results in the death of the mouse embryo.

7

Monosomy and nullisomy of autosomes

7.1 Monosomy

The frequency and developmental effects of autosomal monosomies are much less well studied than autosomal trisomies, both in laboratory mammals and in humans.

7.1.1 Frequency of spontaneous monosomy

Each case of spontaneous monosomy in clinical genetics is an exceptionally rare event. In cytogenetic analysis of almost 1500 early spontaneous abortuses collected during the first month of pregnancy, only a single case has been reported, making the frequency of spontaneous monosomy in humans about 0.06 per cent (Boué and Boué 1974; Boué *et al.* 1976). According to more recent data, the frequency of autosomal monosomies in the earliest human abortuses is only about 0.1 per cent (Hassold, Matsuyama, Newlands, Matsaura, Jacobs, Manuel, and Tsuei 1978) and not a single case of monosomy has been reported among cumulative data from four extensive surveys comprising 3040 abortions (Lauritsen 1982). Most autosomal monosomies in humans are represented exclusively by the smallest G-group autosome (Boué and Boué 1971, 1974; Ohama and Tadashi, 1972; Carr 1969, 1971a,b, 1983). Some exceptionally rare cases of monosomy for chromosome 21 (G-group) are also reported in a few post-natal specimens (Greenwood and Sommer 1971; Franklin and Schmickel 1971; Berger 1972).

Spontaneous autosomal monosomy in humans is thus extremely rare, and encountered predominantly at the earliest stages of gestation. It might be presumed that autosomal trisomies and monosomies, being complementary by their nature, are generated at approximately equal proportions at the onset of pregnancy, but most monosomics are eliminated before the pregnancy is ascertained and hence their loss would be interpreted as irregularities of the menstrual cycle. An earlier suggestion on the preferential elimination of male gametes with autosomal nullisomy (Prokofieva-Belgovskaya 1969; Carr 1967a,b, 1969; Hamerton 1971) is not supported by more recent studies. The normal fertilizing ability of hypohaploid (nullisomic) sperm of experimental rodents, and the significant proportion (2–3 per cent) of hypohaploid sperm in men with normal fertility (see Chapter 4), both provide solid evidence against such a suggestion.

All the available data favour the notion of an almost equal proportion of monosomic and trisomic zygotes in mammals (see Chapter 6), but direct experimental data on spontaneous autosomal monosomy in early mammalian embryos are still very meagre. This may be attributed at least partly to the technical difficulties encountered during chromosomal preparations and analysis at these early stages of embryogenesis. In routine cytogenetic practice, the presence of only one metaphase plate lacking one or a few autosomes in an air-dried cleaving egg would usually be interpreted as a technical artefact. Reliable identification of monosomic embryos should thus be limited only to the rare specimens with at least two distinct metaphase spreads lacking one and the same autosome. The number of these 'reliable' cases of monosomy is very low and this limitation is most probably responsible for the shortage of monosomic embryos in laboratory animals, including rats and mice (see Table 1, also Chapters 4 and 5). In contrast to the early stages of embryogenesis, cytogenetic analysis of post-implantation embryos provides a good number of metaphase plates of high quality and thus allows reliable identification of monosomic embryos. Hence the absence of monosomy in post-implantation specimens of mice, rats, and rabbits (Dyban *et al.* 1971; Baranov and Dyban 1972a; Yamamoto, Endo, and Watanabe 1971, 1973; Ford 1970, 1972) is most probably due to their death at somewhat earlier stages of embryogenesis.

7.1.2 Developmental effects

We are still almost completely ignorant of the developmental effects of monosomies in human embryos and of the actual time of their death. Because of the substantial proportion of mature gametes with nullisomy and disomy of the chromosomes involved in centric fusion (see Chapter 5), mice with Robertsonian translocations have repeatedly proved to be most valuable and in many aspects quite unique experimental tools to study developmental effects of autosomal monosomies in mammals.

Since the early investigations on mice with Robertsonian translocations, it has been shown that both male and female gametes lacking one or two different autosomes are nonetheless capable of fertilization, though the resulting monosomic embryos do not survive after implantation (Ford 1972; Gropp *et al.* 1976). F_1 hybrid mice from *Mus musculus* × *Mus poschiavinus* crosses heterozygous for seven different Robertsonian translocations were used in these studies.

Our preliminary data on developmental effects of autosomal monosomies have been obtained on mice heterozygous for three different Robertsonian translocations: Rb1Iem, Rb1Ald, and Rb163H (males and females) crossed to 'all-acrocentric' mice of CBA strain. Of 342 day-4 embryos studied both cytologically and karyologically, 28 were found to have monosomy for one or two different autosomes (see Table 15, Chapter 4). Twenty-two monosomics were at the late morula–early blastocyst stage and displayed some decrease in blastomere count compared to their control diploid littermates when studied in air-dried preparations (24.7 $\pm$ 3.7 and 32.01 $\pm$ 1.3, respectively; t = 2.4; p < 0.02). Six monosomics of the same group looked abnormal both under the stereomicroscope and in air-dried preparation. The blastomere count in these embryos was only about half that of normal diploids and some of their nuclei looked pyknotic. No banding was carried out on these chromosomal preparations, and thus it remains totally unknown which of the six translocated autosomes involved in the corresponding central fusions (8–17; 6–15; 9–19) was missing in the affected embryos. None the less, these preliminary data indicated the value of this experimental approach.

Special breeding designs similar to those used for the generation of trisomic embryos (see Chapter 6) have also been applied for producing and studying monosomies (see Tables 15 and 21) in the main part of our investigations.

Altogether, 1236 mouse embryos were successfully karyotyped in these rather laborious experiments, with only 129 embryos being identified as monosomics. All studies were carried out on three- to four-day embryos and monosomy for 15 different autosomes were identified. As there were only 129 monosomics the actual number of each type of monosomy was rather limited (see Tables 15, 21). Valuable though rather limited information on the phenotypes of mouse embryos monosomic for autosomes 1, 12, 17, or 19 have been obtained by Epstein's group (see Table 28).

Our cumulative data on the incidence of autosomal monosomies and trisomies in three- to four-day embryos of Robertsonian heterozygotes are summarized in Table 21 (Ms1,2,3,5,6,16,17, and 19) and also partly in Table 15 (Ms9,14, and 15). Of 1236 successfully karyotyped embryos, 955 were classified as normal diploids with balanced chromosome arms count (NF = 40); 267 as aneuploid, with 138 being trisomic (NF = 41), and 129 monosomic for one or rarely for two different autosomes (NF = 39 or 38).

Let us now consider the data for each type of autosomal monosomy in more detail.

Chromosome 1. The absence of this autosome, the largest in the mouse karyotype, has been recorded in the progeny of +/+ female mice crossed to single heterozygote Rb(1.3)1 Bnr/+ males (Baranov 1983a,d). Equal proportions of specimens with Ms1 and Ts1 were found on the fourth day. All day-4 monosomics were represented by normal-looking blastocysts, which could not be distinguished externally with a stereomicroscope or in

Table 21. Frequency of aneuploid pre-implantation and early post-implantation embryos in mice heterozygous for different translocations

Heterozygotes	Sex	Day of pregnancy	Live embryos studied	Embryos karyotyped	Diploids No.	Diploids %	Aneuploids No.	Aneuploids %	Type of expected aneuploidy	Trisomy No.	Trisomy %	Monosomy No.	Monosomy %
Rb(1.3)1Bnr/+	♂	4	147	77	49	63.6	28	36.3	1 3	6 8	7.7 10.3	5 9	6.4 11.6
Rb(2.6)4Iem/+	♂, ♀	4	210	181	151	83.4	30	16.5	2 6	4 6	2.2 3.3	7 13	3.8 7.1
Rb(5.19)1Wh/Rb(5.15)3Bnr	♀	3	180	96	80	83.3	16	16.6	5	5	5.2	11	11.4
		4	135	65	48	73.8	17	26.1	5	10	15.3	7	10.7
Rb(16.17)7Bnr/+	♂	3	230	90	80	88.8	10	11.1	16	4	4.4	6	6.6
T(16;17)43H/+	♀	9	87	61	50	81.9	11	18.0	16	—	—	11	18.0
Rb(9.19)163H/Rb(5.19)1Wh	♂	4	146	83	57	68.6	26	31.3	19	14	16.8	12	14.4
		3	318	131	95	72.5	36	27.5	17	21	16.0	15	11.5
Rb(8.17)1Iem/Rb(16.17)7Bnr	♀	4	192	92	73	79.3	19	20.7	17	21	22.8	—	—

Table 22. Karyotype and blastomere number of day-4 embryos in the male and female intercross progeny of Rb4Iem heterozygotes

Litter No.	No. of embryos studied		Diploid normal		Aneuploid	Trisomy 2, 6		Monosomy 2		Monosomy 6	
	Total	No. karyotyped	No. of embryos	No. of cells	Total embryos	No. of embryos	No. of cells	No. of embryos	No. of cells	No. of embryos	No. of cells
1	6	5	4	83, 58, 74, 57	1	—	—	—	—	1	23
2	13	11	10	60,60, 61, 70, 63, 85, 65, 77, 45, 26	1	—	—	—	—	1	30
3	9	9	7	69, 60, 92, 77, 69, 61, 57	2	1	60	1	16	—	—
4	14	12	9	26, 36, 44, 40, 32, 56, 62, 45, 39	3	1	41	—	—	2	18, 29
5	10	9	7	56, 61, 62, 54, 43, 31, 54	2	—	—	—	—	2	35, 21
6	7	5	4	30, 30, 35, 27	1	—	—	1	7	—	—
7	12	11	8	48, 40, 33, 34, 40, 35, 30, 59	3	2	18, 62	—	—	1	31
8	12	9	7	67, 74, 32, 57, 32, 47, 47	2	—	—	1	17	1	45
9	13	13	10	60, 57, 57, 59, 58, 47, 17, 64, 62, 63	3	2	13, 70	1	13	—	—
10	13	12	9	28, 41, 60, 52, 30, 32, 54, 52, 32	3	1	23	1	17	1	58
11	10	7	4	22, 26, 14, 16	3	1	24	1	10	1	14
12	9	8	5	15, 29, 18, 30, 32	3	2	19, 17	1	14	—	—
13	11	9	6	17, 36, 25, 28, 26, 27	3	—	—	—	—	3	28, 12, 13

air-dried preparations from their normal littermates with balanced diploid karyotype, nor from Ts1 embryos. The mean cell counts in monosomic, trisomic, and diploid embryos was almost identical (41.0 $\pm$ 3.7; 39.6 $\pm$ 2.3; 38.5 $\pm$ 3.9, respectively).

Some Ms1 specimens (three on day 4; five on day 5) were recovered in the progeny of double metacentric heterozygous males (Rb(1.3)1Bnr/Rb(1.10)10Bnr crossed to +/+ females; Epstein and Travis 1979). Equal numbers of trisomic and monosomic embryos were found on the fourth day, whereas a substantial preponderance of the trisomics were reported among five-day embryos (17 per cent and 9 per cent *in vivo* and 17 per cent and 10 per cent after 24 h culture *in vitro*). All Ms1 embryos under external inspection looked like normal blastocysts, indistinguishable from karyologically balanced diploids. However no data on cell counts of monosomic or control embryos were included in this study.

According to Baranov (1983a), at least some Ms1 embryos are capable of implantation but they all die shortly thereafter (see Table 15, Chapter 4).

Thus the absence of chromosome 1 is compatible with pre-implantation survival and normal development of mouse embryos.

Chromosome 2. This group of monosomics consisted of seven day-4 embryos in the progeny of mice heterozygous for Rb(2.6)4Iem translocation. There were only 38 acrocentric chromosomes in the karyotype of one embryo, which was considered as doubly monosomic for the two translocated chromosomes (2 and 6) involved in the Rb4Iem fusion. All monosomics of this group, including this one, revealed a significant lag in their growth by the morula stage (Table 22; see also Figs 26(a) and 27(b)). The mean cell count in Ms2 embryos ranged from 7 to 17 (Table 23) with mean value 14.5 $\pm$ 3.5, compared to 52.6 $\pm$ 3.4 in normal littermates. A good correspondence between the frequencies of monosomy and trisomy witnessed against preferential elimination of monosomic embryos by the fourth day of development. Meanwhile, numerous pyknoses and micronuclei in four out of seven monosomics proved that they had died well before blastulation. It may be concluded therefore, that the deficiency for a single autosome 2 results in a significant lag of cleavage rate and is most probably lethal at the morula stage, but this conclusion requires further experimental support based on more Ms2 embryos of different ages and genetic backgrounds.

Chromosome 3. The absence of this autosome has been recorded in nine out of 77 karyotyped embryos in the progeny of heterozygous males Rb1Bnr/+ (see Table 21). All day 4 monosomics were early blastocysts, with a mean cell number per embryo only slightly reduced compared to normal littermates with balanced diploid karyotype and embryos with Ts3 (33.2 $\pm$ 4.2; 39.6 $\pm$ 2.3; 38.5 $\pm$ 3.9, respectively). The reduction was mainly due to some Ms3 specimens with blastomere number below 20. Some of these showed signs of degeneration (pyknoses, micronuclei). It cannot be excluded

Table 23. Karyotype and blastomere number of day-4 embryos in the progeny of Rb3Bnr/Rb1Wh female double heterozygotes

Progeny No.	Embryos studied		Diploids		Aneuploids					
	Total	Karyotyped	No.	No. of blastomeres	No.	Trisomy 5		Monosomy 5		
						No.	No. of blastomeres	No.	No. of blastomeres	
1	8	5	4	16, 15, 26, 32	1	—	—	1	6	
2	6	4	4	15, 28, 33, 32	—	—	—	—	—	
3	8	6	5	17, 22, 28, 19, 42	1	1	41	—	—	
4	5	3	3	21, 16, 23	—	—	—	—	—	
5	9	6	4	17, 20, 28, 28	2	1	26	1	12	
6	9	7	5	32, 33, 57, 20, 18	2	1	40	1	17	
7	8	5	3	26, 25, 25	2	1	38	1	17	
8	7	6	3	30, 22, 28	3	2	30, 33	1	9	
9	5	4	4	30, 17, 26, 30	—	—	—	—	—	
10	11	6	3	31, 19, 20	3	2	22, 16	1	17	
11	9	8	7	32, 32, 33, 43, 36, 51, 55	1	1	57	—	—	
12	7	5	3	31, 32, 53	2	1	23	1	21	

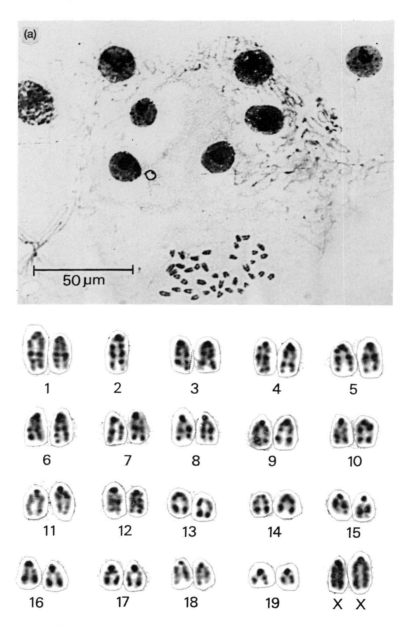

Fig. 26. Total preparations and karyotypes of (a) fourth day embryo (Rb4Iem/+ ♀ × Rb4Iem/+), and (b) third day specimen (Rb7Bnr/+ ♂ × ♀ +/+). Double fixation technique (Dyban and Baranov 1978). Trypsin–Giemsa banding.
  (a) Monosomy 2, (b) monosomy 16 (Baranov 1983a).

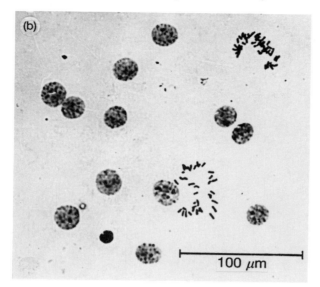

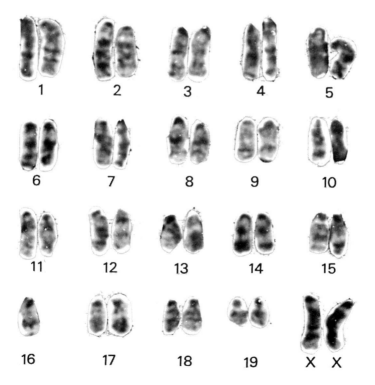

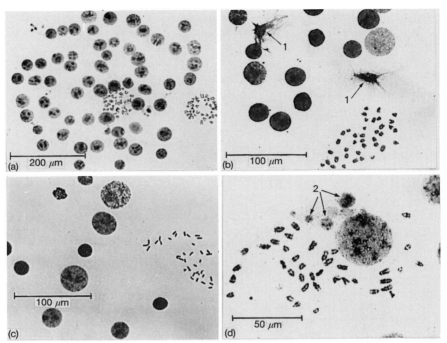

Fig. 27. Total preparations of mouse embryos on the fourth day of development. Double fixation technique, trypsin–Giemsa staining. (a) Balanced diploid karyotype, NF = 40; (b) monosomy of chromosomes 2 and 6, NF = 38 (Rb4Iem/+ × Rb4Iem/+); (c) monosomy of chromosome 5, NF = 39 (+/+ ♂ × Rb1Wh/Rb3Bnr ♀); (d) monosomy of chromosome 19; NF = 39 (Rb1Wh/Rb163H ♀ × ♂ +/+). 1 = nucleus lysis; 2 = micronuclei (Baranov 1983a).

therefore that Ms3 may already be lethal at the late morula–early blastocyst stage, that is during the fifth cleavage division. This suggestion needs further experimental support.

Chromosome 5. Deficiency for this autosome has been observed in day 3–4 embryos from female double metacentric heterozygotes with mono-brachial homology Rb(15.19)1Wh/Rb(5.15)3Bnr crossed to 'all-acrocentric' males (Baranov 1983a,b). There were five Ts5 embryos (5.2 per cent) and 11 Ms5 specimens on the third day. Embryos with Ts5 predominated over Ms5 embryos on the fourth day (10–15.3 per cent and 7–10.7 per cent, respectively). This discrepancy can most probably be attributed to the loss of some Ms5 embryos during days 3 and 4.

All monosomics on the the third day were normal, with a mean blastomere count (4.3 ± 1.6) almost identical to that of normal diploids (4.4 ± 1.3) and to the Ts5 embryos of the same age (4.5 ± 2.8). In contrast, all day 4 monosomics were conspicuously retarded in their growth and still remained at the morula stage (Fig.27(c)). A significant lag in the cleavage rate of monosomics was especially evident within litters (see Table 23). In six out of seven monosomics studied on the fourth day the blastomere number was about half to one-third of that in normal embryos and all showed degenerative changes. The mean blastomere numbers in monosomics and in control littermates at this stage were 14.1 $\pm$ 3.2 and 29.1 $\pm$ 2.4, respectively ($p < 0.05$).

Thus, the absence of autosome 5 appears to result in significant retardation of cleavage rate after the third or fourth division, and in a block of embryonic development beyond the morula stage (12–17 cells).

Chromosome 6. This type of monosomy was recorded in 13 embryos on the fourth day, in the reciprocal crosses of both male and female heterozygotes for Rb(2.6)4Iem translocation (i.e. concomitantly with Ms2) (Baranov 1983a,d). Development profiles of Ms6 embryos varied significantly. Four (Group I) still remained at the morula stage and consisted of 12–18 blastomeres, whereas nine (Group II) looked like normal blastocysts with cell numbers ranging from 22 to 58. Differences in cleavage rate between the control and monosomic embryos were clearly evident within litters (see Table 22). The mean number of blastomeres was almost half of that in the control (27.4 $\pm$ 4.5 and 52.3 $\pm$ 3.6, respectively). Micronuclei, pyknoses, and pathological mitosis were found in six monosomics, suggesting lethality of this aberration at the early blastocyst stage. Whether this is so for Ms6 embryos of specific genetic backgrounds only or is a more widespread phenomenon of Ms6 deficiency remains to be proved.

Chromosomes 9 and 14. Developmental profiles of embryos lacking these autosomes were studied in the progeny of male or female mice, heterozygous for the translocation Rb(9.14)6Bnr. No chromosome banding was used to distinguish between these two types of monosomic. However, chromosomal analysis of the early post-implantation embryos from the same crosses revealed both types of trisomy (see Chapter 6), so two complementary types of monosomies might be expected to occur during early pre-implantation development in the progeny of Rb6Bnr/+ heterozygotes. Indeed on the fourth day of gestation the number of trisomies was only slightly higher than that of monosomies (36 and 25, respectively; see Table 15, Chapter 4). Twenty monosomics lacked one chromosome and five lacked two. All day-4 monosomics were early blastocysts and could not be distinguished by external examination from their karyologically balanced or trisomic littermates. The mean cell count of monosomics was 32.9 $\pm$ 3.1, i.e. within the normal range (see Table 24). Normal nuclei and abundant mitotic figures suggested that survival of these monosomics before implantation was normal, so that they might be expected to show some further development. Three small empty sacs lacking any embryonic derivatives were recovered on

Table 24. The frequency of four-day embryos with different number of blastomeres monosomic for the autosomes 9–14 or 15

Karyotype	No. of embryos	Blastomere number							
		10–14	15–19	20–24	25–32	32–42	43–53	55	Mean count
Balanced diploid (NF = 40)	178	4	12	14	67	48	16	17	32.0 ± 1.3
Monosomies (NF = 39)									
9–14	25	0	4	3	9	4	3	2	32.9 ± 3.4
15	16	1	5	2	5	2	1	0	23.7 ± 3.2

the ninth day and were found to be monosomic for chromosome 14. Whether Ms9 embryos can also survive implantation remains unknown.

Thus, monosomies 9 and 14 do not interfere with cleavage, blastulation, nor most probably with implantation; some Ms14 embryos survive beyond implantation but die soon thereafter.

Chromosome 12. Developmental effects of this monosomy have been studied in the progeny of +/+ mice crossed to the doubly heterozygous males Rb(8.12)5Bnr/Rb(4.12)9Bnr (Epstein and Travis 1979). Of 15 monosomics, 10 were formed on the third day and maintained *in vitro* for a further 24 h, and five were obtained on the fourth day and also cultured for 24 h. The relative frequencies of trisomy and monosomy were virtually the same in both groups studied and all aneuploid specimens both with Ts and Ms were normal looking blastocysts. No data on blastomere count in the monosomic, trisomic or euploid embryos are available. Thus the cleavage rate in Ms12 embryos remains unknown, but Ms12 does not appear to interfere with pre-implantation survival and blastulation.

Chromosome 15. Two different experimental approaches have been used in our laboratory to study developmental effects of Ms15. Initially (Baranov 1976a), 4-day embryos with trisomy or monosomy 15 were obtained in the progeny of +/+ females crossed to Rb(5.15)3Bnr/Rb(6.15)1Ald males. Sixteen monosomics were recovered in total, eight at the morula stage and the remaining eight as early blastocysts (Table 24). The mean cell count of all Ms15 embryos was significantly reduced compared to their genetically balanced littermates (23.7 ± 3.0; 32.0 ± 1.3 respectively). Numerous micronuclei and pyknoses were found in five monosomics. It was concluded therefore that Ms15 affects pre-implantation development in mice (Baranov 1976a).

The same breeding scheme has been utilized to produce Ms15 embryos in the second approach (Severova 1985). Identification of Ts and Ms embryos was carried out at the four-cell stage by cytogenetic study of one microsurgi-

Table 25. Mean cell count in euploid mouse embryos, compared with Ts or Ms 15 embryos (Severova 1985)

Culture conditions	Karyotype	Duration of culture interval			
		42–44 h		66–68 h	
		No. of embryos	Mean cell count	No. of embryos	Mean cell count
Four-cell stage embryos, cultured intact (non-operated)	Euploid	9	26.4 ± 1.9 (17–32)	11	78.0 ± 5.0 (55–107)
	Trisomy	—	—	3	80.0 ± 11.1
	Monosomy	—	—	1	33
Three blastomeres of four-cell stage embryos, cultured after operation	Euploid	48	22.4 ± 0.7 (12–38)	21	49.8 ± 2.9 (26–79)
	Trisomy	15	20.7 ± 0.8 (12–25)	5	41.8 ± 11.4 (20–47)
	Monosomy	11	19.2 ± 1.2 (11–24)	5	31.4 ± 7.7 (19–60)

cally isolated blastomere. The remaining three blastomeres were maintained *in vitro* for a further 44 or 68 h. During the culture period all euploid embryos reached morula–blastocyst stages and were quite similar to non-operated control littermates, apart from a slight reduction in cell number. The great advantage of such an approach is the opportunity that it provides to follow the development of embryos of known karyotype, including Ts15 and Ms15.

Altogether 162 four-cell stage embryos have been treated in this way (see Table 25). Thirty-four were not operated and served as positive controls. Twenty-four were successfully karyotyped at the end of the culture period. A balanced euploid karyotype was found in 20 embryos; three were trisomic and one monosomic for chromosome 15. Successful splitting of four-cell embryos into 3 + 1 parts was carried out in 128 embryos: 105 (82 per cent) were karyotyped, giving 69 diploids, 20 Ts15 and 16 Ms15. This suggests that Ms15 does not interfere with normal compaction and cavitation. However, it does induce a substantial reduction in blastomere count. The mean cell count of monosomic embryos was only 19.2 ± 1.0 compared to 20.7 ± 0.8 in Ts15 and 22.4 ± 4.0 in euploid embryos after 42–44 h in culture. These differences were especially conspicuous when the culture time was extended to 66–68 h (31.4 ± 7.7 Ms-15; 41.8 ± 11.4 Ts15; 49.8 ± 2.9 normal euploids). A reduced blastomere number has also been encountered in some euploid specimens. The mean number of mitotic figures per monosomic embryo was significantly less than per euploid (2.0 and 3.0, respectively), but somewhat higher than in trisomic embryos (0.9). Whether these differences in mitotic index were random, or whether there is some actual block preventing the accomplishment of mitosis in Ms15 blastomeres remains unknown. A significant proportion of aneuploid embryos (both trisomic and monosomic) contained no mitotic figures in air-dried preparations. These would be missed in routine cytogenetic studies if they had not been successfully karyotyped beforehand.

Thus, Ms15 embryos can accomplish cleavage and give rise to normal looking blastocysts, but it seems probable that Ms15 affects cleavage rate and reduces mitotic activity of blastomeres both *in vitro* and *in vivo*.

Chromosome 16. Two different breeding designs have been used to produce embryos deficient for this autosome. Pre-implantation Ms16 embryos have been obtained in the progeny of Rb7Bnr/+ mice (Baranov 1983a). On the third day of gestation all six Ms16 embryos had passed the third cleavage division and were indistinguishable either externally under a stereomicroscope or in air-dried preparations from their normal littermates (Fig.26(b)). Eleven more monosomics were found on the ninth day in the progeny of female mice heterozygous for the translocation T(16;17)43H. In these embryos the absence of practically the whole autosome 16 was combined with a partial trisomy of the CDE segment of autosome 17. In spite of such a heavy genetic imbalance, the monosomic embryos survived implantation but died soon thereafter. Eight of them were represented by small empty vesicles without visible traces of embryoblast derivatives and three had remnants of neural folds and allantois. Whether a similar developmental profile is relevant to ordinary Ms16 embryos without concomitant structural aberrations of autosome 17, or whether they might proceed to even more advanced stages of embryogenesis, remains to be clarified. It seems probable at present that Ms16 does not interfere with pre-implantation development and survival and may be compatible with implantation, but results in death soon thereafter.

Chromosome 17. The developmental effects of Ms17 have been studied in more detail than those of other monosomies. At least three different studies have been devoted to this subject. In Epstein's group, Ms17 embryos have been generated in the progeny of Rb(11.17)1Rma/Rb(8.17)1Iem males crossed to +/+ females and studied on the fourth and fifth days of development. Some Ms17 embryos degenerated before blastulation but at least half remained alive after cleavage and transformed into normal blastocysts. No information on cell number was published (Epstein and Travis 1979; Magnuson and Epstein 1981).

In our group (Baranov, Dyban, and Chebotar 1980), three- or-four-day Ms embryos were recovered in the progeny of doubly heterozygous females crossed to +/+ males. Of 156 three-day-old embryos, Ms17 was found in 15 and Ts17 in 21 embryos. None of the embryos identified karyotypically as Ms17 was found on the fourth day, though the frequency of Ts17 still remained rather high (24 per cent). In spite of significant fluctuations in developmental rate among litters, the cell number in monosomic embryos on day 3 was practically within the normal range. Mean values were 6.6 ± 2.6 for Ms17, 8.3 ± 2.5 for the control, and 8.2 ± 2.5 for trisomics. The most typical trait of Ms17 was the presence of micronuclei, which were found in air-dried preparations of 11 embryos (73 per cent). At least 15 per cent of all the embryos studied (29 out of 192) still remained at the early morula stage on day 4 (i.e. they were composed of 8–16 blastomeres) and contained numerous micronuclei and pyknoses. All attempts to karyotype these degenerating embryos failed, since they had no mitoses and thus their development was totally blocked. Probably these degenerating embryos were actually Ms17. Damage incurred during oogenesis in Rb1Iem/Rb7Bnr females might have a significant influence on the early death of Ms17 embryos (Baranov *et al.* 1980). One may conclude therefore that Ms17 affects embryonic development after a few cleavages and is always lethal by the morula stage.

This conclusion however is justified only for a particular combination of Rb translocations, i.e. it clearly depends on genetic background. In a recent experiment (Severova 1985), Ms17 embryos have again been generated by doubly heterozygous females, but instead of an Rb(8.17)1Iem translocation of laboratory origin (Baranov and Dyban 1971a), a wild mouse derived Rb(8.17)1Sic translocation has been used. Thus, in spite of identical chromosomal rearrangements the genetic background of the females was different.

As with Ms15 (see above), a microsurgical method of identifying Ms embryos was used. Four-cell embryos were microsurgically divided into two unequal parts (3 + 1). One blastomere was used for karyotyping and the rest were cultured for a further 44–48 h (Severova and Dyban 1984). Of 163 embryos cultured *in vitro*, 12 were found to be monosomic and 18 trisomic for chromosome 17. External examination of these aneuploid embryos revealed no damage during cleavage and compaction, but a substantial delay in cavitation was noted in at least some Ms17 embryos.

Cell numbers varied significantly in all groups of embryos studied on air-dried preparations (see Table 26). Though Ts17 embryos could not be distinguished with certainty from their diploid littermates (28.1 ± 4.1 and 33.6 ± 1.4 respectively), the former displayed a reduction in mitotic figures (1.2 per Ts embryo). Most Ms17 embryos had less than half the blastomere count of normal controls (14.9 ± 1.6 and 33.6 ± 1.4, respectively) and their mitotic activity was significantly reduced (0.9 and 2.1 per embryo, respectively). However, at least a few Ms17 embryos contained 30–32 cells, some of which continued to divide mitotically. Thus expression of Ms17 may vary significantly depending on its sex origin (whether it was lost in oogenesis or spermatogenesis), as well as on the parental genotypes. Monosomics with paternal chromosome 17 deficiency have much more chance to survive beyond the morula–blastocyst stage than monosomics of maternal origin

Table 26. Mean cell count in euploid mouse embryos and in specimens with Ts or Ms 17 (Severova 1985)

Culture conditions	Karyotype	Duration of culture			
		42–44 h		66–68 h	
		No. of embryos	Mean cell count	No. of embryos	Mean cell count
Four-cell stage embryos, cultured intact (non-operated)	Euploid	25	54.2 ± 3.7 (29–101)	—	—
	Trisomy	4	41.3 ± 3.4 (35–50)	—	—
	Monosomy	2	28,32		
Three blastomeres of four-cell stage embryos, cultured after operation	Euploid	60	33.6 ± 1.4 (17–67)	12	66.7 ± 2.5 (50–85)
	Trisomy	10	28.1 ± 4.1 (10–53)	4	63.2 ± 5.0 (55–79)
	Monosomy	9	14.9 ± 1.6 (5–21)	1	29

(chromosome 17 lost during oogenesis). Ms17 due to a non-disjunction event in Rb1Sic/Rb7Bnr females might be occasionally compatible with blastocyst formation, though it significantly hampers cleavage (Severova and Dyban 1984; Severova 1985). Ms17 in the progeny of Rb1Iem/Rb7Bnr females is already lethal after a few cleavages and is eliminated by the morula stage (Baranov *et al.* 1980).

Chromosome 19. Monosomy for this autosome is at present being studied in great detail. Besides routine embryological approaches, biochemical, ultrastructural, and experimental embryological methods are being widely used in these studies. All monosomics were obtained by the double metacentric technique in the progeny of 'all-acrocentric' females (C57BL/6J; ICR/Swiss; F_1 CBA × C57BL) crossed to males Rb(9.19)163H/Rb(5.19)1Wh (Magnuson *et al.* 1982; Baranov 1983a) or to males (Rb(9.19)163H/Rb(8.19)1Ct) (Epstein and Travis 1979).

Our studies were based on 12 monosomics picked up on the fourth day. Eleven looked like normal early blastocysts, composed of an average of 35.6 ± 3.7 cells (compared to 34.0 ± 3.1 in control diploids and 29.8 ± 3.7 in Ts19 embryos; see Table 27). One of the monosomics was still at the morula stage and had 13 blastomeres. As shown in Table 27, interlitter comparison of the cell numbers in euploid and aneuploid embryos does not show any significant change in the cleavage rate of embryos with Ms19. However, the presence of numerous micronuclei and pyknoses in four monosomics

Table 27. Karyotype and blastomere number of day-4 embryos in the progeny of Rb1Wh/Rb163H male double heterozygotes

Progeny No.	Embryos studied		Diploids		Aneuploids				
	Total	Karyotyped	No.	No. of blastomeres	No.	Trisomy 19		Monosomy 19	
						No.	No. of blastomeres	No.	No. of blastomeres
1	8	7	5	27, 42, 27, 33, 32	2	1	18	1	29
2	6	5	5	33, 31, 34, 48, 29	—	—	—	—	—
3	7	5	3	23, 31, 23	2	1	11	1	27
4	10	8	5	39, 29, 14, 14, 18	3	1	51	2	32, 13
5	12	7	6	55, 44, 34, 51, 45, 38	1	—	—	1	42
6	6	6	4	33, 32, 35, 40	2	1	35	1	31
7	10	9	6	51, 31, 14, 17, 56, 55	3	1	27	2	57, 53
8	12	10	7	22, 23, 20, 46, 67, 22, 25	3	2	28, 28	1	21
9	10	7	5	17, 25, 25, 35, 30	2	1	48	1	21
10	7	7	4	30, 30, 35, 63	3	2	19, 14	1	25
11	9	8	5	38, 38, 36, 20, 53	3	2	20, 15	1	18

indicates a progressive degeneration (Fig.27(d)). An equal proportion of Ms19 and Ts19 embryos were found at the blastocyst stage but the proportion of monosomic embryos was drastically reduced shortly thereafter. Thus Ms19 causes little damage to developing mouse embryos up to the early blastocyst stage, but it induces embryonic death by the onset of implantation.

This conclusion is in good agreement with the experimental results of Epstein's group. Note that the day of vaginal plug was counted as the first day of gestation (day 1) in our studies but it was day 0 in the studies of Epstein's group. Hence day-4 and day-5 embryos in our work correspond to the three-to four-day embryos in Epstein's publications. Table 28 summarizes their findings, but expresses the stage of the embryos according to our convention (first day of gestation = day of plug).

A similar proportion of trisomic and monosomic embryos (18 and 20 per cent, respectively) were recorded in four-day progeny of female mice (ICR-Swiss or C57BL/6J) crossed to Rb163H/Rb1Wh. A considerable drop in the percentage of monosomic embryos compared to trisomics was however found on the fifth day (2 and 24 per cent, respectively). Death of all monosomic embryos at this stage has been observed both *in vivo* and *in vitro* after 24 h growth in culture. Almost half of all Ms19 embryos proceeded up to the blastocyst stage but most died shortly thereafter, i.e. on the fourth or fifth day (Epstein and Travis 1979). No information on cell counts in Ms19 embryos was given in this early work. These data as well as many others concerned with the early developmental effects of Ms19 are thoroughly discussed by Magnuson (1982).

Step by step comparison of monosomic and euploid embryos has been made throughout the whole pre-implantation period, that is early and late morula, early and late blastocyst stages. As seen in Tables 27 and 28, equal proportions of Ts and Ms embryos (27–21 per cent and 21–16 per cent, respectively) have been found at the 8–16 cell stage. The incidence of aneuploidy was somewhat reduced at the late morula stage, though the relative numbers of both aneuploids (trisomics and monosomics) remained almost constant (11 and 13 per cent respectively). The same correlation was seen at the early blastocyst stage (approximately 17 per cent), but at later stages Ts embryos significantly prevailed (17 per cent for Ts19 embryos and only 3 per cent for Ms19 embryos). Only 7 per cent Ms19 embryos and 22 per cent Ts19 specimens were identified after 24 h *in vitro*.

Thus the incidence of Ms19 is comparable to Ts19 at early stages of embryogenesis, but drops drastically both *in vivo* and *in vitro* at the late blastocyst stage.

Reduction in the cleavage rate becomes evident from the third cleavage division onwards, and gradually increases throughout pre-implantation development. Cell counts of Ms embryos were within the normal range at the early morula stage but became significantly reduced compared to both Ts and

Table 28. Survival capacities and cell numbers in pre-implantation mouse embryos with monosomy and trisomy for chromosome 19 (Magnuson et al. 1982)

Developmental stage	No. of embryos studied Total	Monosomy 19			Trisomy 19			Normal diploids		
		No.	%	Mean cell count	No.	%	Mean cell count	No.	%	Mean cell count
Early morulae (8–16 cells)	132	21	16	13 ± 2	28	21	13 ± 2	83	63	12 ± 2
Morulae	102	11	11	15 ± 2	13	13	21 ± 2	78	76	22 ± 2
Early blastocysts	122	21	17	46 ± 3	21	17	60 ± 3	80	66	56 ± 2
Late blastocysts	77	2	3	?	13	17	?	62	80	?
In vitro from fourth to fifth	161	11	7	?	35	22	?	115	71	?
Inner Cell Mass from 4th day to 6th	58	2	3	?	10	17	?	46	80	?

normal euploid embryos at the late morula stage (15 ± 2; 21 ± 2; 22 ± 2, respectively). Reduction of cell counts of Ms embryos compared to Ts19 and euploid littermates was especially striking at the early blastocyst stage (46 ± 3; 60 ± 3; 56 ± 2, respectively).

Using their original 'twin-technique' (separation of one blastomere for cytogenetic analysis at the two-cell stage), Epstein *et al.* (1978) made possible ultrastructural as well as biochemical investigations of Ms19 embryos. The only ultrastructural defect of Ms19 embryos was an increase in the number of degenerating bodies in the cytoplasm of monosomic cells. But these degenerative changes could be interpreted more readily as a consequence rather than the cause of embryonic death due to Ms19. Comparison of about 500 proteins in monosomic, trisomic, and euploid embryos did not reveal any consistent differences.

In vitro culture of microsurgically isolated inner cell masses of day 4 embryos revealed drastically reduced ability of Ms cells to make cell outgrowths after four or five days. But if aggregated with euploid morulae, Ms19 cells readily participated in genetic chimaeras. All five Ms19 ↔ diploid chimaeras produced in this way were alive and resembled normal day 9 embryos, with about 30 per cent of their cells lacking one chromosome 19. Meanwhile in Ts19 ↔ diploid chimaeras the proportion of Ts cells was usually more than 50 per cent.

Thus, Ms19 is not a true cell lethal and behaves as a lethal mutation at the organismal level only. Ms19 cells can be rescued for a prolonged time and may participate in morphogenetic processes when combined with normal euploid cells. Most Ms19 embryos are eliminated between days 4 and 5 of embryonic development though some may reach an advanced blastocyst stage. On the other hand Ms19 reduces cleavage rate from at least the 12–16 cell stage. Reduced proliferative capacities of Ms19 cells cannot be completely rescued in experimental chimaeras.

4.1.3 Characteristics of autosomal monosomy expression in pre-implantation mouse development

Cumulative data on the developmental effects of autosomal monosomy in mice shown in Table 29 readily demonstrate the insufficiency and fragmentary nature of our knowledge in this area. We are still almost completely ignorant of the embryonic manifestation of Ms4,7,8,10,11,13, and 18, and only a few cases of Ms1,2,3,5,6, and 9 have so far been examined. Autosomal monosomies 15 and 17 have been somewhat better studied, but only Ms19 has been investigated by a range of different experimental methods.

This limited information cannot sustain any fundamental generalizations as to the developmental effects caused by the deficiency of different

Table 29. Characteristics of pre-implantation development in mouse embryos with autosomal monosomy

Autosome No.	Parental karyotype and genotype		Embryos with monosomies studied							Diploid embryos of the same litter		References
	Males	Females	Total no.	Day of pregnancy			Morphological stage	Cell count	Abnormalities	Cell count	Stage	
				3rd	4th	5th						
1	Rb(1.3)1Bnr/+	+/+ F₁(CBA × C57BL)	5			5	Blastocysts	41.0 ± 3.7	No	38.5 ± 3.9	Blastocysts	Baranov 1983a,c
	Rb(1.3)Bnr/ Rb(1.10)10Bnr	+/+ ICR-Swiss	8		3		Late morulae	?	No	?	Morulae	Epstein and Travis 1979
						1	Blastocysts	?	No		Blastocysts	
						4	Late blastocysts, 24 h *in vitro*	?	No	?	Late blastocysts	
2	Rb(2.6)41em/+	Rb(2.6)41em/+	7		7		Morulae	14.5 ± 3.5 (7, 13, 16, 17, 17, 10, 14)	Pyknoses in 4 from 7 embryos	52.6 ± 3.4	Blastocysts	Baranov 1983a,c
3	Rb(1.3)1Bnr/+	+/+ F₁(CBA × C57BL)	9			9	Blastocysts	33.2 ± 4.2	5 embryos (less than 20 cells) pyknoses and degenerations	39.6 ± 2.3	Blastocysts	Baranov 1983a,c
5	+/+	Rb(5.19)1Wh/ Rb(5.15)3Bnr	18		11		Early morulae	4.3 ± 1.6		29.1 ± 2.4	Blastocysts	Baranov 1983a,c
					7		Morulae	14.1 ± 3.2	Micronuclei, pyknoses in 6 embryos			
6	Rb(2.6)41em/+	Rb(2.6)41em/+	13			8	Blastocysts	27.4 ± 4.5	Micronuclei, pyknoses in 6 embryos	53.3 ± 3.6	Blastocysts	Baranov 1983a,c
					5		Morulae	12–18				

Table 29. *Contd*

Autosome No.	Parental karyotype and genotype		Embryos with monosomies studied							Diploid embryos of the same litter		References
	Males	Females	Total no.	Day of pregnancy			Morphological stage	Cell count	Abnormalities	Cell count	Stage	
				3rd	4th	5th						
9 or 14	Rb(9.14)6Bnr/+	+/+	25	25			Blastocysts	32.9 ± 3.1	No	32.0 ± 1.3	Blastocysts	Baranov 1976
12	Rb(8.12)5Bnr/ Rb(4.12)9Bnr	+/+ ICR-Swiss	15		10		Late morulae	?	No	?	Blastocysts	Epstein and Travis 1979
						5	Late blastocysts	?	No	?	Late blastocysts	
15	+/+ F₁(CBA × C57BL)	Rb(5.15)3Bnr/ Rb(6.15)1Ald	16		16		8 early blastocysts 8 late morulae	23.7 ± 3.2	Pyknoses, micronuclei in 5 embryos	32.0 ± 1.3	Blastocysts	Baranov 1976a
	Rb(5.15)3Bnr/ Rb(6.15)1Ald	+/+	16 (*in vitro* since 4 cell stage)		11		Late morulae	19.2 ± 1.2	No	22.4 ± 1.9	Late morulae	Severova and Dyban 1984 Severova 1985
						5	Blastocysts	32.4 ± 7.7	No	49.8 ± 2.4	Blastocysts	
16	Rb(16.17)7Bnr/+	+/+	6	6			Early morulae	4.6 ± 1.5	No	4.8 ± 1.7	Early morulae	Baranov 1983a
	+/+	T(16;17)43H/ –	11			11 (9th day)	Early morulae	?	Empty degenerating vesicles	?	Late morulae	Baranov 1983b
17	Rb(11.17)11Rma/ Rb(8.17)11em	+/+	6		2		Late morulae	?	No	?	Late morulae	Epstein and Travis 1979
						4	Late blastocysts	?	No	?	Late blastocysts	
	+/+	Rb(8.17)11em/ Rb(16.17)7Bnr	21	21			Early morulae	6.6 ± 2.6	Micronuclei and pyknoses in 11 embryos	8.3 ± 2.5	Early morulae	Baranov, Dyban, and Chebotar 1980

					Stage		Abnormalities		Stage reached	Reference
+/+	Rb(8.17)1Sic/ Rb(16.17)7Bnr	19	12	2	Late morulae	28, 32	No	54.2 ± 3.7	Blastocysts	Severova and Dyban 1984
				9	Late morulae	14.9 ± 1.6	No	33.6 ± 1.4	Blastocysts	Severova 1985
				1	Early blastocysts	29	No	66.7 ± 2.5	Late blastocysts	
Rb(9.19)163H/ Rb(8.19)1Ct	+/+ ICR-Swiss		24	21	Late morulae, early blastocysts	?	No	?		Epstein and Travis 1979
				3	Blastocysts					
Rb(9.19)163H/ Rb(8.19)1Ct	+/+ C57BL/6J		10	9	Late morulae, early blastocysts	?	No	?		Epstein and Travis 1979
				1	Late blastocysts					
Rb(9.19)163H/ Rb(5.19)1Wh	+/+ F₁(CBA × C57BL/6J)		12	12	Early blastocysts	35.6 ± 3.2	Micro-nuclei, pyknoses in 4 embryos	34.0 ± 3.1	Early blastocysts	Baranov 1983a
Rb(9.19)163H Rb(5.19)1Wh	+/+ ICR-Swiss, C57BL/6J		67	21	Early morulae	13 ± 2	No	12 ± 2	Early morulae	Magnuson et al. 1982
				11	Morulae	15 ± 2	No	22 ± 2	Morulae	
				20	Early blastocysts	46 ± 3	No	56 ± 2	Early blastocysts	
				3	Late blastocysts	?	No	?		
				11	Late blastocysts (24 h in vitro)	?	No	?		

autosomes. Moreover, all these experiments were based on the same routine approach—comparative cytogenetic and embryological analysis of Ms specimens at successive stages of development, predominantly on days 3 and 4. The absence of monosomy on the fourth day was usually accepted as evidence of its elimination at a somewhat earlier stage, which was considered to be the phenocritical stage for this aberration.

However, such an experimental approach may actually provide only preliminary data on the developmental effects of each monosomy. The main reason for such an interpretation lies in the reduced mitotic activity of aneuploids and especially of monosomics. Because of this, many monosomic embryos studied *in vivo* possess no mitotic figures in air-dried preparations and thus may escape identification in routine studies. Hence a precise determination of the embryolethal effects of monosomies needs both *in vivo* and *in vitro* studies. The latter approach might be especially useful for distinguishing the actual cause of embryonic death, whether it is due to failure of implantation because of decreased blastomere number, or reflects the inability of monosomic embryos to develop beyond some specific embryonic stage.

Although decisive conclusions on exact embryolethal stages of autosomal monosomies cannot yet be made, it seems plausible that the absence of different autosomes, though mostly incompatible with post-implantation survival, has quite specific effects on developing mouse embryos. Monosomies 1,3,6,9,12,14,15,16, and 19 do not interfere with cleavage, compaction, and blastocyst formation, while monosomies 2 and 5 may induce embryonic death by the morula stage, as also does Ms17 if carried on a particular genetic background. The present experimental data indicate an unequal impact of maternal and paternal chromosome 17 loss in monosomic embryo survival. Whether the parental influence on monosomy is a general phenomenon, or a unique property of chromosome 17 remains as yet obscure.

The amount of euchromatin and heterochromatin does not determine phenocritical stages for Ms1,12,17, or 19, and the early death of monosomic embryos is probably due to the hemizygosity of genetic loci on the remaining chromosome rather than to its absolute size (Epstein and Travis 1979; Magnuson, Smith, and Epstein 1982). Ms19 is not a true cell lethal, as monosomic cells can readily be rescued by aggregation with cells of normal diploid embryos. The same is also true for Ms15 (Epstein 1984). Whether other types of monosomies are detrimental predominantly in the affected cells themselves (cell lethals) or whether they behave as lethals at the level of the organism as a whole remains to be studied.

The mechanism of the lethal effects induced by autosomal monosomy is one of the still unresolved and most crucial problems of modern developmental cytogenetics. According to one suggestion launched by Epstein's group, the death or abnormal development of monosomic embryos is provoked by 'alterations in specific cellular processes or structures directly affected by the aneuploid genes'. Morever, 'monosomy results in obligatory states of dosage deficiencies which might have more serious consequences on concentration-dependent cellular processes than trisomy' (Epstein, Epstein, Cox, and Weil 1981; Magnuson *et al.* 1982) According to Baranov (1983a–d), developmental failures of monosomic embryos are mainly due to the functional nullisomy resulting from the partial or complete block of genetic activity of the unpaired odd chromosome during early embryogenesis (see Chapter 13 for details). Both suggestions need solid experimental verification.

Mice with Robertsonian translocations provide an exceptionally useful tool for detailed investigations of the developmental effects of each particular type of monosomy in mice. Further progress in this direction may make an invaluable contribution to our understanding of chromosome activity in the early embryogenesis of mammals.

7.2 Autosomal nullisomy

No cases of spontaneous nullisomy (i.e. complete absence of both homologous chromosomes) have been reported so far in humans or in embryos of experimental animals. Most of our present knowledge on nullisomy and homozygous deficiencies (the loss of both homologous segments) comes from the field of plant cytogenetics. A few cases of partial nullisomy for the distal or interstitial segment of the X chromosome in male children have been reported (Tiepolo, Zuffardi, and Rodewald 1977; Tabor, Andersen, Lundsteen, Niebuhr, and Sardemann 1983).

Much better survival of individuals with partial nullisomy has recently been reported for the Syrian hamster (Breckon and Savage 1982). By crossing two animals that already lacked a small product of the unequal reciprocal translocation T(11.20), two homozygous deficient (i.e. nullisomic) males were obtained. Both animals had only 42 chromosomes instead of the usual 44 and were nullisomic for the centromeric part and short arm of chromosome 11 and for the larger portion of the long arm of chromosome 20. In spite of substantial genetic loss (about 2.5 per cent of the total chromosome length), externally and anatomically these unique animals looked apparently normal, with one of them even being fertile. The missing chromosomal segments fell into the category of late-replicating chromatin and were probably lacking true structural genes.

Autosomal nullisomy provides a unique opportunity to study the actual impact of any particular part of the genome, any chromosome or its segments, in the control of morphogenetic processes, so is of special interest for developmental cytogenetics.

There are a few possible methods of inducing complete or partial nullisomy in experimental mammals. Nullisomy for the X chromosome is efficiently generated in the progeny of XO females or female mice with an inversion of the X chromosome. It may also be produced microsurgically in androgenetic haploids (see Chapter 1).

Production of autosomal nullisomy usually meets with substantial technical difficulties. Several possible ways have been suggested to overcome these difficulties, but none has been successfully realized so far. Ford suggested that it should be possible to produce nullisomic zygotes by intercrossing double Robertsonian translocation carriers, which may yield 3–4 per cent nullisomic zygotes (Ford 1975). This interesting suggestion has still not been realized. One of the main obstacles is the rather low probability of nullisomic embryos on the one hand, and male sterility in double heterozygotes on the other. A much more promising approach has been recently advanced by Searle and Beechey (1982), who suggested generation of a compound of three different Robertsonian translocations with tribrachial homology producing only aneuploid gametes. A compound with Rb(4.6)2Bnr/Rb(4.15)4Rma/Rb(6.15)1Ald could be suggested as an example. We succeeded in producing a similar compound with other Robertsonian translocations—Rb(9.15)3Bnr/Rb(9.19)163H/Rb(5.15)1Wh—but unfortunately, all the males were completely sterile.

Autosomal nullisomy is more readily produced in parthenogenetic embryos (parthenogenones). Even with a normal euploid karyotype, however, parthenogenones display rather poor development compared to normal fertilized eggs (see Chapter 1). This significantly hampers the use of parthenogenetic embryos to study the developmental effects of nullisomy.

Theoretical considerations of this experimental route have been presented elsewhere (Phillips and Kaufman 1974; Kaufman 1975, 1983a). Haploid embryos with nullisomy and disomy of a particular chromosome might readily be produced by artificial activation of ovulated eggs from female mice heterozygous for different Rb translocations. Suppression of second polar body emission by cytochalasin B treatment of recently ovulated eggs in Rb-translocation carriers might produce diploid parthenogenones with nullisomy of the translocated chromosomes. This method has been successfully tested in our laboratory on doubly heterozygous mice, Rb(9.19)163H/Rb(5.19)1Wh, generating a high proportion of eggs with non-disjunction of autosome 19 (see Chapter 1). Chromosome counts have been made in 225 parthenogenetically activated one-cell stages previously diploidized by cytochalasin B treatment (Balakier and Tarkowski 1976). Cytogenetic analysis of air-dried preparations revealed a balanced diploid karyotype in 137 parthenogenones (Dyban and Noniashvili 1985a,b), an aneuploid chromosome set in the remaining 82 (35.5 per cent), 44 tetrasomic and 38 nullisomic (see Chapter 1, Tables 7 and 8). This was the first experimental work to establish the validity of this approach for deliberately producing parthenogenones with nullisomy of a particular chromosome (chromosome 19). The developmental capacities of these nullisomic parthenogenones have not yet been studied. Some preliminary data on the developmental properties of hypohaploid parthenogenones in mice already exist from experiments carried out on mice heterozygous for the unequal reciprocal translocation T(14.15)6Ca (Kaufman and Sachs 1975; Kaufman 1983a). Chromosomal analysis of artificially activated eggs at the first cleavage division revealed haploidy ($n = 20$) in 81 parthenogenones, hypohaploidy ($n = 19$) in 12 (10.9 per cent), and hyperhaploidy ($n = 21$ or 22) in 17 (15 per cent). Significant changes of chromosomal pattern in the parthenogenones were evident on the fourth day of development. Absolute numbers of euploid and hyperhaploid parthenogenones were 59 and 14 respectively, while only one specimen was found with hypohaploidy due to the absence of the small translocation product, chromosome T6. Both normal haploid and hyperhaploid parthenogenones readily proceeded to the morula stage, but the unique hypohaploid parthenogenone consisted of four cells only. Thus nullisomy for a very small centromeric fragment carring NOR of chromosome 15, plus the distal tip of chromosome 14, may not be compatible with development beyond the four-cell stage (Kaufman and Sachs 1975).

Efficient production of hypohaploid parthenogenones may also be achieved by ethyl alcohol treatment of ovulated eggs. This is thought to interfere with normal disjunction of meiotic chromosomes and thus results in parthenogenones with hypo- and hyperhaploidy (Kaufman 1982a, 1983a).

In this experimental series ovulated eggs from mice of normal (all acrocentric) karyotype were activated *in vitro* by transient ethyl alcohol treatment and then used for chromosomal studies (Kaufman 1982a). Of 326 eggs at M-I of the first cleavage division, 23 (7 per cent) were classified as hyperhaploid (21 or 22 chromosomes) and 33 (11 per cent) as hypohaploid (18 or 19 chromosomes). Cytogenetic analysis of more advanced parthenogenones (22.8–13.6 blastomeres) revealed hyperhaploidy in six and hypohaploidy in only one parthenogenone. The cell count in this unique hypohaploid specimen is not recorded, but the observations strongly suggest progressive elimination of hypohaploid parthenogenones during cleavage. This conclusion has gained more support in a recent study by Kaufman (1983a). Alcohol-activated mouse eggs were cultured *in vitro* up to the morning of the fourth day and then used for chromosome counts in air-dried preparations. Of 105 karyotyped parthenogenones, 95 were euploid and 10 aneuploid, including nine hyperhaploid and one hypohaploid ($n = 19$). The cell count of this parthenogenone is not given and the absent autosome was not identified, but it is clear that at least some nullisomic haploids are able to proceed to the morula stage. Unexpectedly, a human embryo totally lacking chromosome 15 in three metaphase plates has recently been discovered among seven or eight cell stages grown from *in vitro* fertilized oocytes (Angell *et al.* 1983). So nullisomy for chromosome 15 seems to be compatible with at least three cleavage divisions in humans.

However, Kaufman's work unambiguously proves the poor viability of hypohaploid parthenogenones in mice, most being eliminated during the second or third cleavage divisions.

We may conclude that while haploid parthenogenones with an excess of one or more autosomes proceed normally up to the morula stage and probably even beyond it, autosomal nullisomy is undoubtedly highly detrimental for pre-implantation development. In exceptionally rare cases, nullisomic haploids may be encountered at the morula stage, indicating clear-cut differences in manifestation of nullisomy for different autosomes. Many more combined cytogenetic and embryological studies are needed in hyper-haploid and hypohaploid parthenogenones to confirm or discount this last suggestion. We shall deal with this problem in Chapter 13, concerning the functional activity of chromosomes in early mammalian development.

One special type of nullisomy concerns homozygous deletions. Experimental analysis of embryological effects caused by homozygous deficiencies may be a fruitful approach to clarify the participation of different segments of particular autosomes in the genetic control of early development.

One of the best studied clusters of homozygous deficiencies are radiation-induced deletions of chromosome 7. Two are known to be expressed prenatally and the other four perinatally. Of special interest is the cytologically identifiable c^{25H} deletion, occupying the middle segment of chromosome 7. Although well tolerated in heterozygotes the deletion in homozygotes (c^{25H}/c^{25H}) blocks early cleavage somewhere between two and six blastomeres. The affected embryos remain alive for the next 24–48 h but then gradually degenerate (Nadijka *et al.* 1979; see also Chapter 12).

Another cluster of radiation-induced deletions is located in the proximal region of autosome 17, in the *t*-locus. These are T^{hp}, T^{Orl}, and T^{Or}. As with the c^{25H} albino deletions, these aberrations are not detrimental in heterozygotes but all are lethal in homozygotes. $T^{hp} T^{hp}$ embryos cease their development soon after the morula stage; $T^{Orl} T^{Orl}$ do not proceed beyond blastocyst stage and $T^{Or} T^{Or}$ die after the egg-cylinder stage. T^{hp} is the largest deletion as it occupies the greater part of the *t*-locus. T^{Orl} is intermediate and T^{Or} is the smallest of the three. Embryological studies of these *T*-deletions in homozygotes have been successfully used (Babiarz 1983) to map the region of chromosome 17, which is crucial for the genetic control of early development. The region includes the *Tcp-1* gene and is responsible for the normal development of mouse embryos up to the morula stage (see Chapter 12 for more details). There are some indications that a special gene or genes present in the proximal part of chromosome 17 close to the *H-2* complex controls the onset and duration of cleavage divisions (Goldbard, Verbanac, and Warner 1982a,b; see also Chapter 12). More studies are needed to put this suggestion on to solid experimental ground.

The developmental effects of combined partial nullisomy and mono-somy for the same chromosome have been studied in our experiments on mice heterozygous for both Robertsonian and non-Robertsonian reciprocal translocations (Baranov 1983b, 1985). Female mice hetero-zygous for the T(16.17)43H translocation (see Chapter 9) were crossed to Rb(16.17)7Bnr/+ males and all day-4 embryos were studied cytogenetically. As a result of meiotic non-disjunction, a substantial number of genetically unbalanced gametes with structural and numerical aberrations of the trans-located chromosomes were generated by both sexes. After fusion, aneuploid zygotes lacking one whole chromosome 17 and carrying substantial deficien-cies in the remaining chromosome 17 were generated.

Of 151 day-4 embryos of known karyotype, four with Ms17 also had a deficiency of the proximal segment of the same chromosome (17AB), which encompasses the genes of the complex t-locus (Forejt *et al*. 1982). Thus the actual aberrations of these unique embryos included Ms17 + nullisomy 17AB. In spite of such crude genetic imbalance, all four embryos were composed of 8–16 blastomeres. Numerous micronuclei, pyknoses, and pathological mitotic figures with prematurely detached chromatids, as well as a significantly reduced cell count, unambiguously pointed to the death of these embryos by the morula stage. The same stage is known to be crucial for the development of pure Ms17 embryos as well as for embryos homozygous for the T^{hp} deletion (see above).

Further studies of the developmental effects due to autosomal nullisomy of whole autosomes or of some chromosomal segments may shed light on the genetic and chromosomal control of early development in mammals (see Chapter 13).

8

Numerical aberrations of sex chromosomes

8.1 Frequency

Nearly two-thirds of the known chromosomal anomalies in liveborn children consist of abnormalities of the sex chromosomes, most of which are numerical aberrations (Sankaranarayanan 1979; Lauritsen 1982; Davidenkova *et al*. 1973; Bochkov, Zakharov, and Ivanov 1984; Hamerton, Canning, Ray, and Smith 1975; Vogel and Motulsky 1979). The XO karyotype has also been recorded in mice, rats, monkeys, cats, and sheep (Russell and Saylors 1960, 1961; Dyban *et al*. 1971; Weiss, Weick, Knobil, Wolman, and Gorstein 1973; Norby, Hegreberg, Thuline, and Finlay 1974; Long, Berepubo, and Amaitar 1980; Zartman, Hinesley, and Gnathowski 1981). The XXY karyotype is known in mice, black rats, Chinese hamsters, cats, and dogs (McLaren 1960; Yosida 1978; Ivett, Tice, and Bender 1978; Clough, Ryle, Hare, Helly, and Patterson 1970; Centerwall and Benirschke 1973, 1975). The XYY karyotype is known in laboratory mice (Cattanach and Pollard 1969; Rathenberg and Müller 1973; Evans, Beechey, and Burtenshaw 1978; Burgoyne and Diddle 1980; Das and Karr 1981). And the XXX karyotype has been recorded in cattle (Rieck, Höhn, and Herzog 1970; Norberg, Rejsdat, Garm, and Nes 1976). A unique mouse with a 42 XXXY constitution has also been reported (Ford 1971). Even more complex numerical sex-chromosome aberrations are encountered in human populations. They are represented by tetra-X, penta-X, etc. (Davidenkova *et al*. 1973; Vogel and Motulsky 1979).

The available information on the frequency of numerical sex-chromosomes aberrations in humans is much more voluminous than in other mammals. In post-natal life, 47 XXY (Kleinfelter's syndrome) and 47 XYY are encountered in 2.1 per 1000 and 1–3 per 1000 people respectively. These rates are almost twice those for triplo-X, 47XXX (1.2 per 1000) and five or six times more then for 45 XO children (Maclean, Harnden, Brown, Bond, and Mantle 1964; Carr 1969; Boué and Boué 1974). The lower birth rate for XO individuals compared to those with an extra sex chromosome (X or Y) is probably due to the massive elimination of the former before birth. 45X is known to be the commonest single chromosome disorder in humans, accounting for almost 19 per cent of all spontaneously aborted embryos in the first trimester of pregnancy. Taking into account the overall frequency of spontaneous abortuses, which constitute over 15 per cent of all conceptuses in humans, this means that roughly 0.75–1 per cent of all human zygotes lack one X-chromosome. Thus the overwhelming majority of XO conceptuses are eliminated during embryogenesis, with only one in 40 reaching term (Carr 1969). The very high frequency of XO conceptuses compared to other sex-chromosome abnormalities in humans and probably in other mammals may be due to sex-vesicle loss during spermatogenesis (Ashley 1983).

The XO karyotype is an even more common aberration in laboratory mice. It has been recorded in 0.7–0.9 per cent of all conceptuses in one strain and in almost 1.7 per cent in other strains of mice (Russell 1962, 1964). A very low frequency of the XXY karyotype has been reported for laboratory mice compared to the identical abnormality in humans (a tenfold difference; Russell 1962). An extremely low incidence of XXY anomalies has also been recorded in cats (Centerwall and Benirschke 1973, 1975). No reliable data on the frequency of sex-chromosome aberrations in other mammalian species are yet available.

8.2 XO karyotype

8.2.1 Effects on human development

Developmental arrest in XO conceptuses usually occurs at 6 weeks of age (Boué and Boué 1974, 1976). Almost two-thirds of all early XO abortuses are represented by closed sacs containing the remnants of macerated embryonic tissues. The remaining one-third of XO conceptuses are less severely affected, but macerated embryos of a developmental age below 6 weeks are often encountered. Pathological changes show that the death of these XO embryos is secondary to placental defects. A few XO embryos develop beyond 7–12 weeks, but they are exceptionally rare among second trimester abortuses. These advanced XO conceptuses usually manifest some common morphological characteristics such as generalized oedema, hygroma on each side of the neck, and horse shoe kidneys (Singh and Carr 1966; Carr 1971a,b). Lymph-filled sacs in the neck region and horseshoe kidneys are two very common features of XO fetuses. Other abnormalities include heart damage, malformations of skeletal and urinary systems, nodular amnion, and atresia of the umbilical vessels (Kuliev 1972; Kulazenko 1976). One of the most conspicuous features of XO monosomy is gonadal dysgenesis. The ovaries of XO individuals are histologically normal in 45 per cent of embryos for up to 2 months of intrauterine life and contain numerous dividing oogonia and follicle cells. The germ cell number declines progressively during the second half of pre-natal life. The ovarian picture at term is highly variable. The infant may have streak gonads completely devoid of germ cells, or on the other hand the ovary of an XO infant may be normal (Carr, Haggar, and Hart 1968). A few normal looking primordial follicles were encountered histologically in most XO ovaries. Degeneration of oocytes not surrounded by follicle cells during the initial stages of folliculogenesis was suggested to be the major cause of sterility in XO women (Conen and Glass 1963; Hodel and Egli 1965). As a rule the number of ova was greatly reduced and only about 1.5 per cent of follicles were judged to be normal.

Thus, the XO anomaly does not seem to affect germ cell migration to the gonadal ridges or their proliferation or entry into meiotic prophase, but all or almost all oocytes of XO constitution invariably die thereafter. A dramatic loss of XO oocytes occurs at a stage when many XX oocytes are undergoing atresia. According to one early suggestion, the inability of follicle cells to form normal primordial follicles might be a primary reason for the massive death of XO oocytes (Singh and Carr 1966). According to a more recent view, however, the reduced level of some X-coded gene products in the XO oocytes renders them highly susceptible to degeneration (Epstein 1981b).

Adult XO individuals are always sterile due to the total loss of germ cells from their atrophic 'streak' gonads. Numerous other phenotypic peculiarities of XO adults are outlined in great detail elsewhere (Davidenkova, Verlinskaya, and Tysiachnuk 1973, 1974; Ohno 1967; Hamerton 1971; Vogel and Motulsky 1979). Nonetheless, up to now the exact reasons for the congenital defects and abnormal development of XO individuals in humans remains obscure. The high mortality among XO individuals is likely to be correlated in some way with the X chromosome inactivation mechanism operating during early development. Incomplete inactivation of the heterocyclic X chromosome and the developmental significance of two X chromosomes during oogenesis and early embryogenesis are suspected as the most likely causes of the early death and inherited abnormalities in XO individuals (Epstein 1981b, 1983a,b).

Thus the XO anomaly is the most common single chromosome disorder in humans. A very high proportion of XO embryos is encountered during the first trimester. Only 2 per cent of XO embryos go to term and these display serious congenital defects involving the lymphatic, cardiovascular, and urinary systems as well as some other abnormalities, comprising the so-called Turner's syndrome.

8.2.2 XO mice

8.2.2.1 Experimental production. The spontaneous incidence of XO mice was found to be within the range 0.1–1.7 per cent according to stock, with an average incidence of 0.7 per cent (Russell 1962, 1964). Abnormal sex chromosome constitutions may be readily induced by irradiating female mice soon after insemination, or by irradiating males prior to copulation. The former was found to be much more efficient.

Total body irradiation of female mice with only 100 R between 8.30 and 11.00 a.m. on the day of mating yielded about 3–3.5 per cent XO monosomy in the progeny. Sensitivity rapidly decreased, showing that the completion of the second metaphase and the onset of pronuclear formation were the most vulnerable stages for induced X chromosome loss. When X-irradiation is applied at the pronuclear stage, more paternal than maternal X chromosomes are lost. Comparison between the induction of paternal sex-chromosome loss in irradiated spermatozoa and in the male pronucleus shortly after sperm entry indicates the latter stage to be more sensitive by approximately one order of magnititude. The induction rate of X chromosome loss after spermatozoa irradiation with 600 R does not exceed 1.2 per cent (Russell 1964, 1965).

The existence of coat-pattern mutations carried on the X chromosome, including the semidominant tabby (*Ta*) mutation, makes XO individuals easy to recognize and maintain in stocks carrying appropriate markers, without additional cytogenetical control (Cattanach 1962).

8.2.2.2. Developmental manifestations. In contrast to the abundant information that we have on XO mice post-natally we have until recently been almost completely ignorant as to the developmental effects of an XO karyotype before birth. Since the number of XO embryos in the progeny of XO mice is somewhat more at the first than at the second cleavage division, elimination of at least some XO embryos during early cleavage has been suspected (Luthardt 1976). Cleavage rate and blastocyst formation were found to be retarded in XO embryos compared to their XX or XY littermates (Burgoyne and Biggers 1976). The retardation of XO embryos was even more marked during later stages of embryogenesis (Burgoyne, Evans, and Holland 1983a). The lower viability of XO embryos both *in vivo* and in *vitro* could be attributed not primarily to the XO karyotype but rather to the reduced developmental potential of eggs produced by XO mice. Poor pre-implantation survival *in vitro* of chromosomally normal embryos of XO mothers seems consistent with such an interpretation (Burgoyne and Biggers 1976.) More direct evidence of the deleterious developmental effects of the XO anomaly has been obtained in female mice heterozygous for a paracentric inversion of the X chromosome, In(X)1H. Dosage effects for X-coded genes were normal in the oocytes of these mice but because of regular meiotic non-disjunction, 25 per cent of their daughters were XO (Burgoyne *et al.* 1983a; Burgoyne, Tam, and Evans 1983b). Egg cylinder growth was substantially retarded in XO embryos at 7.5 days *post coitum*. Growth delay of XO embryos was somewhat less evident during major organogenesis (11–12 days) but became more marked later. The placental weight of pre-term XO fetuses

was somewhat greater than that of XX fetuses, indicating a specific deleterious effect of the XO anomaly in embryonic cells proper (Burgoyne *et al.*
1983b). It should be remembered that only the maternal X chromosome
remains active in the ectoplacental cone and chorionic plate—that is in two
major constituents of the chorio-allantoic placenta, while maternal and
paternal X chromosomes are randomly inactivated in the embryonic tissues
proper (Papaioannou and West 1981; see also Chapter 13). Thus the absence
of a paternal X chromosome seems not to be harmful for the trophectodermal and primary endodermal derivatives, but is unfavourable for the
growth of the embryonic tissues proper, where inactivation of the X chromosomes is at random.

Since the deleterious effects of XO monosomy are already evident during
pre-natal development in both human and mouse embryos, XO mice provide
a useful experimental tool for the study of developmental abnormalities due
to the lack of one X chromosome. These model experiments can be fruitfully
extended to post-natal life.

It has been known for a long time that an XO karyotype is compatible with
post-natal survival in mice (Russell *et al.* 1959; Welshons and Russell 1959).
Early observations suggested that the growth and post-natal development of
XO mice was not retarded but was normal or 'fairly normal' (Welshons and
Russell 1959; Cattanach 1962). More detailed analysis of the growth of XO
offspring of either In (X)1H or XO females are not consistent with such a
conclusion (Deckers and van der Kroon 1981; Burgoyne *et al.* 1983b).

The growth rate of XO mice during the first 5 weeks of life was shown to be
significantly lower (15 per cent) than that of normal XX mice (Deckers and
van der Kroon 1981). The interpretation of these findings was hampered
however by the presence of the tabby mutation which itself is unfavourable
for post-natal growth. This difficulty was circumvented in mice heterozygous
for In(X)1H (Burgoyne *et al.* 1983b). The mean weight of XO newborns was
significantly less than that of normal XX mice in the same litter. Weight
differences increased up to 15 weeks, when the mice were karyotyped.
Reduced growth rates of XO mice showed a strong positive correlation with
lowered body temperature and thyroid activity, and were therefore thought
to be due to hypothyroidism and a lower basal metabolic rate (Deckers, van
der Kroon, and Douglas 1981).

Thus growth retardation should be considered as a constant feature of the
XO anomaly, common to both human beings and mice.

The reproductive capacities of XO mice have been subjected to especially
intensive studies for many years. XO mice were found to be fertile, but their
reproductive lifespan was little more than half that of normal XX mice
(Cattanach 1962; Morris 1968). Mean litter size of XO mice was only 55 per
cent of that of XX females and the total number of offspring generated by the
former did not exceed 34 per cent of the control (Lyon and Hawker 1973). It

has been suggested therefore that the markedly reduced reproductive life of XO mice was due to premature exhaustion of the oocyte supply (Lyon and Hawker 1973). According to a more recent study, the reproductive capacity of XO mice reaches its maximum and minimum earlier than in normal mice (Deckers *et al.* 1981).

The short reproductive lifespan of XO mice can be attributed to a dramatic loss of oocytes immediately after birth (Burgoyne and Baker 1981). As a result the ovaries of XO females contain only half as many oocytes as do normal XX ovaries (Burgoyne and Baker 1981). The whole ovary as well as the number of maturing follicles are substantially smaller in XO mice than in XX controls. This was tentatively attributed to a reduced activity of the oestrogen–progesterone system (Deckers *et al.* 1981).

Thus the lower productivity of XO mice might be caused by a diminished secretion of gonadotrophic hormones together with a smaller number of primordial germ cells.

Interspecies differences in the reproductive capacity of XO individuals are probably quantitative in nature and rely on the time scale characteristics of each species. XO mice reach puberty before the X gene-dosage deficiency in oocytes or in some extragonadal tissue becomes severe, while XO women reach puberty after all their oocytes have degenerated (Burgoyne 1978).

The small litter size of XO mice cannot be attributed to the ovary defect as the number of ovulated ova in XO females does not differ significantly from that in control XX mice (Morris 1968). The ovulated oocytes of XO females may be successfully fertilized but many are eliminated both before and after implantation (see above). All YO embryos and some XO are subject to elimination during pre-natal development, resulting in a mean litter size of little more than half the control.

The selective elimination of chromosomally unbalanced embryos cannot be the sole factor responsible for the large excess of XX over XO females in the progeny of XO mice (Cattanach 1962). Direct cytogenetic studies of ovulated oocytes from XO females found a substantial preponderance (almost twofold) of oocytes with 20 chromosomes (X-bearing gametes) over those with 19 chromosomes (X-deficient gametes) (Kaufman 1972; Luthardt 1976). Abnormally low segregation of nullo-X gametes was initially suggested by Morris (1968). This meiotic drive seems to be the only explanation for the excess of sex-chromosome balanced embryos (XX or XY).

Abnormal growth of XO embryos, developmental retardation before term and after birth, and underdevelopment of ovarian tissues including oocytes suggested only quantitative and not qualitative differences between XO mice and Turner's syndrome women, making XO mice a valuable experimental tool for studying the detailed mechanisms of developmental failure due to X chromosome monosomy in humans.

Our preliminary studies with co-workers from the Behavioural Genetics

Department of Moscow State University have revealed a reduced ability for extrapolation and conditioned reflex production in adult XO mice compared to normal XX mice, indicating a somewhat reduced mental capacity in XO individuals.

8.2.3 XO karyotype in embryogenesis of other mammalian species

Data on the developmental effects of the XO anomaly in other mammals are still very fragmentary. XO embryos of the Norway rat were still alive on the twelfth day of gestation (Butcher and Fugo 1967) and some were recovered after birth (Dyban *et al.* 1971). The reproductive capacities of XO rats as well as other phenotypic characteristics of these individuals remain unknown. XO rhesus monkeys revealed a gross gonadal dysgenesis and hormonal disturbances similar to those of XO women (Weiss *et al.* 1973).

All available data on ovarian morphology and fertility in other mammalian species with an XO karyotype are consistent with the hypothesis that quantitative rather than qualitative differences between species are manifested due to differences in their developmental timescale (Burgoyne 1978). XO individuals from species with a short generation time (cats, black rats, mole rats, wood lemmings, field mice, etc.) are usually fertile, while those from species with much longer generation times (human, horse, rhesus monkey) are sterile (see Burgoyne 1978 for more details).

Assuming an absolute homology for the gene content of X chromosomes in all placental and marsupial mammals (so-called Ohno's Law—Ohno 1979), the similarities of X monosomy phenotype in different mammalian species are to be expected.

One of the most thoroughly studied karyological phenomena of early mammalian embryos is X chromosome inactivation. The principal cytological manifestations of this phenomenon include the appearance of a sex-chromatin body (in interphase nuclei), late replication of one of the two X chromosomes (in metaphase plates) as well as some other minor changes in morphology of the inactivated X chromosome (Lyon 1972, 1974). Let us follow briefly the timing and pattern of X chromosome inactivation in mammalian embryos.

No sign of a sex-chromatin body is evident in blastomere nuclei of cleaving embryos. In most of the mammals studied so far sex-chromatin and late replication of the X chromosome appear first at the blastocyst stage, just at the onset of or soon after implantation (Lyon 1966, 1972), a little earlier in the trophoblast cells than in the embryo proper (Park 1957). Sex-chromatin identification, however, cannot be considered as a reliable or sensitive method for estimating the time of X chromosome inactivation. The embryonic time of X chromosome differentiation may be more precisely measured by autoradiography, or by some differential staining technique for metaphase chromosomes. For instance in the rabbit, late replication of the X chromosome appeared 24 h earlier than did sex chromatin (Issa, Blank, and Atherton 1969).

Cytogenetic studies on both mice and rats with marker X chromosomes have shown that there is preferential paternal X chromosome inactivation in the chorion and in the yolk sac of early post-implantation embryos. At the same stage no preponderance of either parental X chromosome inactivation could be detected in embryonic cells (Takagi and Sasaki 1975; Wake, Takagi, and Sasaki 1976). Among post-implantation mouse embryos a single brightly fluorescent X chromosome was found in blastocysts consisting of more than 40 cells, implying that differentiation of the X is initiated by the blastocyst stage. The heterochromatic X chromosome became bright over its entire length, resulting in an indistinct banding pattern. Contrary to the corresponding data in rabbits (see above), cytological indications of X chromosome inactivation in mice are evident at least one cell cycle earlier than is delayed initiation of replication. The allocyclic X chromosome is often more contracted than its normal homologue, though this correlation is not the same for all cells. Direct cytological evidence for preferential inactivation of the paternally derived X chromosome in the mouse blastocyst was obtained in mice heterozygous for Cattanach's translocation (Takagi, Wake, and Sasaki 1978).

The time of X inactivation estimated by chromosome staining methods is usually more than 24 h earlier than that estimated by any other available methods, including genetic and biochemical ones. There is strong genetic evidence, based on injection of single inner cell mass (ICM) cells into host blastocysts, that both X chromosomes are still active in the ICM at 3.5 and 4.5 days *post coitum*, i.e. much later than the cytologically detectable X chromosome appears (Gardner and Lyon 1971). Cytological changes in the heterochromatic X chromosome revealed by a special technique (Kanda 1973) precede biochemical indications of X chromosome inactivation in epiblast derivatives of early post-implantation mouse embryos (Rastan 1982).

Thus at the present time it is not easy or even possible to reconcile the data on the precise time of X chromosome inactivation obtained by different methods. However, there is no doubt that the biochemical approach provides a major contribution to our knowledge of X chromosome differentiation in XX mammals. These studies have been made almost exclusively on mouse embryos heterozygous for various allelic isozymes of the phosphoglycerate kinase (*Pgk-1*) gene, carried on the X chromosome (Nielsen 1977). The other highly sensitive biochemical approach was based on determinations of the ratio of activities of the X-linked enzyme hypoxanthine phosphoribosyl transferase (HPRT) to an autosome-coded enzyme, adenine phosphoribosyl transferase (APRT) (Monk 1978).

According to combined cytogenetic, biochemical and genetic data, the general scheme of X chromosome differentiation in XX mouse embryos can be outlined as follows. Both maternal and paternal X chromosomes are active during cleavage, until the early blastocyst stage. The X chromosome in trophectoderm appears to differentiate by the middle or late blastocyst stage (Monk and Kathuria 1977; Kratzer and Gartler 1978) and that of primary endoderm by the sixth day of gestation (Monk and Harper 1979). It is predominantly or even exclusively the paternal X chromosome that is inactivated in these cells. The randomness of X chromosome expression is correlated with embryonic cell lineage. This means that the maternal X (X^m) is almost exclusively expressed in the trophectoderm derivatives (ectoplacental cone, chorionic plate, primary giant cells) and in that of primitive endoderm (parietal and visceral layers) while both X^m and X^p (paternal) are randomly expressed in the derivatives of primitive ectoderm (allantoic and visceral yolk sac mesoderm; amnion and tissues of embryo proper) (Takagi et al. 1978; Papaioannou and West 1981). Non-randomness of X chromosome inactivation was shown not to result from cell selection or secondary reversal of a primarily active X to an inactive one, but was probably due to the preferential inactivation of the paternally derived X chromosome in the trophectoderm and primitive endoderm (Harper, Fosten, and Monk 1982). X chromosome inactivation is completed before the differentiation of yolk sac mesoderm, the three germ layers, and the germ line. The question of whether these tissues 'bud off' from the epiblast according to the stem-line postulated by Monk's hypothesis (Monk and Harper 1979; Monk 1981) still awaits an answer. The earliest time of X inactivation in the epiblast (primary ectoderm) cells is calculated from the biochemical data to be between 4.5 and 5.5 days *post coitum*. All embryonic tissues proper are derived from the same pool of epiblast just after X inactivation. The size of the pool was roughly estimated at 47 cells (McMahon, Fosten, and Monk 1983). Random inactivation of the parental X chromosomes in female germ cells of fetal mice makes it highly unlikely that germ cells are derived from the yolk sac endoderm, where as mentioned above the paternally derived X chromosome is preferentially inactivated. By their pattern of X chromosome inactivation germ cells evidently resemble other epiblast derivatives (McMahon, Fosten, and Monk 1981). It seems probable from X chromosome inactivation studies that the germ line stems from epiblast not as a group of few cells but as a pool of some appreciable number of cells roughly estimated at 193 cells (McMahon et al. 1983). The successive stages of X chromosome progression have been thoroughly studied in the germ cells of various mammals including mice and humans, and very similar results were obtained. The presence of sex chromatin and heterochromatinization of one X chromosome is reported in female germ cells of human embryos after they have reached the germinal ridge (Semyonova-Tian-Shanskaya and Patkin 1978). The reactivation of the previously genetically silent X chromosome in the female germ cells occurs at the onset of meiosis. A close relationship between reactivation of the heterochromatic X chromosome and initiation of the meiotic process has been suggested (Monk and McLaren 1981).

Thus the X chromosome is the first genetic element of the mammalian karyotype whose differentiation, both cytological and biochemical, can be so intimately linked with morphogenetic processes during gametogenesis and early embryogenesis. Whether X chromosome inactivation–reactivation can be considered as a genetic trigger that initiates differentiation, or whether it is one of the manifestations of early differentiation processes, remains unknown.

8.3 YO karyotype

Not a single case of YO karyotype has ever been recovered in spontaneous abortions even at the earliest stages of human development (Therkelsen, Jensen, Jonasson, Lamm, Launtsen, Lindsten, and Petersen 1973; Boué and Boué 1974; Boué et al. 1976; Kuliev 1974; Vogel and Motulsky 1979; Sankaranayanan 1979; Lauritsen 1982). Our knowledge of YO manifestation in laboratory mammals is very poor indeed and limited exclusively to laboratory mice.

In the progeny of XO mice, an equal proportion of XX, XY, XO, or YO embryos might be expected However, not a single YO karyotype (i.e. nullisomic for all X chromosome coded genes) was ever found after birth, or at the late morula–early blastocyst stage among embryos from XO mice (Cattanach 1962; Morris 1968). However almost 20 per cent of all 4-day embryos among the progeny of XO females still remained at the two-blastomere stage and were degenerating. A YO karyotype was suggested for these degenerating embryos, though no direct cytological proof was available. The presence of YO embryos in the progeny of XO females was proven directly on chromosomal preparations from embryos of the first and second cleavage divisions (Kaufman 1972). A significant difference was observed in the number of YO embryos recovered at the first and at the second cleavage, indicating a substantial loss of YO embryos between these stages (Luthardt 1976). Some indirect evidence from *in vitro* experiments supports the notion that YO embryos survive up to the eight-blastomere stage (Burgoyne and Biggers 1976). These data are not yet confirmed karyologically. Only one YO embryo was identified among 129 four-cell stages recovered from XO mice, suggesting continued selection against YO embryos at this developmental stage (Sinow and Luthardt 1978).

Thus a YO karyotype, that is nullisomy for all genes coded by the X chromosome, is already evident during early cleavage and is inevitably lethal, at least in mice, by the eight-blastomere stage. Thus a YO karyotype is not compatible with the accomplishment of cleavage, compaction, and blastocyst formation.

8.4 Sex-chromosome trisomy (karyotypes XXY, XXX, XYY)

Karyotypes with an extra sex chromosome are rarely found among abortuses, in contrast to their high frequency in newborns (approximately 1 per 1000 male (XXY, XYY) or female (XXX) births; Sankaranarayanan 1979). This discrepancy could be partly due to misidentification of the Y chromosome if banding procedures that allow reliable identification of the Y chromosome and autosomes 21 are not used. On the other hand it may be that an extra sex chromosome is rarely lethal for developing mammalian embryos, owing to the effective X chromosome dosage compensation mechanism, so that extra copies are compatible with prolonged survival and post-natal life. No data on extra sex-chromosome effects on embryogenesis of experimental animals have been reported so far, but numerous cases of extra sex chromosomes have repeatedly been reported in post-natal animals of various mammalian species (see Section 9.1).

The most constant trait of the XXY karyotype is spermatogenic failure, found in men with Klinefelter's syndrome as well as in mice, cats, and dogs (Russell 1961; Centerwall and Benirschke 1973, 1975; Clough *et al.* 1970). All known XXY specimens were sterile due to complete absence of meiosis. Testicular biopsies in some XXY men revealed two types of tubule. Most lacked spermatogonia, but occasional fertile tubules were encountered, probably as a result of local repopulation by XY germ cells generated from XXY cells by a non-disjunctional accident (Burgoyne 1978).

Failures of spermatogenesis are much less common in XYY men, though some suffer oligoozospermia of varying extent. However, most develop as normal males and become inconspicuous in the general population (Tettenborn, Schwinger, and Gropp 1970; Baghdassarian, Bayard, Borgaonkar, Arnold, Solez, and Migeon 1975; Skakkebaek *et al.* 1973). The loss of one Y chromosome in the germ cells of XYY men with subsequent normal spermatogenesis has been reported (Hulton and Pearson 1971). The causes of such Y chromosome loss before meiosis remain unknown (Benirschke 1974).

Reports on XYY karyotype in other mammals have so far been confined to laboratory mice. Data on XYY mice indicate a severe impairment of spermatogenesis, leading to sterility at an early age (Evans, Lyon, and Ford 1969; Cattanach and Pollard 1969; Rathenburg and Müller 1973; Evans *et al.* 1978). The sterility was associated with a much reduced testis weight and spermatogenic blockage. Detrimental effects of the XYY constitution start early and progress through the seminiferous cycle (Evans *et al.* 1978). The maturation processes, however, were not inhibited completely and six out of seven males produced some sperm. Only one XYY male had epididymal sperm, and then only 1 per cent of the normal count. This unique male proved fertile early in life but soon became sterile. Another case of a probably fertile XYY male has been recently reported. The testis weights of this individual were 84 and 104 mg and the epididymal sperm count was about half the control value (6.3 $\times$ 10^6 ml^{-1} and 10.07 $\times$ 10^6 ml^{-1} respectively, Das and Karr 1981). Meiotic analysis of XYY males has revealed various sex chromosome complements in proportions that were rather specific for each individual. Most of the M-I cells (20–82 per cent) had XY bivalent + Y univalent; 5–19 per cent had YY-bivalent + X univalent; 3–33.6 per cent had XYY trivalent and in some M-I cells all sex chromosomes were unpaired. At least some of the sex chromosome complements were probably technical artefacts, as all available data on the second metaphase stage argue in favour of random segregation of the sex chromosomes from M-I (Rathenberg and Müller 1973; Evans *et al.* 1978). Thus, an XYY constitution in mice imposes considerable sterility, although like in humans the degrees of spermatogenic impairment can vary substantially so that fertile XYY mice might perhaps be found (Cattanach and Pollard 1969; Das and Karr 1981).

8.5 Conclusions

Numerical aberrations of the sex chromosomes are encountered both in pre- and post-natal life of many mammalian species. 45 XO monosomy is known as the most common single chromosome disorder in humans. The very high frequency of XO compared to other sex chromosome disorders in humans and in other mammals may be due to sex vesicle loss during spermatogenesis (Ashley 1983). No more than 2 per cent of all human XO zygotes reach term, with most suffering developmental arrest at 6 weeks of intrauterine life. XO conceptuses reaching the second or third trimester of pregnancy reveal a very specific phenotype, including progressive decrease of the germ cell count in their gonads. The reasons for the early death, congenital defects, and abnormal development post-natally of XO individuals are mainly unknown, though some mechanisms have been postulated (Epstein 1981a,b, 1983a,b). XO monosomy in mice can be readily produced by irradiation of female mice soon after fertilization (completion of the second meiotic division, start of pronuclear formation) by crossing female mice heterozygous for In(X)1H or by crossing XO mice with tabby markings to normal XY males. Irrespective of their origin, cleavage rate, blastulation, and early post-implantation development were affected to a varying degree in most XO embryos. Some XO mice are eliminated before term but most survive birth. Post-natal effects of XO monosomy in mice are confined to retardation of growth, lowered body temperature, some behavioural disturbances, reduced thyroid activity, and impaired reproductive capacity. The reduced litter size and unusually short reproductive life are the most typical features of XO mice. The lower productivity of XO mice is probably caused by diminished secretion of gonadotrophic hormones, together with a reduced number of primordial follicles owing to the dramatic loss of oocytes in a short period of time immediately after birth (Burgoyne and Baker 1981; Deckers *et al.* 1981). An abnormally low segregation of nullo-X gametes was suggested, and was later proved by direct chromosomal analysis of M-II oocytes of XO mice. All available data on ovarian morphology and fertility for different mammalian species are consistent with the hypothesis that the quantitative features of XO karyotype expression are due to differences of the developmental time scale; XO females from species with a short generation time are usually fertile, while those from species with much longer generation times are sterile (Burgoyne 1978). XO mice should be considered a valuable experimental tool for the study of mechanisms of developmental failure due to X chromosome monosomy in humans. The YO karyotype, i.e. nullisomy for all genes encoded by the X chromosome, is probably lethal by the eight-cell stage.

Sex chromosome trisomies are rarely encountered pre-natally but are rather common after birth, at least in humans. Spermatogenic failure is a constant feature of the XXY karyotype in humans (Klinefelter's syndrome), but only moderate impairment of spermatogenesis is encountered in XYY males. The reproductive capacity of XYY male mice varies from complete sterility in most cases to partial transient fertility in some. Assuming homology of the gene content in X chromosomes of all mammalian species, the similarities in manifestation of sex chromosome imbalance seem quite logical and offer a unique opportunity for developmental biology and medical genetics.

9

Structural chromosomal aberrations

9.1 Frequency of spontaneous aberrations

The early cytogenetic studies of spontaneous and induced abortions showed that the incidence of spontaneous structural aberrations in humans is low compared to the incidence of aneuploidy and triploidy (Carr 1969, 1971b). The introduction of new chromosome staining techniques had little influence on the number of structural anomalies detected in animals before or after birth, though it greatly extended our knowledge of this type of chromosomal aberration in humans. The frequency of structural aberrations varies substantially, depending mostly on the staining technique employed as well as on the number of translocation carriers in the group of selected patients. Of 921 abortuses with abnormal karyotype, structural anomalies were observed in 3.8 per cent with routine staining techniques (Boué and Boué 1976) while only one abortus out of 154 analysed by G-banding had a structural aberration (Andrews, Dunlop, and Roberts 1984). In other studies a very high incidence of spontaneous structural aberrations (6.8 per cent) has been reported for 639 spontaneous abortuses (Kajii, Ferrier, Niikawa, Takahara, and Oharma 1980). Of 11 structural aberrations in this study, seven were found to be inherited and only four were sporadic.

The impact of chromosome banding techniques in revealing many new types of structural chromosomal aberration can hardly be overestimated. Numerous new partial trisomies, deletions, inversions, translocations, etc. have been detected both in spontaneous abortuses (Therkelsen *et al.* 1973; Tadashi, Koso, Niikawa, Arlanc, and Sugandhi 1973) and in newborns with malformations (Vogel and Motulsky 1979). Numerous new syndromes due to partial trisomies or to minor autosomal deficiencies have been described (Lewandowski and Yunis 1975), while some forms of inherited disease initially suspected to be single gene mutations or multifactorial in origin were actually found to result from small deletions. The Prader–Willi syndrome and some forms of de Lange's syndrome are good examples of the present situation in modern medical genetics (Ledbetter, Riccardi, and Airhart 1981; Lazyuk, Lurie, and Cherstvoi 1983). An unusually high incidence of spontaneous structural aberrations (3.3 per cent of total 8.5 per cent) has recently been reported for mature human spermatozoa karyotyped using a method for direct visualization of the chromosome constitution (Martin 1983, 1984).

Thus our view of the actual impact of spontaneous structural aberrations in pre- and post-natal pathology in humans gradually changes in parallel with advances in our technical capacity for more and more precise analysis of chromosomal structure. The introduction of prometaphase chromosome techniques, with extremely high resolution of banding patterns in individual chromosomes, may substantially facilitate progress in this field of medical genetics.

Data on the frequency of spontaneous structural aberrations in laboratory mammals are rather meagre. Analysis of meiotic I spermatocytes suggests that the frequency is extremely low in laboratory mice. Not a single spontaneous translocation, inversion, or other structural aberration inducing abnormal meiotic chromosome pairing was detected in male mice of various strains (Ohno *et al.* 1969; Muramatsu 1974). Male mice of advanced age (2–3 years) revealed spontaneous translocation in 0.8–1.2 per cent of their meiotic cells (Schleiermacher 1972; Muramatsu 1974). Approximately 35 000 meiotic cells from diakinesis to the first metaphase were examined in control male mice. Nine of these cells, obtained from the same specimen, included a quadrivalent configuration, indicating an exceptional low frequency of spontaneous structural aberrations (Ford 1970). Cytogenetic data from several laboratories on 3531 adult mice yielded a higher frequency of spontaneous translocations of one in 1200 mice tested. A post-meiotic origin for new translocations in male gametes or a predominance of new translocations in the female germ line was therefore suggested (Ford 1970). However, not a single multivalent configuration has yet been reported in M-I oocytes of mice (Henderson and Edwards 1968; Luthardt *et al.* 1973; see also Chapter 5), though one of 273 zygotes has been found to carry a structural chromosomal anomaly (Martin-Deleon and Boice 1983).

Thus the frequency of spontaneous structural aberrations in mice is most probably very low, though more work is needed to clarify the relative contribution of male and female gametes in the origin of these aberrations.

Besides chromosomal translocations other types of structural rearrangements, including deletions, duplications, and isochromosomes, have been repeatedly reported. All of these aberrations except isochromosomes were however recorded in the progeny of mice heterozygous for reciprocal translocations or treated with various mutagens (see Chapter 4).

The isochromosome is a common structural aberration in laboratory mice of normal karyotype or in mice with Robertsonian translocations (Vickers 1969; Baranov and Dyban 1971b; see also Tables 1 and 11 in Chapter 4).

9.2 Effects on development

Data on the developmental effects of structural aberrations in humans are scarce, as their relative contribution to fetal wastage is rather low (see above).

Since the classical work of Snell on mice heterozygous for chromosomal translocations, it is generally accepted that most structural aberrations arising during gametogenesis behave as dominant lethals in various mammals, including humans (Snell *et al.* 1934; Snell 1941, 1946; Russell 1962, 1964; Ford and Clegg 1969). Sterility or semi-sterility of adult males is therefore taken as an indication of chromosomal rearrangements in the germ cells.

The introduction of methods for direct chromosomal studies during pre- and post-natal embryogenesis provided direct experimental evidence for this suggestion, and thus opened up a new field in developmental cytogenetics with the study of the phenotypic expression of genetic imbalance due to duplication or deficiency of particular segments in the translocated chromosomes. Some theoretical implications of these studies are more throughly discussed in Chapter 5 (see also Ford and Clegg 1969).

9.2.1 Pre-natal period

Developmental effects of partial trisomies or deletions during embryogenesis have so far been studied, with only a few reciprocal translocations. These are T(14;15)6Ca; T(16;17)43H; T(7;14)2Iem; T(9;17)138Ca; T(2;8)26H. Mice heterozygous for these translocations regularly produce a proportion of chromosomally unbalanced gametes, which result in zygotes with duplications and deficiencies of the chromosomes involved in the translocations.

9.2.1.1 T(14;15)6Ca. This is an unequal reciprocal translocation with one breakpoint at 14E5 and the other at 15B3. The small product of translocation (designated 15^{14}) can be easily distinguished with any routine chromosomal staining procedure, while its large product, 14^{15}, can be identified only with banding techniques. Mice heterozygous for this translocation produce gametes with duplications and deficiencies of the translocated chromosomes, as depicted in Fig 11. The phenotypic effect of these aberrations has been studied soon after implantation and during major organogenesis (Baranov and Dyban 1968, 1970; Oshimura and Takagi 1975; Baranov 1976a). The number of embryos with different types of aberration involving translocated chromosomes 14 and 15 are listed in Table 30.

(a) *Duplication of the distal part of chromosome 15 + deletion of the distal tip of chromosome 14 (Dp(15C–15F), Df(14E5)).* This has been studied in 59 embryos soon after implantation (6.5–7.5 days of gestation) and in 27 embryos during major organogenesis, after 9–12 days of gestation. The mean size of implanted embryos with these aberrations was only half that of normal littermates (Oshimura and Takagi 1975). Their further development

Table 30. Incidence of embryos with structural aberrations of chromosome 15 at different stages of embryogenesis in mice heterozygous for T(14;15)6Ca translocation

Type of aberration	Chromosome number	Days of pregnancy		
		6.5–7.5[a]	9–12[b]	16.5–18.5[b]
Dp(15C–15F) + Df (14E5)	40	59	27	—
Ts(15^{14})6Ca	41	31	58	3
Dp(15C–15F) + Ts(15^{14})6Ca	41	24	12	0
Df (15C–15F)	40	4	—	—
Df (15C–15F) + Df (15A–15B) = Ms 15	39	—	—	—

[a] Oshimura and Takagi 1975.
[b] Baranov 1976a.

varied, however. By 11–12 days of gestation two morphologically distinct groups of abnormal embryos with identical karyotype changes could be distinguished (Baranov 1976a). At least two-thirds had severe general retardation most evident in the head region, and numerous abnormalities of the heart and internal organs. The allantoic stalk failed to provide proper vascularization in the fetal portion of the placentae and this abnormality was probably the principal cause of their death after 10–12 days of gestation. One of these specimens is depicted in Fig. 28.

The embryos of the second group were only slightly smaller than their normal littermates, but they showed almost identical abnormalities of the head region. Anophthalmia in conjunction with very small or even absent maxillary processes, and severely underdeveloped hyoid arches and mandibular rudiments were the most conspicuous features of this syndrome (see Fig. 29). Regular heartbeats, properly vascularized yolk sac, and allantoic placenta witnessed their viability. Histological studies revealed moderate to severe hydrocephaly, impaired differentiation of all brain vesicles, absence of eye rudiments and tongue, and maldevelopment of the liver (see Fig. 30). Not a single embryo of this kind was ever recovered at later stages of gestation, either in our studies or in Oshimura and Takagi's experiments, showing that they were eliminated long before term.

The reasons for the different developmental pathways taken by embryos with apparently identical chromosomal abnormalities remain obscure. It must be remembered that besides duplication 15CDEF all the specimens of both groups carried a very small deletion of the most distal part of chromosome 14 (14E5). The absence of this small segment (which cannot yet be demonstrated cytologically; Nesbitt and Francke 1973) may account for the earlier elimination of these specimens compared with embryos trisomic for the whole of chromosome 15 (see Chapter 7). However it cannot provide a

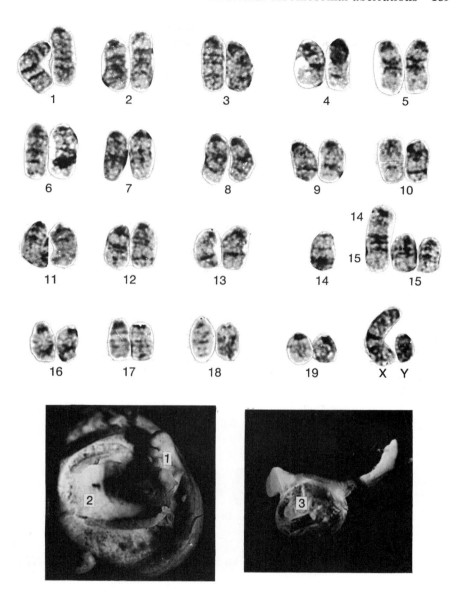

Fig. 28. Duplication of distal part of chromosome 15 and deletion of distal part of chromosome 14 (Dp(15C–15F) and Df (14E5)); T6Ca/ + ♀ × +/ + ♂; twelfth day embryos fixed in Bouin's fluid. General severe retardation; deformed head region, expanded pericardium. 1 = embryo, 2 = placenta, 3 = heart.

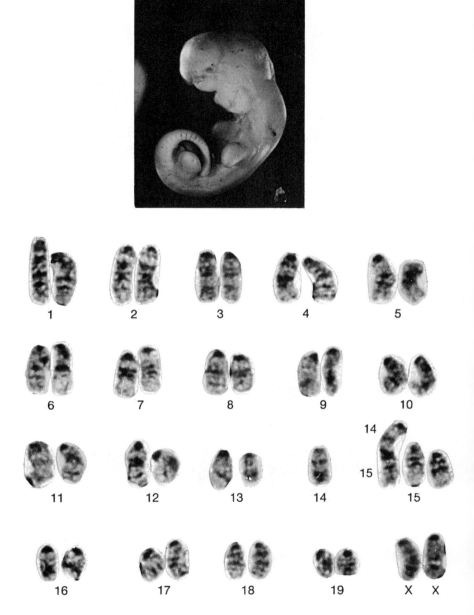

Fig. 29. Moderate retardation, hydrocephaly, the absence of hyoid and mandibular arches; Dp(15C–15F) and Df (14E5); T6Ca/ + ♀ × + / + ♂. Twelfth day monse embryo, Trypsin–Giemsa chromosome staining.

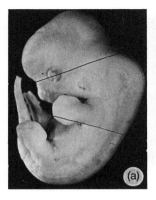

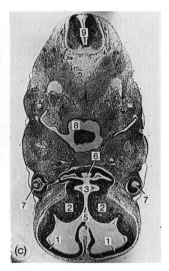
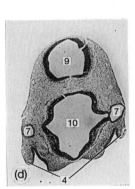
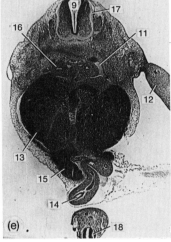
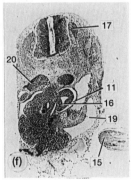

Fig. 30. External view (a,b) and transverse sections at the head (c,d) and thoracic (e,f) regions of twelfth day mouse embryos in the progeny of T6Ca/ + ♀ × +/+ ♂. Bouin's fixative, Mayer's haemalum and eosin. (a,c,e) Partial Ts(15¹⁴)6Ca; (b,d,f) Dp(15C–15F) and Df (14E5).

1 = telencephalon, 2 = diencephalon, 3 = third ventricle, 4 = maxillary arches, 5 = foramen of Monro, 6 = Rathke's pocket, 7 = eye, 8 = pharynx, 9 = spinal cord, 10 = diocoel, 11 = pleural cavity, 12 = forelimb buds, 13 = liver, 14 = intestine, 15 = allantoic vessels, 16 = lungs 17 = spinal ganglion, 18 = tail, 19 = peritoneal cavity, 20 = cardial vein.

clue for the existence of two morphologically different groups of embryos bearing the same genetic imbalance.

(b) *Duplication of the distal part of chromosome 15 + partial trisomy for chromosome 15^{14} (T6) (Dp(15C–15F) + Ts15^{14})*. The extra genetic content in the 36 specimens of this group (see Table 30) amounted to almost the whole of chromosome 15, but their developmental profile was different from that of Ts15 embryos (see Chapter 6). All 26 (Dp(15C–15F) + Ts15^{14}) embryos recovered after implantation 6.5–7.5 days and all 12 specimens detected during major organogenesis were very similar and almost identical to the severely affected embryos with Dp 15 CDEF only (see above). As might be inferred from external inspection the embryos with structural aberrations Dp(15C–15F) + Ts15^{14} were very stunted, often disorganized, and generally looked much more severely affected than the embryos with Ts15 in the progeny of Robertsonian translocation carriers (see also Chapter 6). Whether these obvious discrepancies were due to some unfavourable positional effects of rearranged parts of chromosome 15 or whether they stem from some strain-specific genetic peculiarities of this chromosome remains unknown as yet.

(c) *Partial trisomy 15 ^{14} (Ts 15^{14} 6Ca)*. [Duplication for the centromeric region of chromosome 15 (Dp(15A–15B)) and for telomeric region of chromosome 14 (Dp(14 E-?))]. Non-disjunction of the small translocation product, marker chromosome 15^{14}, is one of the most common chromosomal aberrations in the gametogenesis of T6/ + female mice (but not in males). The karyotype of these specimens contains 41 chromosomes with one very small marker product distinguished also by its secondary constrictions—15^{14} (chromosome T6). Ninety-two Ts15^{14} embryos have been studied at post-implantation stages (Table 30). All of them looked quite normal soon after implantation and during major organogenesis. Their phenotypic normality has been proven at term and moreover all of them were found to be capable of post-natal development (Fig. 31; Cattanach 1967; Baranov and Dyban 1970; Eicher 1973; see also Section 9.2.2.)

(d) *Partial deletions of chromosome 15 (Df(15C–15F); Df(15A–15B))*. Deletions for the major distal part of chromosome 15, i.e. Df(15C–15F) as well as for its minor proximal region, Df(15A–15B) were found in eight embryos in total, soon after implantation. All of these specimens were most severely retarded at examination (Oshimura and Takagi 1975) and none of them has been ever found at more advanced stages.

Thus, the survival of mouse embryos deficient for the various parts of chromosome 15 is somewhat better than for the corresponding full mono-somics (Ms15). Nonetheless, all are inevitably lethal during implantation or soon thereafter. The most unexpected finding of this study was the very poor survival of embryos deficient for a relatively small part (about 37 per cent) of chromosome 15. The presence of a ribosomal gene cluster and probably of

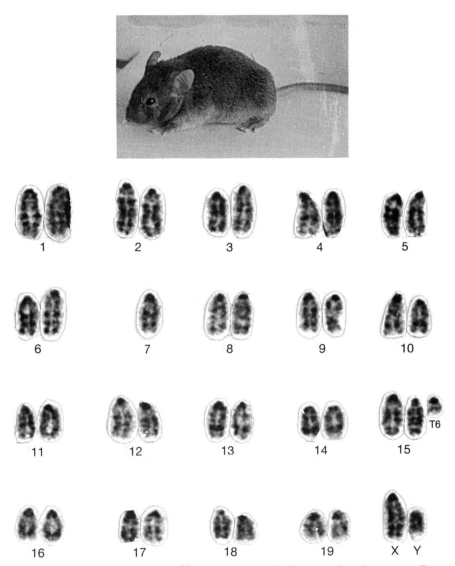

Fig. 31. Partial trisomy, Ts(15^{14})6Ca. Phenotypically normal male mouse. Bone marrow biopsy, Udalova's method (Udalova 1970). Trypsin–Giemsa staining of chromosomes.

other structural genes involved with the early stages of embryogenesis in mice may be suspected in the AB region of chromosome 15.

9.2.1.2 T(7;14)2Iem. This is an unequal reciprocal translocation, very similar to the T6 translocation already discussed. One of its break points is located in the very distal part of chromosome 7 (presumably in bands 7F3–4) and the other is very close to the centromeric region of chromosome 14 (14B). Both translocation products are easily identified in metaphase plates without banding (Baranov 1979, 1982b). Two basic types of genetically unbalanced

Table 31. Frequencies of embryos with structural aberrations of autosomes 7 and 14 before and after implantation in the progeny of mice heterozygous for translocation T2Iem

	Day of pregnancy			Total examined
	3	14	9–11	
Total no. of embryos	336	208	93	637
No. scored[b]	109	97	78	284
Karyotype analysis[a]				
Genetically balanced				
No.	51	49	54	154
%	46.7	50.5	69.2	—
M^+m^+	34	27	26	87
M^+m^-	17	22	28	67
Genetically unbalanced				
No.	58	48	24	130
%	53.3	49.5	30.7	—
Euploid				
No.	49	41	19	109
%	45.0	42.3	24.3	—
M^+m^-	29	10	0	39
M^-m^+	20	31	19	70
Aneuploid				
No.	9	7	5	21
%	8.3	7.2	6.4	—
Trisomy				
M^+m^+	2	2	2	6
M^+m^-	1	1	0	2
M^-m^+	1	0	2	3
Monosomy				
M^-m^-	5	2	1	8
M^+m^-	0	0	0	0
M^-m^+	0	2	0	2

[a] M, a major marker 7^{14}; m, a minute marker 14^7.
[b] Karyotype determination possible.

gametes have been recorded in the progeny of T2/+ males or females. Development of the two groups was very different (Table 31).

(a) *Deficiency for the distal part of chromosome 7 and duplication of chromosome 14 (Df(7F4), Dp(14B–14E))*. This abnormality has been recorded in 39 pre-implantation embryos, 28 on the third and 10 on the fourth day of gestation. All chromosomally abnormal embryos were at the morula stage on the third day (mean cell count 10.2 ± 2.3 compared to 13.1 ± 1.8 in the control littermates) and at the early blastocyst stage on the fourth day (mean cell count 34.1 ± 2.3 and 33.5 ± 4.3, respectively). However, numerous pyknoses and micronuclei, indicating progressive degeneration, were present in six out of 10 embryos of this group. Taking into account that the relative frequency of these specimens was almost 2.5 times higher on the third day than on the fourth, these data are consistent with the suggestion that chromosomal imbalance due to Df(7F4), Dp(14B–14E) affects blastulation and is already lethal by the early blastocyst stage. The cytologically identified c^{25H} deletion in the middle of chromosome 7 (Miller and Miller 1972) is lethal during cleavage (Gluecksohn-Waelsch 1979). The breakpoint of T2Iem in chromosome 7 is located in the most distal region, probably 7F3-4, so that it is unlikely to share a common segment with a c^{25H} deletion. The genetic map of the translocated segment 7F3–F4 is almost completely unknown as yet (Eicher and Washburn 1978), but our results indicate that the genetic content of the distal segment of chromosome 7 is very important for the control of early development in mice. Whether the lethal effect is due to the expression of lethal genes in hemizygous form or is caused by defects of genes with regulatory functions remains unknown (Table 31).

(b) *Duplication of the distal part of chromosome 7 + deficiency of chromosome 14 (Dp(7F4), Df(14B–14E))*. A total of 70 embryos of this karyotype have been studied before and after implantation. The proportion was almost equal on the third/fourth and on the ninth/eleventh days of pregnancy, indicating little or no elimination of this unbalanced genotype up to the major organogenesis stage (see Fig. 32). However, 10 of the 19 post-implantation embryos of this group were represented by small empty yolk sacs containing no traces of embryoblast derivatives. The other nine, though very small and pale on external examination, revealed remnants of neural folds. Maternal blood in the implantation chamber and the absence of the allantois showed that these embryos were degenerating. The prolonged survival of these abnormal embryos, lacking almost the whole of chromosome 14 except for its small centromeric region, is in good agreement with the recovery of some Ms14 embryos after implantation (see Chapter 7). Thus, all information gained so far on the developmental effects of chromosome 14 aneuploidy can be taken as evidence of a very modest part played by its genes in the early embryogenesis of mice.

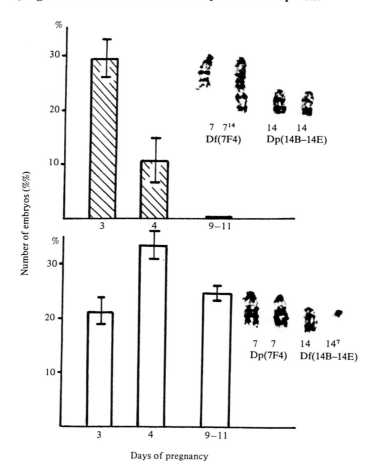

Fig. 32. Percentage frequencies of the embryos with structural aberrations of chromosomes 7 and 14 on third, fourth, and ninth to eleventh days of pregnancy in mice, heterozygous for T(7;14)2Iem translocation (T2Iem/ + ♂ × +/+ ♀).

9.2.1.3 T(16;17)43H. One breakpoint disrupts the centromeric hetero-chromatin of chromosome 16, and the other is located at the distal end of band 17B (Searle, Beechey, and Evans 1978), just distal to *tf* (Lyon, Evans, Jarvis, and Sayers 1979). The latter break physically separates the two major genetic complexes *T-t* and *H-2*, with the *T-t* complex left in the large trans-location product 17^{16} and *H-2* in the small 16^{17}. Both translocation products are easily distinguished in C-banded chromosome plates. The large 17^{16} is dis-tinguished by double blocks of centromeric heterochromatin separated by a

Table 32. Frequencies of embryos with structural aberrations of chromosomes 17 and 16 before and after implantation in the progeny of T43H/+ females crosses to +/+ or Rb4Iem/+ males

Day of pregnancy	4	8–10
No. of females	37	20
Total no. of implants	217	147
Karyotype possible	82	81
Karyotype genetically balanced	43	45
Karyotype genetically unbalanced		
Total no.	39	36
Type of aberration (nos.)		
Df(17C–17E) + Dp(16)	7	—
Dp(17C–17E) + Df(16)	12	10
Dp(17A–17B)	7	14
Df(17A–17B)	13	12

Giemsa-negative region, while the small 16^{17} bears a very small block of C-heterochromatin (see Fig. 33(a)).

Four main groups of genetically unbalanced karyotypes have been identified in the progeny of T43H/+ females (see Table 32, also Baranov 1983b).

(a) *Deficiency for the distal part of chromosome 17 + duplication of chromosome 16 (Df(17C–17E), Dp(16))*. All seven specimens were found on the fourth day, that is before implantation, and none were recovered later. Morphologically all these rare embryos looked abnormal and demonstrated a conspicuous lag in growth by the morula stage (Fig. 33a). The mean cell number of embryos lacking the 17CDE region was two or three times lower than in embryos of genetically balanced karyotype from the same cross (12.1 ± 2.6 and 31.2 ± 1.8 respectively). Numerous micronuclei, pyknoses, unusually large and lightly stained nuclei with metaphase chromosomes separated prematurely into paired chromatids, indicated advanced degeneration and death of these embryos by the morula stage. Thus the survival profile of these embryos is comparable with that of Ms17 embryos (see Chapter 7).

(b) *Duplication of the distal part of chromosome 17 + deficiency of chromosome 16(Dp(17C–17E), Df(16))*. All 12 embryos recovered on the fourth day looked like normal early blastocysts. On days 8–10 all 10 specimens of the same karyotype revealed gross abnormalities and were degenerating. Thus, deficiency for almost the whole chromosome 16 apart from its very small centromeric fragment does not seem to be very harmful before implantation. These data are in agreement with the developmental profiles of Ms16 embryos in mice heterozygous for Rb translocations (see Chapter 7).

(a)

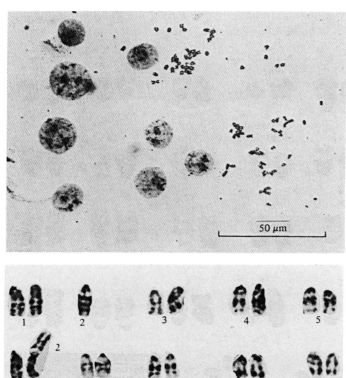

Fig. 33. (a) Deletion 17 CDE + Trisomy 16. Total preparation and trypsin–Giemsa banded karyotype of the fourth day embryo (Rb4Iem/+ ♂ × T43H/+ ♀). Double fixation method. Greatly reduced blastomere number.

(c) *Deficiency for the proximal part of chromosome 17 (Df(17A–17B))*. Almost equal proportions of such specimens have been reported before and after implantation (Table 32), providing evidence for their survival during the initial stages of implantation and neurulation (Fig. 33b). All looked severely abnormal at later stages. Nine out of twelve 8–10-day embryos were

(b)

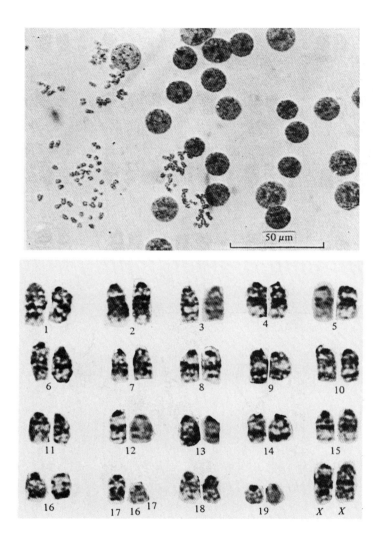

(b) Deletion 17 AB. Total preparation and trypsin–Giemsa banded karyotype of the fourth day embryo (+/+ ♂ × T43H/+ ♀). Double fixation method. Normal cell count, no sign of degeneration (Baranov 1983b).

represented by small trophoblastic vesicles with no embryoblast derivatives visible under a stereomicroscope; three looked like abnormal neurulae (Fig. 34a). However, the complete absence of embryonic remnants in at least half of all implantation sites on days 9–10 proved that embryonic death occurred soon after implantation.

(a)

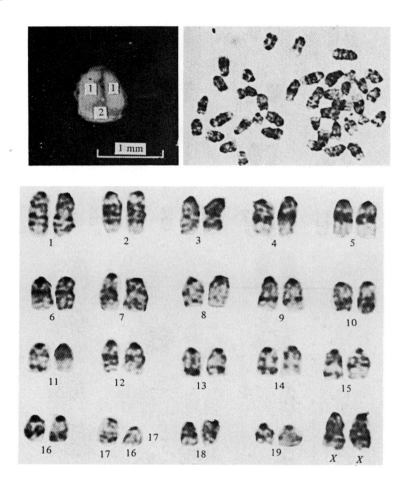

Fig. 34. (a) Deletion 17 AB. External view under stereomicroscope, metaphase plate, and trypsin–Giemsa banded karyotype of the ninth day embryo (+/+ ♂ × T43H/+ ♀). 1 = Neural folds, 2 = allantois. Greatly reduced in size malformed neurula.

(b) Deletion 17 ABCD + Duplication 9 AB. Total preparation and trypsin–Giemsa banded karyotype of the fourth day embryo (T138Ca/+ ♂ ×/+ ♀). Double fixation method. Reduced blastomere number. 1 = pyknotic nuclei, 2 = abnormally large nuclei with transparent nucleoplasm (Baranov 1983b).

(b)

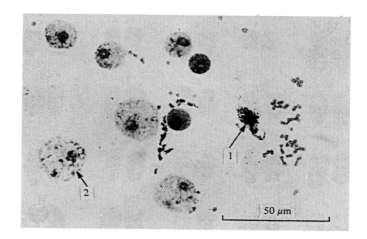

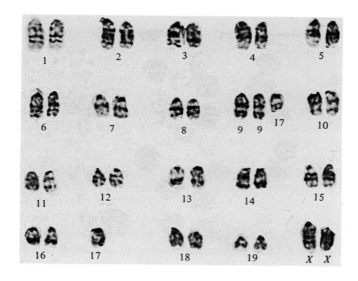

(d) *Duplication of the proximal part of chromosome 17 (Dp(17A–17B))*. Both before and after implantation almost all specimens looked normal and were identified only on chromosomal preparations. Two embryos were slightly retarded and one showed malformations. This genetic imbalance has been found to be compatible with post-natal development (Forejt, Capkova, and Gregorova 1980). At birth, however, the trisomic embryos were smaller than normal and some died. More details on their post-natal development are given later in this chapter (Section 9.2.2).

Table 33. Frequencies of embryos with structural aberrations of chromosomes 17 and 9 before and after implantation, in the progeny of T138Ca/ + males crossed to + / + females

Day of pregnancy	4	8–10
No. of females	13	25
Total No. of implants	96	120
Karyotype possible	55	70
Karyotype genetically balanced	29	52
Karyotype genetically unbalanced		
Total No.	26	18
Type of aberration (Nos.)		
Df (17E) + Dp(9C–9F)	10	9
Dp(17E) + Df (9C–9F)	4	3
Df (17A–17D) + Dp(9A–9B)	6	0
Dp(17A–17D) + Df (9A–9B)	6	6

Thus, deficiency for the proximal part of chromosome 17 (Df(17A–17B)) or duplication of any part of this chromosome (AB or CDE segments) is not detrimental before implantation, but deficiency for the CDE segment of the same chromosome is inevitably lethal during cleavage.

9.2.1.4 Translocation T(9;17)138Ca. Break points are located in the D segment of chromosome 17 and in the B segment of chromosome 9 (Miller and Miller 1972). The average frequency of the chromosomally unbalanced embryos on the fourth day was almost twice that on days 8–10 (47.3 per cent and 25.7 per cent, respectively), indicating strong embryonic selection at implantation (Baranov 1983b).

Four major groups of karyologically different embryos with visible chromosomal imbalance have been identified (Table 33). One group, deficient for the proximal segment of chromosome 17 (Df(17A–17D), Dp(9A–9B)) was recorded exclusively before implantation. All six embryos of that group were composed on the fourth day of 6–12 blastomeres (mean cell count 8.7 ± 3.6 compared to 31.5 ± 1.8 in controls), some of which were pyknotic or abnormally large with a very light transparent nucleoplasm (Fig. 34b). These embryos were dead by the morula stage. Of 18 chromosomally unbalanced embryos on the eighth to tenth days, nine were deficient for the distal part of chromosome 17 (Df(17E)) and had an extra copy of most of chromosome 9 (Dp(9C–9F)). These aberrations did not interfere with cleavage, blastulation, and implantation, but embryos died soon after implantation. The development of chromosomally unbalanced embryos of the two other groups (see Table 33), Dp(17E), Df(9C–9F) and Dp(17A–17D), Df(9A–9B), was morphologically similar to Df(17E), Dp(9C–9F).

This study of T43H and T138Ca heterozygotes provided evidence that

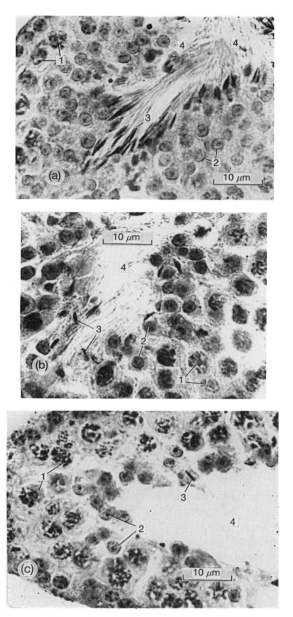

Fig. 35. Transverse sections of seminiferous tubules at the 7–9 stages of spermato-genesis. PAS–Shiff haematoxylin–eosin. 1 = spermatocyte I at pachytene stage, 2 = spermatids at the 'cap' phase; 3 = spermatozoa; 4 = lumen. (a) C3HA, $2n = 40$; (b) C3HA/CBA-T6($2n = 40$)-T6 + ; (c) C3HA/CBA-T6($2n = 41$); Ts(15^{14})6Ca.

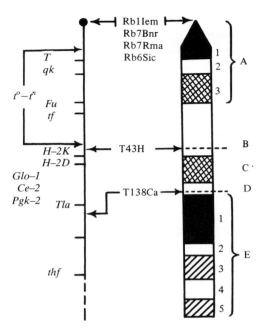

Fig. 36. Regional assignment of genetic map (left) to G-bands of chromosome 17 (right) by means of T43H and T138Ca translocation break points (arrowed). Gene symbols in normal chromosome 17 are brachyury, *T*; quaking, *qk*; fused, *Fu*; tufted, *tf*; *H-2K*, *H-2D*; glyoxalase, *Glo-1*; catalase, *Ce-2*; phosphoglycerate kinase, *Pgk-2*; Thymus antigen, *Tla*; thin fur, *thf*.

deficiencies for different parts of chromosome 17 impaired the development potential of mouse embryos to varying extents. Absence of the distal part (17E1–17E5) or of the proximal region (17A1–17A3 B) of this chromosome (Fig. 36) does not affect cleavage, blastulation, or implantation, and some of these embryos survived up to neurulation, though all of them looked abnormal. Thus, hemizygosity for the genes of the *T-t* complex located in the AB region of chromosome 17 (Lyon *et al.* 1979) is not detrimental to embryonic survival before implantation.

On the other hand, the effects of deletions 17 ABCD and 17 CDE are already evident during cleavage and they completely block embryonic development by the morula stage. So, the damaging effects of these deletions are almost identical with and are expressed at almost the same stage of development as monosomy for the whole autosome 17 (Baranov, Gregorova, and Forejt 1981; Baranov 1983a; see also Chapter 7). The deletions (17 ABCD and 17 CDE) have in common the region of chromosome 17 enclosed between the breaks T43H and T138Ca (see Fig. 36). As deletions for the proximal (AB) or the distal (E) segments of chromosome 17 do not affect pre-implantation development, we may assume that absence of the middle segment, located between exchange points of T43H and T138Ca, is responsible for the early lethal effects of partial deletions 17 CDE and 17 ABCD as well as monosomy of chromosome 17. This suggests that region CD of this chromosome contains genes controlling the initial stages of mouse embryogenesis. The genetic map of the central part of chromosome 17, encompassed by the breakpoints T43H and T138Ca, is known in detail. Besides the complex *H-2* locus it carries DNA for a number of other structural genes such as catalase (*Ce-2*), phosphoglycerate kinase (*Pgk-2*), glyoxalase (*Glo-1*), etc. (Womack 1978). Expression of some *H-2* genes only from the late blastocyst stage can be reliably demonstrated by immuno-fluorescent methods (Searle, Sellens, Elson, Jenkinson, and Billington 1976; Magnuson and Epstein 1981), so the imbalance for these particular genes is unlikely to be responsible for the death of CD-deficient embryos at the morula stage. On the other hand at least some *H-2* associated genes (*Ped*— pre-implantation embryonic development) affect the time of the first cleavage division and the rate of subsequent development (Goldbard, Verbanac, and Walker 1982a,b). Though the real nature of *Ped* genes as well as their gene products remains unknown, they might be considered as main candidates for the so-called 'early' genes, controlling the initial stages of embryogenesis in mammals.

9.2.1.5 Translocation T(2;8)26H. The translocation chromosome 2^8 is longer than the original chromosome 2 and the translocation product 8^2 is shorter. The exact positions of the relevant breakpoints are not known (de Boer and Maar 1976). Histological analysis at day 6 and day 8 revealed three different groups, which could be reconciled with the expectations of the three classes of abnormal karyotypes, based on cytological studies. The survival capacity of the genetically unbalanced embryos was in direct proportion to the extent of deficiency observed. Unfortunately, the absence of direct chromosomal studies make it impossible to conclude which chromosomal deficiency is responsible for abnormal early development of the affected embryos.

9.2.2 Post-natal period

9.2.2.1 Chromosomal translocations. Genetically balanced chromosomal rearrangements quite compatible with otherwise normal post-natal development in humans and in other mammals might nonetheless lead to substantial failure of reproductive function. Male translocation carriers are usually sterile or semi-sterile, while both male and female translocation hetero-zygotes are at high risk of having progeny with structural or numerical chromosomal aberrations related to meiotic non-disjunction. Problems related to translocation sterility and reproductive failure in humans are thoroughly discussed elsewhere (Lurie and Savenko 1978; Lazyuk and Lurie 1979; Chandley, Christie, Fletcher, Franckiewicz, and Jacobs 1972; Chandley, Edmond, Christie, Cowans, Fletcher, Franckiewicz, and Newton 1975; Chandley 1979; Vogel and Motulsky 1979; Michels, Medrano, Venne, and Riccardi 1982; Joseph and Thomas 1982).

The impact of Robertsonian and non-Robertsonian (reciprocal) transloca-tions on reproductive impairment of mammals and humans is not identical (see Chapter 5; also Gropp and Winking 1981; Searle 1981). Reciprocal trans-locations are usually distinguished by more severe impairment of gameto-genesis and reproduction than Robertsonian rearrangements (Searle 1981; Gropp and Winking 1981; Searle and Beechey 1982; Baranov 1983a,b 1980). Substantially reduced reproductive capacity is common in both male and female translocation heterozygotes, though the exact reasons for repro-ductive failure may be identical. Sterility or semi-sterility in male hetero-zygotes is predominantly due to a meiotic block in spermatogenesis, while the reduced fertility of the females is primarily caused by non-disjunction of the translocated chromosomes, with subsequent elimination of genetically unbalanced embryos (Carter, Lyon, and Phillips 1955; Baranov and Dyban 1968). For example, in mice heterozygous for T(14;15)6Ca, mean testis weight of T6/+ males was only 40 mg, less than half that of +/+ C57BL/6J males (Baranov and Dyban 1968). Histological examination of the testis revealed massive death of primary spermatocytes at the diakinesis–metaphase I stages. Direct cell counts of spermatocyte I and resulting spermatids suggested that almost 86 per cent of the cells in C57BL/T6 hetero-zygotes died, compared to only 16–20 per cent cell loss in CBA-T6T6 or C57BL/6J males (Baranov 1976a).

Unlike Robertsonian translocations, some reciprocal translocations turn out to be completely sterile even in simple heterozygotes. These are T(5;12)31H, T(6;12)32H, T(X;11)38H, T(X;4)37H, T(11;19)42H, T(1;7)40H, T(16;17)43H, and some others (see Searle et al. 1978). One of these male-sterile translocations has been successfully 'cured' by the introduction into the already heterozygous karyotype (T43/+) of another translocation, this one of Robertsonian type, involving the same autosomes 16 and 17 (Forejt et al. 1980). Curiously enough, these doubly heterozygous males, Rb7/T43H, turned out to be fertile and made it possible for the first time to get viable T43/T43 homozygotes (Forejt et al. 1980). However, the combination of several different reciprocal or Robertsonian translocations sharing one common chromosome often results in complete sterility of the male (see Chapters 4 and 5; also Searle 1981; Gropp and Winking 1981; Evans 1976).

9.2.2.2 Duplications, partial trisomies, and deficiencies. All these aberrations may arise spontaneously during gametogenesis, but more often they are generated as products of meiotic non-disjunction in reciprocal translocation carriers.

Numerous different syndromes due to structural anomalies of autosomes, and thus primarily to genetic imbalance caused by duplications or deficiencies, are known in humans (Vogel and Motulsky 1979; Lazyuk and Lurie 1979; Lazyuk, Lurie and Cherstvoi 1983).

If a duplicated region of any particular chromosome possesses its own centromere and thus behaves as a supernumerary product of chromosome breakage in an otherwise normal karyotype, the aberration is often termed 'partial trisomy' (Patau, Therman, Smith, Inhorn, and Picken 1961). Such aberrations are especially common in laboratory mice (Griffen and Bunker 1964; Cattanach 1964; Griffen 1967). Most partial trisomies known in mice are produced through meiotic non-disjunction in unequal reciprocal translocation carriers, and are often termed in the literature 'tertiary trisomies' (see Chapter 4). Lyon and Meredith (1966) were the first to describe and confirm cytologically tertiary (partial) trisomic mice in the offspring of females heterozygous for translocations T194H and T158H. Since that time a good many reciprocal translocations have been tested for their ability to generate viable offspring with partial trisomies or small deletions. Most of these translocations are listed in recent comprehensive reviews by Searle (1981) and Epstein (1984). In view of the opportunities provided to the developmental biologist by these natural models of genetical imbalance capable of prolonged survival post-natally, we will briefly consider some of these partial trisomies.

(a) *Ts(15^{14})6Ca.* It has already been mentioned (Section 9.2.1.1(c)) that an extra copy of the small translocation product of T(14;15)6Ca does not interfere with embryonic development but proceeds safely to term (Cattanach 1967; Baranov and Dyban 1970). The marker chromosome T6 includes approximately 37 per cent of chromosome 15 and probably only the telomere of chromosome 14. Viable progeny with 41 chromosomes (including the 15^{14} element) are produced by female, but not by male translocation carriers (Eicher 1973). All trisomic mice looked normal both as adults (Cattanach 1967) and during the early stages of post-natal development (Fedorov, Dyban, Baranov, Dimifriev, and Udalova 1973), and had no visible malformations. Ts15^{14} females were fully fertile and had the expected 50 per cent trisomic offspring. Most however were bad mothers and often killed their offspring. Both males and females with Ts15^{14} showed significant impairment of their mental activity and had behavioural abnormalities as well (see Chapter 10). Ts15^{14} males are completely sterile, with a mean testis weight only half that of normal males (45 $\pm$ 3 and 87 + 6, respectively). No mature sperm was visible in the highly expanded lumen of their seminiferous tubules or (Fig. 35, p. 203) in air-dried meiotic preparations. Spermatocytes of Ts15[14] males were able to enter meiosis but all of them were eliminated soon after (Fedorov *et al.* 1973; Baranov 1976a). The viability of the trisomic mice does not seem to be seriously affected.

(b) *Ts(1[13])70H*. The duplicated segment is formed by the centromeric region (about 13 per cent) of autosome 1 and by the distal part (also about 13 per cent) of chromosome 13. This partial trisomy was most thoroughly studied by Peter de Boer and his colleagues (de Boer 1973; de Boer and Groen 1974; de Boer, de Hoeven, and van der Chardon 1976; de Boer and van Beek 1982; de Boer and Speed 1982). The trisomic mice carry the short marker 1[13] in addition to a normal set of chromosomes. Some of the partial trisomies showed a characteristic pattern of skull bone malformations, which included deficiency of the anterior skull bones, bent nasals, and abnormal growth of upper and lower incisors (de Boer 1973). Both sexes of Ts1[13] mice were fertile, though the average litter size was very low (3.3 compared to 8.45 in the control Swiss mice). The low and very variable reproductive performance of the tertiary trisomic males was caused mainly by lowered sperm production (de Boer and Groen 1974). Many sperm heads were morphologically abnormal irrespective of their actual haploid karyotype. About 35–46 per cent of male Ts1[13] offspring were also found to be partially trisomic; their number fell to about 24 per cent by weaning (de Boer and Groen 1974). Detailed studies of meiotic chromosomes in Ts males suggested that proximity of the interstitial chiasma of chromosome 13 to the centromere affects the frequency of adjacent 2-disjunction and consequently affects disjunction of those centromeres within the multivalent (de Boer and van Beek 1982). During pachytene, a firm association of the 1[13] univalent with the sex-chromosome complex inside the sex vesicle has been discovered (de Boer and van Beek 1982). The significance of this non-homologous pairing, and its association with male sterility both in the tertiary trisomics and in translocation heterozygotes, are discussed later in this chapter.

(c) *Ts (17[16])43H*. The duplicated segment is represented by the proximal part (about 37 per cent) of chromosome 17. Cytologically it corresponds to the AB region of this chromosome and thus incorporates all the genes, including those of the *T-t* complex, from the centromere up to the *H-2[k]* complex (Forejt *et al.* 1980). Unlike most other partial trisomies, the excess genetic material in T43H is not free, but is translocated to another chromosome (chromosome 16), resulting in a large marker 17[16], easily distinguished in metaphase plates by two prominent blocks of C-heterochromatin, one occupying the centromere region and the other lying interstitially (Fig. 36). Since virtually all of chromosome 16 but its centromere is present in the 17[16] marker, the phenotypic effect of Ts43H may be attributed solely to the duplicated segment of chromosome 17. The viability of Ts43H embryos was first shown by Forejt and his colleagues

(Forejt *et al.* 1980). At birth and during the first two weeks of post-natal life, all trisomics were smaller than their normal sibs and some died before weaning. Later, however, the difference disappeared and Ts17[16]43H mice were phenotypically indistinguishable from normal diploids (Forejt 1981), although they were still somewhat smaller (Gregorova *et al.* 1981). A normal proportion of trisomic embryos in the progeny of Ts17[16]43H males (Baranov *et al.* 1981) during major organogenesis and up to the eighteenth day of gestation suggests some moderate lethal effect of Ts43H soon after birth (Gregorova *et al.* 1981). Post-natal viability and growth of Ts mice are to a great extent related to their genetic background. The female trisomics were fertile too, but even fewer partial trisomics were recovered in their progeny than in the progeny of Ts17[16] males.

Unexpectedly, the mental capacities of Ts17[16] mice, both males and females, were found to be negligibly impaired and their ability to solve standard behavioural tasks was found to be even superior to that of animals homo- and heterozygous for this translocation (see details in Chapter 10).

Thus, an excess of about 37 per cent of chromosome 17, including the *T-t* gene complex responsible for some basic steps in embryonic development and spermatogenesis, is compatible with prolonged post-natal survival and does not inhibit fertility.

(d) *Ts(5[12])31H; Ms(5[12])31H.* These were recovered with frequencies of 16.1 and 4.5 per cent respectively in the progeny of heterozygous females T(5;12)31H (Beechey *et al.* 1980). The duplicated (or deficient) chromosomal segment 5[12] is composed of the terminal part (approximately 11 per cent) of chromosome 12 attached to the proximal segment 9 (about 27 per cent) of chromosome 5. Surviving monosomics of both sexes were stunted at birth and tended to have unusual skeletal fusions. The trisomics were also somewhat smaller than the normal sibs at weaning. The trisomic males were sterile but the females generated the expected 50 per cent trisomic offspring. Tertiary monosomic males were also sterile but displayed less severely affected spermatogenesis.

Other unequal reciprocal translocations generating viable partial trisomies or monosomies are T(1;17)90Ca (Lyon *et al.* 1978), T(X;11)38H (Beechey and Searle 1979), T(-;-)158H and T(-;-)194H (Lyon and Meredith 1966), T(10;13)199H and T(2;4)1Sn (Washburn and Eicher 1975), and T(10;-)Ho (Hollander and Waggie 1977). The cytogenetic descriptions of these translocations and the phenotypic characteristics of the corresponding partial trisomies or monosomies have been discussed in several comprehensive reviews (Searle 1982; Epstein 1984), as well as in the above-mentioned original publications.

One of the most puzzling problems concerning both partial trisomies and translocated heterozygotes is the mechanism underlying male sterility. This sterility (or semi-sterility) was earlier attributed to the following causes:

(a) the selection of genetically unbalanced gametes during post-meiotic stages of spermatogenesis (Koller 1944);

(b) impairment of the functions of many of the genes involved in spermatogenesis and scattered throughout the genome (Ford 1969a);

(c) breakage of quadrivalents into trivalents and univalents (Lyon and Meredith 1966);

(d) position effect inactivation of some translocated genes or chromosomal segments (Cacheiro, Russell, and Swartont 1974)

(e) unequal length of translocated products, interfering with their attachment to the nuclear membrane of spermatocytes (Searle 1974).

A very interesting finding that may throw new light on the nature of 'chromosomal' sterility in mice and probably in other mammals as well has been recently reported by Forejt in Prague. He noted an abnormally high frequency of C-band contacts between the XY-chromosome pair and the translocation configuration in a number of mouse translocations affecting male sterility (Forejt 1974, 1979). This curious phenomenon proved to be true both for numerous male-sterile reciprocal translocations (T199H;T43H;T145H;T6Ca) (Forejt and Gregorova 1977) and for double Robertsonian translocations (Forejt 1979). It was also established for male-sterile tertiary trisomies (de Boer and Branje 1979) and for male-sterile partially overlapping inversions (Forejt et al. 1981). The existence of such associations has recently been shown as early as the pachytene stage, when the synaptonemal complexes of the sex chromosomes usually lie close to or are visibly attached to those of the translocation configuration (Forejt et al. 1981). In cases of tertiary trisomies, an extra chromosomal segment is often found lying inside the sex vesicle, closely attached to the XY synaptonemal complex (de Boer and van Beek 1982). A very similar close association of the XY-bivalent and a trivalent has been very recently reported for the human sterile chromosomal translocation 13q, 14q (Luciani, Guichaoua, Mattei, and Morazzani 1984).

The intimate association of the euchromatic translocated chromosomes or some extra-chromosomal segments with the X chromosome could interfere with X-inactivation, which in turn would be incompatible with normal spermatogenesis (Lifschytz and Lindsley 1972). An alternative hypothesis suggests that information for inactivation spreads from the already inactivated X chromosome to the attached autosome, switching off part of the genetic information carried in the translocated chromosome (Forejt and Gregorova 1977). The latter suggestion however is not consistent with the sterility of partial (tertiary) trisomies. Moreover, crucial evidence for a causal relationship between the non-random X-autosome contact and spermatogenic failure is still lacking (Forejt 1981). Nevertheless, pairing failure of the translocated chromosomes, with a resulting shift in balance between these and the sex chromosomes, should be considered as the most likely hypothesis at present to account for male sterility in translocation heterozygotes and in tertiary trisomics (Forejt 1979; de Boer and Searle 1980; Forejt *et al.* 1981). Many more experiments are needed to substantiate this interesting suggestion, and to clarify the actual sequence of events underlying the mechanism of male sterility caused by chromosomal rearrangements.

9.3 Conclusions

Chromosome banding techniques made a substantial impact on our understanding of the incidence and the type of structural chromosomal aberrations both in humans and in other mammals. Numerous new malformation syndromes due to partial trisomies or to minor autosomal deficiencies have been described. An unusually high incidence of spontaneous structural aberrations (3.3 per cent) has been reported in human spermatozoa. The incidence of spontaneous structural aberrations in early human abortuses varies from 0.6 to 6.8 per cent. The frequency of spontaneous chromosomal aberrations in laboratory mice is probably very low during spermatogenesis, early embryogenesis, and post-natally. Isochromosome, and especially spontaneous Robertsonian translocations are rather often encountered in laboratory and in feral stocks of mice.

The developmental effects of partial trisomies or deletions were studied most thoroughly in mice with the unequal reciprocal translocations T(14;15)6Ca, T(7;14)2Iem, T9(16;17)43H, T(9;17)138Ca, and T(2;8)26H. The role in early embryonic development of the different parts of the translocated chromosomes has been elucidated in mice heterozygous for these translocations. The data on developmental profiles of mouse embryos with duplications and deficiencies of chromosomes 2,7,8,9,14,15,16, and 17 are highlighted. Partial deletions are generally more detrimental for embryonic survival than partial trisomies. In experiments with two different translocations (T43H and T138Ca) dividing chromosome 17 into almost three equal parts, the region B–C, corresponding to the *H-2* complex, was found to be of the utmost importance for embryonic development in mice.

Genetically balanced chromosomal rearrangements are quite compatible with normal post-natal development in various animals and in humans but they may substantially affect reproduction both in male and in female carriers. Males heterozygous for many unequal reciprocal translocations, or for two Robertsonian translocations with monobrachial homology, are either partly or completely sterile.

Duplications (partial trisomies) and minor deficiencies are often compatible with normal post-natal survival in mice. Many reciprocal translocations have been tested for their ability to generate viable offspring with partial trisomy or monosomy. The characteristics of embryogenesis and post-natal life in mice with Ts(15^{14})T6Ca, Ts(1^{13})T70H, Ts(17^{16})T43H, Ts and Ms(5^{12})T31H have been briefly outlined.

Special attention is paid to the theories of male sterility mechanisms, operating in individuals heterozygous for different chromosomal transloctions or carrying partial trisomies or monosomies. Pairing failure of the translocated chromosomes with a resulting shift in balance between the autosomes and the sex chromosomes is now the favoured hypothesis to account for male sterility of 'chromosomal' origin.

10

Higher nervous activity in animals with chromosome anomalies

Much clinical evidence exists of mental and psychological abnormalities in humans due to karyotype aberrations (Pratt 1967; Davidenkova *et al.* 1973; Kessler and Moos 1973). However, having summarized the literature on this problem we concluded that in humans only the XO karyotype is accompanied by specific behaviour impairment of space orientation and several personality peculiarities. All other cases of karyotype anomalies are characterized in general by rather similar behavioural changes, e.g. different degrees of mental retardation (Kessler and Moos 1973; Polani 1977).

Although there is no doubt that brain development and functional maturation are under the control of numerous loci, one has to question whether all chromosomes are equally involved in these complicated and multistaged processes. Clinical evidence is not able to answer this question because of the extreme complexity of human psychology and its strong dependency on environmental influences, rearing conditions, and many other social causes. The answer may be found in animal behaviour experiments (Fuller and Thompson 1960; Lindszey and Thiessen 1970; Ehrman and Parsons 1981).

Changes in behavioural character are known to be very sensitive indicators of mutational phenocopies induced by teratogens (Barlow and Sullivan 1975). It therefore seemed reasonable to investigate the behaviour and higher nervous activity of animals with various chromosomal anomalies, in order to understand their influence on the phenotype in cases where no morphological indications of such anomalies were present.

Evidence of behavioural alterations in mice with chromosomal anomalies is described below and the possible uses of these animal models in developmental and behavioural genetics are discussed (see also Dyban 1974, 1976a,b).

The formation of conditioned reflexes in mice with the reciprocal translocation T(14;15)6Ca was investigated in the Embryology Department of the Institute of Experimental Medicine, Academy of Medical Sciences USSR, and in the Laboratory of the Genetics of Higher Nervous Activity at the I.P.

Pavlov Institute of Physiology, Academy of Sciences USSR (Fedorov *et al.* 1973). Later, the influence of other karyotype rearrangements on behaviour in mice was investigated in a joint project with the Laboratory of Physiology and Genetics of Behaviour at Moscow State University (Krushinsky, Dyban, Baranov, and Poletaera 1976, 1978, 1981, 1982).

When these investigations started, no indications of behavioural alterations in mice with chromosomal aberrations existed in the available literature. It seemed that our work was the first attempt of this kind. Later it was demonstrated that the chromosomal translocation T(5;13)264Ca influences the development of mouse pups' sensitivity to their mother's odour (Koski, Dixon, and Fahrion 1977). In other rodent species (e.g *Spalax ehrenbergi*; Guttman, Naftali, and Nevo 1975; Nevo, Naftali, and Guttman 1975) possessing several different karyotype patterns (Robertsonian rearrangements), differences in aggression were also demonstrated. The authors consider that these differences are connected with reproductive isolation of chromosomal forms that can be regarded as incipient species or subspecies.

The main results of our investigations are given in the following sections, together with an evaluation of other chromosomal rearrangements in mice that can be used in solving some of these problems.

10.1 Conditioned reflex peculiarities in mice with T(14;15)6Ca translocation

Special crosses of mice with T(14;15)6Ca made it possible to obtain animals that were trisomic for the fragment of autosome. Trisomic mice can be found at the final stage of a two-step breeding experiment. Sexually mature females heterozygous for T(14;15)6Ca were mated to normal males. In the F_1 hybrids, karyotypes were analysed using bone marrow biopsies, and several trisomic mice were found. Among F_2 animals, trisomics were present as well as mice heterozygous for T(14;15)6Ca. Biopsies were carried out in 546 mice: 22 had trisomy of the marker autosome, 220 were heterozygous for the translocation, and 304 had a normal karyptype.

Mice with partial trisomy for autosome T6 showed shivering, muscle jerking, and low muscular tonus, but in different animals these characteristics were displayed in different degrees and were not observed in all trisomics (Cattanach 1967; de Boer 1973; Eicher 1973). In the mice that we studied, these changes were not displayed.

Two-way avoidance conditioning in a shuttle box was investigated in partially trisomic mice, in mice heterozygous for the translocation, and in control littermates with normal karyotype. A light served as the conditioned stimulus, an electric shock (0.3–0.5 mA) being delivered after 6 s of light exposure. Each experiment included a block of 20 trials with a 30-s intertrial interval. In the first series of experiments each animal was subjected to 10

20-trial blocks with an interval of 24 h between blocks.

The level of active avoidance learning was measured according to the following indices:

1. The total number of active avoidance responses during learning in experiments 1–5 and 6–10, and their mean latencies.

2. The increase in number of avoidance responses during experiments 1–5 and 6–10 and during experiments 1–10 as a whole, as well as changes in latency. Linear regression was used to calculate the gradient of change.

In the first series of experiments five trisomics, 21 heterozygotes, and 36 normal karyotype mice were used. The mean number of avoidance responses and the mean latencies in the three groups differed significantly from one another. Trisomic mice acquired the conditioned response far more slowly than mice with a normal karyotype (8.8 ± 2.3 and 26.0 ± 2.5 avoidance reactions in 10 experiments for trisomics and normal mice, respectively). The mean reaction latencies in trisomics and normal mice were 8.5 ± 0.09 and 7.0 ± 0.06, respectively. The behaviour of heterozygous mice was intermediate between the trisomic group and normal karyotype mice.

Analysis of the development of conditioning during the whole 10 experiments in three groups of animals revealed that the regression coefficient (number of avoidance responses on experimental days) was maximal in mice with normal karyotype, minimal in trisomics, and intermediate in heterozygous mice: the regression coefficients were 0.49 ± 0.04 for normal animals, 0.30 ± 0.05 for heterozygous mice, and 0.11 ± 0.02 for trisomic mice.

The calculation of regression coefficients for response latencies on experimental days demonstrated similar differences. The shortening of response latencies in the course of active avoidance learning was most pronounced in normal karyotype mice and least in trisomics.

In the second series of experiments (four trisomics, 22 heterozygous mice, and 25 mice with normal karyotype) results similar to those described above were obtained as well as some new information on the memory span of these animals.

The experimental schedule was modified in such a way as to permit investigation of memory trace formation over 30-s and 60-min intervals. The rate of increase of conditioned avoidance reactions during the first experimental day (20 trials) was 0.165 ± 0.001 for normal mice, 0.140 ± 0.110 for heterozygotes, and 0.034 ± 0.011 for trisomic mice. The data for experimental days 1–5 were 1.964 ± 0.094, 1.163 ± 0.62, and 0.810 ± 0.106, respectively. As the 40–60-min period after a learning experiment can be considered as a time when memory trace consolidation takes place, our data on learning with 60-min intervals between trial blocks can be considered as revealing memory trace consolidation differences in our animal groups. Thus, the low linear regression coefficient in trisomic mice may serve as an indication of memory consolidation deficits in trisomics and heterozygous mice.

These facts demonstrated that in mice with translocation T(14;15)6Ca, a substantial impairment of higher nervous activity takes place. The impairment is most marked in mice with partial trisomy for the T6 autosome.

Later it was also demonstrated that trisomy and translocation T(14;15)6Ca are accompanied by changes in the density of neurons in the neocortical areas and the hippocampus, as well as by shifts in RNA content in neural and glial cells in the same brain regions (Hoffman, Eremeiev, Gluschenko, and Dmitrieva 1979; Hoffman, Dmitrieva, and Lopatina 1980).

It is worth noting that trisomy for the T6 autosome induces a genetic imbalance of a very peculiar type, consisting of extra genetic material from chromosome 14 and centromeric heterochromatin from chromosome 15 (Nesbitt and Francke 1973).

At this stage of the investigation it was not yet established to what extent the learning deficit in mice with T6(14;15)6Ca is determined by an imbalance of chromosome 14 gene loci. It is clear that the specificity of these changes in behaviour needs to be proved by investigations of reciprocal translocations of other karyotype regions.

Further steps in the analysis of behaviour in aneuploid mice were made in a joint research project with the Laboratory of Physiology and Genetics of Behaviour (Moscow State University). The behavioural reaction chosen for this investigation needs some comment. The ability of animals to learn certain conditioned reactions or to achieve some more complicated habits is generally accepted as the main index of behavioural adaptivity and plasticity. At the same time a large volume of evidence exists suggesting that learning ability is not the only factor in behavioural adaptation to rapid environmental changes (Krushinsky 1977).

Krushinsky (1977) suggests that animals may use elementary reasoning ability as a means of rapidly adapting to environmental changes. Elementary reasoning is the ability to comprehend simple empirical laws operating between objects and phenomena of the external world and the capacity to apply this knowledge in new situations when no rapid learning could occur.

The reasoning ability of many vertebrate species has been investigated (Krushinsky 1977). The ability of mice to extrapolate the direction of a food stimulus movement was studied in a special experimental chamber (Krushinsky *et al.* 1976, 1978). Ability to extrapolate is considered to be one of the indices of reasoning ability in animals. Our experimental chamber took the form of a plastic box, the front wall of which had three openings, one in the centre, one on the right, and one on the left. The food bait (milk) was poured into a small cup. This cup slid behind the front wall of the chamber and could be placed in front of any opening. There was also another cup which slid each time in the opposite direction, unseen by the animal and serving to balance the odour stimuli.

The experimental procedure was as follows. Animals were deprived of food and water for 24 h. The food bait was placed in front of the central opening. As the mouse began to drink milk the cup moved to the right (or to the left) and disappeared from view. Thus the animal could see the moving bait only for 1.5–2 cm of its trajectory. After the bait disappeared from sight the mouse could obtain food only if it moved to the opening on the appropriate side. Such a response counted as a correct solution. If the animal chose the opposite opening the response counted as incorrect. Each animal had to solve the task six times, the direction of bait movement being alternated in a quasi-random order. The results were presented as the percentage of correct choices and also as the percentage of animals which solved the task correctly. Laboratory and inbred mice, with few exceptions (but see below) possess no extrapolatory ability, the percentage of correct solvings being non-significantly different from 50 per cent chance level.

Experiments of this type were performed using mice with chromosomal anomalies.

Two types of translocation and trisomy were investigated. Homo- and heterozygosity for T6(14;15)6Ca was the first type of anomaly, while the second was the translocation T(16;17)43H and respective trisomy, where an additional part of chromosome 17 is attached to the centromere region of chromosome 16 (Forejt *et al*. 1980). This type of chromosomal anomaly was described earlier (see Section 9.2.3.4). Mice with this kind of trisomy were obtained from the Institute of Molecular Genetics of the Czechoslovak Academy of Sciences, Prague (Dr J. Forejt and Dr S. Gregorova). The results of extrapolatory task-solving by mice with the above chromosomal anomalies are presented in Table 34. In the case of T(14;15)6Ca, mice with all three types of chromosomal rearrangement were completely unable to solve the task. The groups were small, but it is clear that no correct choices occurred in the trisomics or in animals homo- and heterozygous for trans- locations. Unfortunately no normal karyotype littermate control mice were available.

The results of this test in T(16;17)43H mice were quite different. We found that trisomic mice were superior in this task to animals homo- and hetero- zygous for the translocation. The results of the task-solving at its first presen- tation in trisomic and homozygous mice were significantly above chance level, while heterozygous animals proved to be incapable of solving the task. The summarized data on six task presentations were similar. The increased level of extrapolatory ability in T(16;17)43H trisomics has two important implications. First, this fact means that the chromosomal rearrangements tested in our experiments appear to have specific effects on certain kinds of behavioural reaction. Two types of translocation involving fragments of

Table 34. Extrapolatory task-solving in mice with reciprocal translocations and with two different types of trisomy

Genetical groups	Translocation T(14;15)6Ca		Trisomy for T6	Translocation T(16;17)43H		Trisomy for the fragment of chromosome 17
	Heterozygous	Homozygous		Heterozygous	Homozygous	
Number of subjects	6	8	5	30	27	12
Correct task solvings at first presentation						
No. Per cent	3 50.0[b]	3 37.5[b]	3 60.0[b]	17 56.7[b]	20 74.1[a]	11 91.7[a]
Statistical significance	$p>0.5$	$p>0.5$	$p>0.5$	$p>0.5$	$p<0.01$	$p<0.001$
No. of task presentations	27	43	30	169	142	101
Correct task solvings at 1–6 presentations						
No. Per cent	15 55.5[b]	23 53.5[b]	18 60.0[b]	95 52.2[b]	85 60.0[a]	72 71.3[a]
Statistical significance	$p>0.5$	$p>0.5$	$p>0.5$	$p>0.5$	$p<0.05$	$p<0.001$

[a] Statistical significance of being above 50 per cent chance level.
[b] Non-significantly different from 50 per cent chance level.

different autosomes and inducing partial trisomy of different chromosomal regions are also accompanied by different behavioural features. Trisomic mice with an extra region of chromosome 17 demonstrate a high level of extrapolatory ability, which is unusual for mice. Mice with T6 trisomy show no such ability.

Second, the elevated level of extrapolatory ability in T(16;17)43H trisomics could be treated as an indirect indication that loci on chromosome 17 may be involved in the process of brain development.

It is not yet established whether the extra chromosome 17 region is expressed genetically, so the cause of the behavioural peculiarity associated with this karyotype anomaly is as yet unknown. However, these results constitute the first indication that brain development may be differentially affected by different parts of the genome.

10.2 Robertsonian translocations: their influence on mouse behaviour

Robertsonian translocations (Rb) differ significantly from reciprocal translocations. The number of chromosomes is reduced but the number of chromosome arms remains constant and no loss or gain of genetic material occurs. The new metacentric chromosome obtained as the result of Robertsonian fusion contains two linkage groups, which belong to two different acrocentrics (Matthey 1965; Hsu and Mead 1969).

Centric and tandem fusions are considered to be one of the factors influencing reproductive isolation and thus promoting speciation. It is widely accepted that the main mechanism of reproductive isolation induced by Rb is the low reproductive rate in heterozygous animals, which prevents wide hybridization between normal animals and Rb carriers. At the same time there may be other Rb influences; changes in spatial configuration of interphase chromosomes in cases of Rb could also influence genome function (White 1975).

Mice with different Rbs were tested for their capacity to solve the extrapolatory task. As was mentioned in the previous section, inbred mice and outbred mice and rats are in general unable to solve this task at all, the number of correct solutions at the first task presentation being non-significantly different from the 50 per cent chance level. Yet in several genetic groups of mice (for instances families of C57BL/6J-Lac-Sto mice, feral mice) and in wild brown rats this capacity is obviously present (Poletaeva and Romanova 1977; Krushinsky, Astaurova, Kuznetzova, Ochinskaya, Poletaeva, Romanova, and Sotskaya 1975).

Our experiments revealed that mice with an Rb(8.17)1Iem translocation in their karyotype are capable of extrapolating the direction of movement of a food stimulus. This Rb, discovered in the laboratory colony of outbred mice in the Embryology Department of the Institute of Experimental Medicine

(IEM) in 1971 (Baranov and Dyban 1971a,b; Baranov and Udalova 1975) was described earlier (see Chapter 5).

In Table 35 several indices of extrapolatory capacity in mice with different Rbs and in inbred laboratory mice are presented. It is obvious that mice possessing Rb(8.17)1Iem in their karyotype are superior to other mice in their test performance. Mice possessing two Rbs, one of them being Rb(8.17)1Iem, also demonstrate an increased capacity for extrapolation.

Mice with Rb(8.17)1Iem, used in the early stages of our experiments, belonged to groups with diverse and heterogeneous genetic backgrounds. In the course of further breeding a more homogeneous group of carriers of Rb(8.17)1Iem was formed. These animals also possessed a heightened capacity for extrapolation. Our aim was to clarify the relative role that chromosomal rearrangement *per se* and genetic background played in determining the elevated level of correct task solving in Rb(8.17)1Iem mice. For this purpose eight generations of repeated backcrosses of Rb(8.17)1Iem carriers on to an inbred CBA background were performed. After this series of backcrosses was completed, mice heterozygous for Rb(8.17)1Iem were intercrossed and homozygous individuals from their progeny were bred by means of brother–sister mating. After four or five generations of inbreeding the new CBA-Sto substrain with Rb(8.17)1Iem in its karyotype was formed.

In Table 36 the results of extrapolatory task-solving by these mice are presented, as well as data from the different backcross generations. Mice with Rb(8.17)1Iem on a CBA genetic background demonstrated a statistically significant level of extrapolatory ability as compared to control CBA normal-karyotype animals. Mice heterozygous for Rb(8.17)1Iem derived by back-crossing showed a higher level of extrapolatory ability than their littermates with normal karyotype, but CBA Rb(8.17)1Iem mice are less good at the task than mice with this translocation on the initial genetic background. This suggests firstly that some influence of Robertsonian fusion of eight and 17 chromosomes probably does exist, and secondly, that the general genetic background also exerts some influence on the behavioural capacity tested. The decrease of extrapolatory ability in inbred substrains is unlikely to be due to the influence of homozygosity, as unpublished data on C57BL mice show that a significant level of extrapolatory task-solving ability was detected in these highly inbred mice. The influence of genetic background on the expression of a behavioural character is not very surprising, since polygenic determination of many behavioural traits is a well-established fact. Much more interesting in our opinion is the finding that the chromosomal rearrangement Rb(8.17)1Iem improves behavioural performance. This finding has been strengthened in the course of further work. Mice with the identical but independently derived Robertsonian translocation Rb(8.17)6Sic also possess an elevated level of extrapolatory ability (see Table 36). These

Table 35. Extrapolatory task-solving in mice with normal karyotype and with Robertsonian translocations

Genetic group (2n = chromosome number)	Inbred strains			Laboratory mice			
	CBA 2n = 40	C57BL/6J 2n = 40	AKR with Rb(6.15)1Ald 2n = 38	AKR with Rb(5.19)1Wh 2n = 38	AKR with Rb(8.17)1Iem 2n = 38	AKR with Rb1Iem and RbAld homozygous 2n = 36	AKR with Rb1Iem and RbAld heterozygous 2n = 38
No. of subjects	155	60	27	23	193	26	27
Correct task solvings at first presentation							
No.	74	35	11	13	141	17	15
Per cent	47.7	58.3	40.7	56.5	73.0	65.4	55.6
Statistical significance[a]	NS[b]	NS	NS	NS	$p < 0.001$	NS	NS
Number of task presentations	649	361	199	131	1041	151	153
Correct task solvings at 1–6 presentations							
No.	326	183	95	65	684	98	95
Per cent	50.2	50.6	47.7	49.6	65.7	64.9	62.1
Statistical significance[a]	NS	NS	NS	NS	$p < 0.001$	$p < 0.001$	$p < 0.01$

[a] Statistical significance of being above 50 per cent chance level.
[b] Non-significantly different from 50 per cent chance level.

Table 36. Extrapolatory task-solving in mice from backcrosses and inbreeding and in mice with Rb(8.17)6Sic

Genetic groups (2n = chromosome number)	F_2–F_5 generations of backcrosses		Inbred animals (CBA strain)		Laboratory mice with Rb(8.17)6Sic
	With Rb(8.17)1Iem	Normal karyotype	With Rb(8.17)1Iem CBA	Normal karyotype CBA	
	(2n = 38)	(2n = 40)	(2n = 38)	(2n = 40)	(2n = 38)
No. of subjects	37	23	37	40	14
Correct task solvings at first presentation					
No.	28	14	25	22	14
Per cent	75.6	60.9	67.6	55.0	100
Statistical significance[a]	$p < 0.01$	NS[b]	$p < 0.05$	NS	$p < 0.001$
No. of task presentations	195	152	207	226	75
Correct task solvings at 1–6 presentations					
No.	121	84	123	122	63
Per cent	62.0	55.2	59.4	53.5	84.0
Statistical significance[a]	$p < 0.001$	NS	$p < 0.01$	NS	$p < 0.001$

[a] Statistical significance of being above chance level of 50 per cent.
[b] Non-significantly different from 50 per cent chance level.

mice were kindly presented to us by Professor A. Gropp (Institute of Pathology, Lübeck, FRG).

Thus fusion of chromosomes 8 and 17 appears to influence the development of the central nervous system in a way that enables these animals to solve more complex behavioural tasks than their karyotypically normal counterparts. It is not yet possible to establish the chain of events from the level of chromosomal function to brain physiology. A possible role can be ascribed to some locus on chromosome 17, especially in view of our data on mice with partial trisomy and an extra fragment of autosome 17.

Animal behaviour is considered to be one of the factors determining microevolutionary changes in populations (Mayr 1963). If Robertsonian translocations can really change an animal's capacity to perform some complicated behavioural adaptations in the wild, this factor could augment the viability of their carriers and thus favour these translocations in spite of their adverse effect on fertility.

10.3 Other biological models that can serve the aims of cytogenetic–behavioural analysis in mice

Many types of chromosomal aberration are not compatible with post-natal development in mice (see Chapters 6 and 7) and thus could not be utilized for behavioural genetics studies. At the same time several karyotypic anomalies do not influence mouse viability. It is therefore surprising that carriers of these anomalies are not yet used in behavioural investigations.

10.3.1 X0 mice

The loss of one X chromosome does not affect viability (see Chapter 8).

XO mice are phenotypically females. They can produce offspring, although their fertility is reduced. Some of their progeny possess the same karyotypic anomaly, which allows the maintenance of a stable XO colony. Behavioural parameters of XO mice have not yet been investigated at all. Our preliminary observations of XO mice on the extrapolation problem showed that they lack this capacity, their level of correct solvings being non-significantly above chance level. At the same time these animals showed an extremely poor ability to adapt to our experimental conditions. They required much more time to learn the location of food in the central opening than do normal mice. This suggests the existence of some deficit in recent memory and in the formation of orienting reactions in XO mice. We could not find any reports concerning other behavioural investigations in these mice. According to Ohno's Law, rather distantly phylogenetically removed animals are likely to show similarities in genetic loci carried on the X chromosome (Ohno 1969). Some specific space orientation disturbances have been reported in XO women (Kessler and Moos 1973; Polani 1977), so it seems important to investigate more closely the behavioural peculiarities of animals with this karyotypic anomaly, using various conditioning procedures and neurological as well as behavioural tests.

10.3.2 XYY mice

There exist several descriptions of mice with an extra chromosome (Russell 1962). These mice are sterile and each finding of such an animal is a rare event. It is still a pity that mice with XYY karyotype have not been investigated physiologically. It is worthwhile remembering here the hypothesis of Court-Brown (1969) that the presence of an extra Y chromosome in the human karyotype causes increased aggression. This hypothesis was criticized and in the face of a large amount of experimental evidence was finally discarded, although Y-chromosome genes have been shown to be involved in the determination of aggressive behaviour in mice (Maxon, Platt, Shenker, and Trattner 1982). This makes the problem of behavioural investigations in XYY mice still more interesting.

10.3.3 C-heterochromatin polymorphism

One of the characteristic features of the mouse karyotype is the presence of large blocks of heterochromatin in the centromeric regions of all chromosomes, with the exception of the Y chromosome. In these regions satellite DNA is present, which involves long repeated sequences, so they contain genes with numerous copies.

The role of repeated sequences of DNA is not fully known. It is suggested that this DNA is transcribed but not translated (Georgiev 1971; Samarina, Lukandin, and Georgiev 1973).

C-heterochromatin polymorphism in mice has been described in different strains, involving homologous chromosome differences (Forejt 1972, 1973; Dev *et al.* 1973), and occurs in interstrain as well as inter-individual differences (Dyban and Udalova 1974).

It is interesting to investigate whether behavioural correlates exist when differences occur in the size of heterochromatin regions in two chromosome homologues. As behavioural changes are the most sensitive indices of phenotypic change, they may serve as a means to assess the effects of heterochromatin polymorphism in mouse chromosomes.

10.3.4 Genetic chimaeras: brain structures (allophenic mice)

Experimental embryology possesses well-elaborated methods that allow us to obtain genetic mosaics and chimaeras (McLaren 1976b,c).

Nesbitt (1978) analysed several behavioural parameters in 36 chimaeric mice, which were produced using the morula aggregation method. Mice of two inbred strains (C57BL/6J or C57/BL10J and A/J7) were used. Previously it had been demonstrated that these strains possessed differences in behaviour. It was suggested that study of behaviour in chimaeras would show whether any of the behavioural reactions chosen were governed by single cells or single-cell clones. Four behavioural categories were investigated, chosen on the basis of ease of testing and the magnitude of differences between the two strains: open field activity (two tests), cricket attacking, and rope climbing. If a single clone or single cell controls a particular behaviour, the chimaeras must behave either like C57BL mice or like A/J mice, because the cell or clone will be of C57BL or A/J origin, and not mixed. On the other hand, in cases where the behaviour is controlled by a multiclonal cell constellation, chimaeras will behave in ways not wholly characteristic of one or the other parent strain. It was concluded that these behavioural reactions were not controlled by a clonal cell population.

More interesting are the cases when chimaeric mice are produced using normal and mutant embryos. In the series of investigations performed by Mullen and Herrup (Mullen 1977; Herrup and Mullen 1979; Mullen and Herrup 1979), chimaeras of normal mice and mice with the staggerer mutation were studied. These experimental chimaeric mice were useful in deciding whether the effect of a mutant gene action in a given cell type is direct (intrinsic) or indirect (extrinsic). Analysis of staggerer–wild type chimaeras showed an absence of the locomotion disturbances that are characteristic of *st/st* animals, and also provided a direct demonstration that the Purkinje cell is a primary site of *st*-gene action. The correlation of the Purkinje cell's morphological phenotype and its nuclear genotype that led to this finding was made possible through the use of enzyme activity variants of β-glucuronidase. Histochemical visualization of enzyme activity serves as an independent marker of the Purkinje cell's genotype. Analysis of the *st*-gene involvement in the phenotype of granule cell death has also been carried out (Herrup 1983). In this investigation the ichthyosis mutation has been used as an independent cell marker. This mutation produces no known neurological defects, but does cause a characteristic clumping of heterochromatin, especially in small cells. Analysis of two staggerer–ichthyosis chimaeras led to the conclusion that granule cells that would have died in the mutant are rescued in the chimaeric cerebellum. The death of granule cells in the homozygous staggerer mutant is thus an indirect epigenetic consequence of direct gene action on other cells.

The investigation of chimaeric brains also makes possible clarification of the neural–glial relationships.

The implantation of chimaeric blastocysts in the uterus of a female recipient is a necessary stage of chimaera production. The problem naturally arises of a possible pre-natal influence on higher nervous activity and behaviour, since the maternal organism differs genetically from the embryo.

The technique of transplanting ovaries may be used to assess the relative importance of pre-natal and post-natal maternal effects. Ovaries from inbred donors may be transplanted into hybrid recipients, so that it is possible to obtain inbred offspring of a particular strain that have been carried by females of a different genotype. In the investigation by DeFries (1967), mean open field behavioural scores (activity and defecation levels) were compared in offspring carried by inbred BALB/c (B), C57BL/6 (C) or hybrid (H) mothers (B/H symbolizes inbred BALB/c offspring carried by hybrid mothers, etc.). Comparison of B/B versus B/H and C/C versus C/H offspring indicated little or no maternal effect on open field behaviour but did demonstrate the effect of maternal environment on body weight.

The problem of pre-natal influences on behaviour and higher nervous activity is a very important one, and its investigation requires the use of various experimental methods as well as a knowledge of the consequences of harmful agents affecting the organism during its development. These questions are too wide to be considered in this book. It is worth noting that the use of animals of particular genotypes carried by genetically different mothers is a potentially powerful tool in the investigation of maternal influences on behavioural development.

10.4 Conclusions

Karyotype aberrations in humans sometimes lead to mental abnormalities. However, this kind of clinical evidence is poorly suited to a more precise study of the participation of different chromosomal regions in nervous system development and function.

Although experiments with animal behaviour using a cytogenetic approach are still in their early stages, their results show promise.

Mice with chromosomal anomalies may serve as good models for this purpose. Behavioural investigations of mice with reciprocal and Robertsonian translocations are discussed, as well as studies of mice with XO and XYY karyotypes. The reciprocal translocation T(14;15)6Ca in mice is accompanied by a profound deterioration in active avoidance conditioning, which is especially prominent in cases of partial trisomy with an additional T6 chromosome. The ability of mice to extrapolate the direction of food movement was studied in the case of the latter chromosome anomaly as well as T(16;17)43H and trisomy for the fragment of autosome 17. Mice with T(14;15)6Ca anomalies were as unable to solve this task as are many laboratory and inbred mice, but mice with the T(16;17)43H aberration were found to perform significantly above the chance level. A functional inequality in the influence of different chromosomal rearrangements on the development of brain function is suggested. Mice with the Robertsonian translocation Rb(8.17)1Iem are able to extrapolate, whereas mice possessing fusions of other chromosomes or with normal karyotypes are not. These results may implicate Robertsonian translocations in the processes of speciation in mammals. Animal behaviour experiments may serve as a very sensitive method of investigating the phenotypic expression of chromosomal disorders.

11

Comparison of karyotypic anomalies in human and animal embryogenesis

Some comparative aspects of developmental cytogenetics in laboratory animals and in humans have already been outlined in the previous chapters dealing with each particular type of chromosomal aberration. Features in common as well as some basic species-specific traits will be briefly outlined in this chapter.

11.1 Equivalent stages of development

Only animals that have been well studied both genetically and embryologically may be considered as adequate tools for investigations in developmental cytogenetics. Of more than 4000 species that comprise the class Mammalia (White 1978), only a few meet these strict demands. The laboratory mouse *Mus musculus* is undoubtedly the first in this rank, with the laboratory rat *Rattus norvegicus* and the rabbit *Oryctolagus cuniculus* sharing second place. Embryos of other species including pigs and hamsters are rather rarely used for these purposes.

The laboratory mouse possesses some decisive advantages compared to other mammals as a basic tool in developmental cytogenetics. The gametogenesis and embryogenesis of the mouse have been very thoroughly studied, by all available methods of modern biology. Its genetics, though less well analysed than *Drosophila*, is none the less much ahead of other mammals, including humans.

Rabbits and hamsters have also been well studied as embryological objects, but available information on their genetics is still very meagre. These animals are potentially of great value for developmental cytogenetics but have rarely been used in this field as yet.

Two main conclusions may be drawn. First, it is quite evident that up to now only a few mammalian species have been the basic sources of information in developmental cytogenetics. It might not be unexpected to find some developmental peculiarities in other mammalian species that significantly distinguish them from murine rodents and humans. Second, the urgent necessity arises to determine whether cytogenetic data obtained on murine rodents and human embryos are species-specific and to what extent they are common for all mammals. The re-evaluation of these cytogenetic data from the point of view of equivalent stages of embryonic development is therefore of special importance. Unfortunately such a comparative approach is not always in use in developmental cytogenetics. Human embryos of the second month correspond to the stage of major organogenesis and are considered as rather early in clinical cytogenetics. On the other hand, in murine rodents the same stage embryos are usually regarded as much more advanced in development.

Equivalent ages of human embryos and of laboratory animals are surveyed in a number of original articles and reviews (Otis and Brent 1954; Nishimura and Yamamura 1969; Dyban, Puchkov, Baranov, Samoshkina, and Chebotar 1975; O'Rahilly 1979; Schneider and Norton 1979; Donkelaar, Geysbevts, and Dederen 1979). Strictly speaking a scheme of equivalent ages should be prepared for each embryonic anlage separately. However, this approach is hardly justified for routine embryological studies. Therefore a number of rather simple traits, visible by external inspection, is used to arrange intrauterine development into definite stages. There are 24 stages for rat, mouse, golden hamster, and rabbit embryos according to the equivalent staging scheme elaborated in the Department of Embryology at the Institute of Experimental Medicine, Leningrad. It includes a detailed description of the main traits for each of these stages, which are compared to appropriate embryonic staging outlined by other authors. Human embryogenesis is usually subdivided into so-called 'horizons' named after G.L. Streeter (O'Rahilly 1979).

The embryos of different mammalian species are never exactly the same at any stage of their development. Embryos of various species that share some basic features in common might simultaneously reveal significant discordance in other features.

Let us look briefly at the duration of the main stages of embryonic development in humans and laboratory mice.

As can be seen from Table 37, pre-implantation development in the mouse lasts for 4.5–5 days and in humans for 5.5–6.5 days. Thus it proceeds at approximately the same rate in both species. However, there are delays in subsequent human development, while in the murine rodents the rate of development steadily increases. For instance, gastrulation in human embryos as well as in other higher primates lasts about 10 days, i.e. 3.5 times longer than the same stage in the laboratory mouse and rat. Total duration of the neurula stage and active (major) organogenesis in human embryos is about 40 days. The same stages in the mouse occupy less than 7 days. The most conspicuous difference concerns the fetal period, which is almost 40 times longer in humans than in the mouse. The total duration of the period between

Table 37. The duration and concise description of equivalent stages in embryonic development of humans and the laboratory mouse *Mus musculus*

Stages of pre-natal development	Timing (days)		Duration (days)		Stages	
	Mouse	Human	Mouse	Human	Murine rodents (after Dyban *et al.* 1975)	Streeter developmental horizons
1. Pre-implantation period (cleavage, blastocyst formation)	1–4.5	1–6	4.5–5.5	5.5–6.5	I–VII	I–III
2. Implantation	5.5–7	6–14	1.5	8	VII–IX	IV–VI
3. Gastrulation						
First phase	4.5–6	7–14	1.5	7	VI–VIII	IV–VI
Second phase	6.1–7.5	15–18	1.4	3	IX	VII–VIII
4. Neurulation, axial organ formation	7.5–8	19–21	0.5	2	X	IX
5. Organogenesis	8–14	20–56	7	36	XI–XVIII	X–XXIII
Early (major)	8–10	20–27	3	7	XI–XIII	X–XII
Late	11–14	27–56	4	29	XIV–XVIII	XIII–XXII
6. Fetal period	15–19	56–280	5	224	XVIII–XXIII	XXIII–birth

conception and puberty is 9 per cent of the lifespan in mice, 13 per cent in rabbits, and 18 per cent in humans (Nishimura and Yamamura 1969). Consequently, the developmental rates of intrauterine growth in mice are many times higher than those in humans.

The comparison of various stages in the embryonic development of humans and murine rodents suggests the existence of a striking similarity in their basic developmental pattern before implantation, with respect both to the morphogenetic processes preceding nidation and to their duration. Moreover, the available evidence indicates that with only a few exceptions cleavage and blastocyst formation proceed in a similar way in all mammals. Thus in studying the causal factors that operate in early embryogenesis, the use of mouse embryos may provide valuable information on cleavage, blastulation, and cell interactions in pre-implantation mammalian embryos (see also Blandau 1961; Biggers and Borland 1976) and may help to estimate the impact of chromosome abnormalities in producing disturbances of these crucial embryological processes (Dyban 1974).

As mentioned above, after the onset of implantation the development of humans and of higher primates becomes substantially different from that of all other mammalian species. These peculiarities concern not only the duration of each successive embryonic stage but primarily the morphogenetic processes underlying them.

Embryogenesis in humans is distinguished by the very early formation of the nutritional organs, i.e. chorion and allantoic placenta. During nidation the extra-embryonic mesenchyme spreads outward from the blastodisc, lines the blastocyst cavity, and takes part in the formation of the chorion. Numerous chorionic villi are formed, which invade the deeper uterine tissue and proceed to break down blood vessels until they are bathed in maternal blood. This type of placentation is called haemochorial. The definitive chorio-allantoic placenta in murine rodents is also haemochorial in nature but it belongs to the so-called 'labyrinthine' type and differs significantly in its structure from the human placenta.

Another distinguishing trait of species specificity confined almost exclusively to higher primates and humans is the prematurity of extra-embryonic tissues formation. In these species the development of extra-embryonic parts proceeds somewhat quicker than that of embryoblast derivatives and to some extent does not depend on the latter. Chorionic villi remain alive for some time even after embryonic death and this may result in so-called 'empty embryonic sacs', which are retained in the uterus for some weeks after complete degeneration of the embryo itself (Dyban 1959). The mechanism of implantation in murine rodents is quite different from that in humans and does not include the formation of typical chorionic villi. The rodent trophectoderm differentiates into giant cells and cells of the ecto-placental cone, and this depends on specific interactions between the inner cell mass and trophectoderm at earlier stages of development (Gardner 1972; Gardner and Papaioannou 1975). Moreover, not only implantation but formation of the yolk sac and chorio-allantoic placenta in murine rodents are most intimately correlated with morphogenetic processes in the embryo proper and strictly depend on them. Therefore the disturbances in the development of embryoblast derivatives in murine rodents usually interfere with differentiation of trophectodermal cells and inevitably result in pathology of implantation and placentation with consequent degeneration of the extra-embryonic membranes. This sequence of pathological events is common not only for injuries induced by some harmful external agent, but also for many types of chromosome abnormality. As indicated in Chapter 7, these primarily affect inner cell mass (ICM) differentiation and interfere with morphogenetic processes in the epiblast and its derivatives including egg membranes, resulting in early embryonic death. Unlike murine rodents, the well-developed chorionic villi of the human embryo usually provide prolonged retention of an already dead or even macerated fetus in the uterus.

One more difference in the embryology of murine rodents and humans concerns the provisional functions of the yolk sac. Soon after the first phase of gastrulation in mice and rats, the yolk sac acquires the functions of a main provisional organ. In murine rodents the yolk sac constitutes the principal part of the so-called 'omphaloic' or 'yolk sac' placenta, which provides embryonic nourishment during gastrulation, formation of axial organs, and neurulation, i.e. up to the eleventh day of embryonic development in mice and the twelfth day in rats.

At this time there occurs a gradual transition of the main nutritional functions from the yolk sac placenta to the true chorio-allantoic placenta. The crucial stage of definitive placental development in rodents is an attachment of the allantoic stalk to the inner surface of the chorionic plate. The process of the growth and fusion of the allantois is especially vulnerable to noxious environmental agents and makes a major contribution to the death of rodent embryos with chromosome abnormalities (see Chapter 7).

There is no typical omphaloic placenta in human embryos and the yolk sac does not carry out any nutritive functions. These discrepancies in the mechanisms of yolk sac formation and its function in humans and murine rodents may explain some differences in response of these species both to external agents and to chromosome abnormalities during implantation and placentation.

The murine rodents are also distinguished by the mode of inner cell mass (ICM) differentiation, which initially gives rise to extra-embryonic endoderm and to the unusually inverted germ layers, which acquire their normal position only during neurulation and the separation of the body from the extra-embryonic membranes. Pre-implantation stages of human embryos are not so thoroughly studied as those in murine rodents. Nevertheless, there is no doubt that human embryos do not have inversion of the germ layers and that they lack some morphogenetic processes typical for rodents at equivalent stages of embryogenesis.

The early stages of development in rabbit embryos resemble more closely those of human than those of murine rodents, but some species-specific features can be seen already at these stages of rabbit embryos too. For instance, implantation in rabbits occurs exceptionally late, that is already after the completion of neurulation and the first heartbeats.

Thus considerations of cytogenetic data obtained in human embryos and in the embryos of laboratory animals should be primarily based on equivalent stages of embryogenesis, which are the main clue for an appropriate comparison of experimental and clinical data. Such an approach might be considered as the most reliable guide for understanding morphogenetic mechanisms underlying expression of chromosome abnormalities during embryonic development in humans and in various animal species.

A comprehensive comparison of the developmental cytogenetics of mice and humans has recently been made by one of the pioneers in these studies, Professor Alfred Gropp (Gropp 1981a,b, 1982). The usefulness of mouse models for studying the effects of aneuploidy and human chromosomal disorders has also been thoroughly reviewed by Epstein (Epstein 1981a,b, 1984). The emphasis on Robertsonian translocations as very useful tools in experimental biology and medical genetics has been repeatedly stressed in some of our publications (Dyban 1974; Baranov 1980, 1983a).

11.2 Species-specific features

These include differences in the quality and quantity of spontaneous chromosomal aberrations; some characteristics of survival capacity and retention in the uterus of chromosomally abnormal embryos as well as some very marked differences of abnormal genotype expression during pre- and post-natal development.

11.2.1 Frequency of spontaneous chromosomal aberrations in humans and in animals

The actual incidence of various types of chromosomal abnormality in human pre-implantation stages of development has yet to be established. Direct studies of that sort are hardly possible because of numerous ethical and practical reasons which can hardly ever be overcome. However, it seems clear even now that the incidence of heteroploidy in early human embryos is exceptionally high. This is indicated indirectly by the pronounced reproductive wastage in the human, as well as by direct cytogenetic studies of gametes and aborted embryos.

Earlier studies suggest that at least half of all conceptuses are eliminated before or soon after implantation (Dyban 1959), with almost 30 per cent being eliminated so early that women carrying them were unlikely to have a delayed period (Hertig 1967; Nishimura 1975). The results of recent studies based on detection of HCG production during early pregnancy suggest a post-conception loss rate before 20 weeks of about 43 per cent (Miller, Williamson, Glue, Gordon, Grudzinskas, and Sykes 1980). The major part of this conceptus loss is due to chromosomal aberrations. Chromosomal disorders of various types have been detected in about 60 per cent of all spontaneous abortions in the first trimester (Boué and Boué 1976; Boué *et al.* 1976). An estimate of the total rate of lethal chromosomal aberrations in humans at 9 per cent (Gropp 1981a,b) is probably a lower limit. The incidence of chromosomal abnormalities in human sperm has been found to be about 8–9 per cent (Martin *et al.* 1982, 1983b). Numerous chromosomal aberrations have already been detected in human eggs after cytogenetic studies of *in vitro* fertilized eggs. (See Chapters 6 and 7.)

On the other hand, reproductive wastage in murine rodents, both mice and rats, rarely exceeds 20–30 per cent, with only a very small proportion of blighted ova showing chromosomal anomalies (see Chapters 6 and 7). Cytogenetic studies of post-implantation conceptuses in rats and mice suggest that the highest level of spontaneous chromosomal aberrations (12 per cent) is to be found in empty extra-embryonic membranes on the eleventh day after conception (Dyban *et al*. 1971). According to equivalent stages of embryonic development in mammals (see also Dyban *et al*. 1975), these pathological embryonic specimens in rats are comparable to 'blighted ova' in humans at 3–4 weeks of age. Most of these human conceptuses, as has already been mentioned, are chromosomally abnormal, but only a few carried detectable chromosomal disorders in rats. Similarly, less than 18 per cent of abnormal mouse embryos of comparable age display heteroploidy. It therefore looks quite probable that during early post-implantation stages at least, the incidence of heteroploidy in laboratory animals is almost five or six times lower than in human conceptuses of equivalent age. Pre-natal mortality in murine rodents is probably provoked by non-chromosomal factors (both genetic and non-genetic; Dyban 1974), while the major impact of hetero-ploidy on reproductive wastage in human conceptuses is almost beyond doubt.

The factors responsible for inter-species differences in the frequency of spontaneous heteroploidy, and especially for the unusually high incidence of aneuploidy in humans, remain mostly unknown. Some presumed mechan-isms of chromosomal non-disjunction during gametogenesis and early embryogenesis in humans are discussed in recent surveys (Vogel and Motulsky 1979; Carr 1983) and are reviewed in Chapter 4. Hormonal imbalance in women, intrafollicular overripeness of ova, and the ageing of gametes in the female reproductive tract before gamete fusion are generally accepted now as the most plausible reasons for the high frequency of spontaneous chromosomal aberrations in humans.

A new suggestion is that the high incidence of heteroploidy in man is induced by the mutation of special gene(s) regulating chromosomal dis-junction in meiosis (Juberg and Davis 1970). Mutations of this kind might be similar or even identical to known 'mei'-mutations in *Drosophila melanogaster* (Golubovskaya 1979). These mutations should be eliminated from highly inbred strains of murine rodents but may be abundant in wild populations of mammals. One might therefore expect an unusually high incidence of heteroploidy in wild rodents as compared to inbred animals. Unfortunately no data on the incidence of spontaneous chromosomal aberrations in early embryos of wild rats or mice are available so far. Moreover, the very low incidence of spontaneous chromosomal aberrations in embryos of non-inbred animals (see Chapter 4) seems to contradict the suggestion of the existence of 'mei'-mutations in mammals, though it does not refute it completely.

11.3 Qualitative differences in chromosome anomalies of human and animal embryogenesis

Data on the frequency of various types of spontaneous heteroploidy in humans and in other mammalian species are shown in Fig. 37. It is seen that each mammalian species studied cytogenetically in early pre-natal life possesses its own unique pattern of chromosomal abnormalities. Qualitative differences of spontaneous heteroploidy between murine rodents and humans are evident. The most common type of spontaneous chromosomal anomaly in humans is aneuploidy, taking its origin mainly from meiotic non-disjunction during gametogenesis of both sexes. The overall frequency of trisomy among the most extensive studies in medical genetics varies from 44 to 57 per cent with a mean of 50–51 per cent of total anomalies. The second most frequent type of chromosome anomaly in aborted human fetuses is

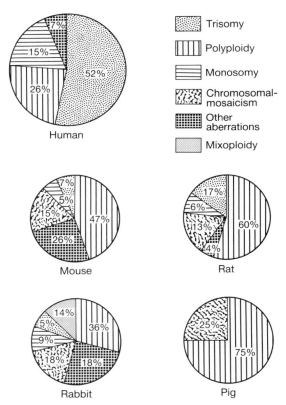

Fig. 37. Relative incidence (in per cent) of different types of chromosomal aberrations in early embryos of humans, mice, rats, rabbits, and pigs.

monosomy OX (18–19 per cent). The next most frequent is triploidy (15–16 per cent), with the remaining 15–16 per cent encompassing all the other less common chromosomal aberrations such as tetraploidy, mosaicism, translocations, and so on (Sankaranarayanan 1979; Lauritsen 1982; Carr 1983).

In laboratory rodents, on the other hand, triploidy provides the major contribution to spontaneous chromosomal aberrations, while the incidence of aneuploidy in both rats and mice is very modest (see Chapters 4 and 6). Triploidy or mixoploidy but not aneuploidy are the commonest anomalies in pigs, rabbits, and hamsters.

Obviously much more information must be gathered before firm conclusions can be drawn regarding inter-species characteristics of spontaneous chromosomal aberrations. Nonetheless all the data available so far indicate that clear-cut quantitative and even more pronounced qualitative differences exist between human beings and laboratory rodents. The mechanism underlying the high level of meiotic non-disjunction in humans remains obscure.

11.4 Time difference in the effects of karyotypic imbalance on human and animal embryogenesis

A comparison of clinical and experimental data on the survival capacity of heteroploid mammalian embryos is consistent with the notion of relatively more prolonged retention of chromosomally abnormal embryos in humans than in murine rodents. It has already been pointed out (Chapters 2 and 6) that most autosomal trisomies as well as all triploid embryos are eliminated in mice within a few days, during placentation and major organogenesis. Most of these affected embryos display severe morphological changes, both in the embryonic tissues proper and in the fetal membranes, especially in the yolk sac and the chorio-allantoic placenta. Extensive haemorrhages in the implantation chamber readily separate abnormal embryos from the surrounding decidual tissue, and provoke intensive resorption and the expulsion of 'blighted ova' in murine rodents. Pathological findings in human abortuses are somewhat different. They usually reveal severely affected macerated or 'nodular form' remains of the embryo proper, while the embryonic membrances and chorionic villi are vascularized, normal in size and shape, and may provide extremely prolonged retention of affected embryos in the uterus. Premature differentiation of extra-embryonic ectodermal and mesodermal derivatives, giving rise to the numerous villi penetrating the decidual wall as well as maternal blood vessels, and thus firmly anchoring the implanting conceptus in the uterus (see Sections 11.2 and 11.3), may be taken as one though probably not the sole explanation for the prolonged retention of chromosomally abnormal human embryos.

Inter-species differences in phenotypic expression of chromosomal abnormalities may also reflect some basic peculiarities of genome function. The experimental data on autosomal trisomy and monosomy in laboratory mice (Chapters 6 and 7) may be taken as an indication that all autosomes and many gene loci are expressed during implantation (Dyban 1974). In humans these early stages of embryogenesis are probably under the control of only a few autosomes and a few gene loci, which would make human embryos less vulnerable to chromosome imbalance during the initial stages of embryogenesis (Dyban 1973a,b, 1974). This hypothesis, though not substantiated at present, is amenable to experimental test. Some of its implications will be more thoroughly discussed in Chapter 13.

11.5 Other species-specific differences of developmental cytogenetics in humans and mice

In contrast to humans, not a single type of autosomal trisomy in mice is compatible with prolonged post-natal survival (see Chapter 7). A few types of trisomy can be recovered at term, but only trisomy 19 on a particular genetic background allows survival for a few days after birth. Most other mouse trisomies are eliminated during major organogenesis.

Pathological changes and survival capacities in human embryos tend to be correlated with the overall size of the extra chromosome. In laboratory mice, however, the situation may be quite opposite. Excess or deficiency for the largest autosome 1 is much less detrimental for embryonic survival than imbalance of the much smaller chromosome 17. (Examples of similar discordance can also be found in humans: Carr 1983.) The absence of any clear-cut correlation between the size of chromosomes and their impact on the genetic control of pre-natal development probably reflects the unequal distribution in the mammalian genome of so-called 'early genes', providing essential information for the early stages of embryonic development (see Chapters 7 and 13).

Not only trisomies but also triploidy can be compatible with rather long survival of human embryos, some of which proceed to term. However, triploidy in mice and rats is inevitably lethal by mid-gestation.

In contrast, XO monosomy seems to be much less harmful for embryonic survival in laboratory mice than in human conceptuses. This chromosomal anomaly is a pre-natal lethal in over 99 per cent of human embryos, while most XO mice readily reach term, look normal on external inspection, and are fertile (see Chapter 8).

Robertsonian translocations gradually accumulating in the karyotype of mice have no detrimental effect on the reproduction and embryonic wastage in simple or even complex heterozygotes, provided they do not share monobrachial homology in translocated chromosomes. Similar chromosomal rearrangements often correlate with spermatogenetic block and serious reproduction impairment in men.

11.6 Basic similarities in developmental cytogenetics of mammals

The expression of chromosomal aneuploidy in humans and mice is basically similar, as shown by the following experimental findings:

1. The phenotype of most trisomics displays a syndrome of non-specific changes, which include general retardation in development (the so-called 'runting syndrome' of Gropp), especially evident in the cranial region of the neural tube and brain, cranio-facial malformations, and heart anomalies. Very similar non-specific changes of the embryo itself and of extra-embryonic membranes are also common for human abortuses (Boué *et al.* 1976; Carr 1983).

2. Trisomy for at least some autosomes is not infrequently associated with rather specific malformations such as exencephaly (trisomies 12 or 14), cranio-facial hypoplasia with or without holoprosencephaly in trisomy 1, pathological outgrowth of neural tissue in the spinal cord region (trisomy 15), and cleft palate in trisomies 13, 18, and 19 (see Chapter 6). As almost all these malformations tend to be encountered in some types of trisomy, the term 'semi-specificity' of malformations seems to be appropriate (Gropp 1981). Numerous 'specific' or 'semi-specific' traits have also been repeatedly reported in chromosomally abnormal human specimens before and after birth. Some of the known 'specific' anomalies in man have their phenotypic counterparts in laboratory mice and vice versa. For instance, holopro-sencephaly and cyclopia typical for trisomy 1 in mouse embryos are also quite common in humans with Patau syndrome (Ts13). Patterns of specific, semi-specific, and non-specific alterations of phenotype in chromosomally abnormal mouse embryos correspond to chromosomal syndromes known in medical genetics.

3. Translocation carriers usually display a substantial increase in genetically unbalanced gametes. Male translocation heterozygotes often show a partial or complete meiotic block, resulting in their sterility or semi-sterility. The non-disjunction rate and the number of aneuploid embryos are usually higher in female than in male translocation heterozygotes. Similarly, ageing effects on chromosome non-disjunction are much more pronounced in females than in males.

4. Both in mice and in humans, the incidence of aneuploidy due to Robertsonian translocations has been found to depend on the specific

properties of the autosomes involved in the fusion, on the number and type of other chromosomal rearrangements in the karyotype, and on the sex of the translocation heterozygotes (see Chapter 4).

Because of these basic similarities in the intrinsic genetic mechanisms underlying meiotic maturation in translocation heterozygotes and pathological changes induced by chromosomal disorders in mammalian zygotes, experimental animals and laboratory mice in particular provide natural biological models for the study of chromosomally induced congenital malformations, reproductive failure, and embryonic wastage.

11.7 Laboratory mice as models for experimental studies in medical cytogenetics

The use of mouse trisomies and mouse translocation carriers for the purposes of experimental biology and medicine has been thoroughly reviewed in a number of recent surveys (Gropp 1981a,b; Baranov 1980; Epstein 1981a,b, 1984; Searle 1981; Searle and Beechey 1982). This section deals with a few of the problems in the field of developmental cytogenetics of man that are amenable for experimental study in the laboratory mouse, to draw the attention of clinical and experimental cytogenetics to the possibilities that exist.

1. One of the most crucial problems of modern medical genetics concerns the identity of the factors responsible for the high rate of meiotic non-disjunction in humans. Laboratory mice and especially those carrying Robertsonian translocations have become the animals of choice for these studies (see Chapter 5). Ageing gametes, univalents, chiasma frequency, and aneuploidy, prezygotic selection of genetically abnormal gametes, the causes of sterility (semi-sterility) in translocation heterozygote carriers—all these and many other related problems are already within the scope of intensive experimental studies. Such studies have revived interest in the problem of mechanisms operating against chromosomally abnormal sperm (Martin-Deleon and Boice 1982; Martin 1983); they have provided new evidence that biological rather than chronological age predisposes to meiotic non-disjunction in females. In good agreement with experimental data (Forejt 1974, Chapter 9), pachytene analysis of infertile men with reciprocal translocations has revealed non-random association of the trivalent with the sex vesicle (Luciani *et al.* 1984). It seems probable therefore that further studies in mice with chromosomal translocations may provide a clue to the intrinsic mechanisms of hybrid sterility in males, and thus might help to devise a rational strategy for its effective treatment.

2. No definite answer can be given at present to the simple question, 'Why does the extra gene dosage caused by the triplication of any particular

chromosome have such damaging effects on early development in mammals and humans?' The existence of mouse models for autosomal trisomies provides a unique opportunity for the detailed experimental analysis of this problem. The decisive advantages of mouse models for these studies, as pointed out by Epstein (1981a, 1984), are the following:

(a) these models facilitate the developmental analysis of the pathogenesis of abnormalities;

(b) *in vitro* and *in vivo* investigations of cells and tissues other than blood elements and fibroblasts become feasible;

(c) pre-implantation and early post-implantation stages of development inaccessible to direct analysis in humans can be readily studied in aneuploid mice;

(d) the effects of genetic and environmental factors on the phenotype of the aneuploid state can be assessed;

(e) chromosomally unbalanced animals facilitate gene dosage studies;

(f) aneuploidy of all parts of the genome can be investigated in a systematic manner;

(g) model test systems for proposed therapeutic approaches in chromosomally abnormal patients may ultimately be developed.

Many of these theoretical expectations have already received substantial experimental support (see Chapter 6). One of the most important achievements in this field includes the artificial production of viable Ts $\leftrightarrow$ *2n* chimaeras by aggregation of early embryos or by injection of trisomic cells into otherwise normal diploid adult hosts. Other advances are the following: dosage studies for known gene products; demonstration of genetic factors influencing the phenotype of post-natal trisomics; developmental analysis of malformations induced by autosomal trisomies (see Chapter 6).

The benefit of animal models stems from the assumption that the mechanisms by which chromosomal imbalance produces developmental disorders are quite similar in mouse and in humans, and it is not hampered by the observed differences in phenotype and genotype in aneuploids of the two species (Gropp 1981a,b).

This assumption cannot however be considered as absolutely justified in the present state of our knowledge of mouse and human genetics. Comparative gene mapping in several mammalian species including the best studied ones, mouse and man, provide strong evidence in favour of a high degree of stability in the structure of the mammalian genome. About 60 per cent of autosomal linkages are conserved in humans and mice, several with homologous segments longer than 20 cM (Nadeau and Eicher 1982). Moreover, about 28 syntenic (carried on the same chromosome) groups of genes have been identified in the genomes of mice and humans. Thus, though it is impossible to find in laboratory mice a biological model entirely identical to certain types of trisomy in humans, the knowledge of syntenic groups of genes permits 'the construction of a mouse in which several genes present on a human chromosome of interest are similarly unbalanced in animals' (Epstein 1984). Ts16 in the mouse might be a good candidate for such 'genetically similar' biological models. This assumption is based upon the conservation and synteny of at least three different gene loci [superoxide dismutase (SOD-1); interferon receptor protein (IFRC); phosphoribosyl glycinamide synthetase (PRGS)] in a distal part of chromosome 16 in mice and of chromosome 21 in humans. The genetic homology between the two chromosomes facilitates the acceptance of Ts16 in mice as a model for Ts21 Down's syndrome in humans (Epstein *et al*. 1981). Unfortunately Ts16 is not compatible with post-natal survival, though it is often recovered at term and induces numerous heart malformations characteristic of patients with Down's syndrome (Gropp *et al*. 1983). The next step towards the creation of a 'genetically identical' model of Down's syndrome in mice has recently been undertaken by Epstein's group (Cox, Smith, Epstein, and Epstein 1984). They have made viable Ts16 ↔ diploid aggregation chimaeras, which may provide new insights into the mechanisms by which Ts21 leads to developmental abnormalities in humans.

3. Studies in mice may provide invaluable information on developmental cytogenetics which could hardly ever be achieved by clinical trials. The fate of monosomic embryos is a good example. Embryos deficient for a whole chromosome have only exceptionally rarely been reported in spontaneous abortuses, and thus the fate of nullisomic gametes in humans remains for the most part unknown. Our systematic studies of monosomy in mice provide arguments in favour of viability of nullisomic gametes, their participation in fertilization, and early death of monosomic embryos in pre-implantation stages (see Chapter 7).

An unusual application of biological models in developmental cytogenetics may stem from recent studies of complete hydatidiform moles (extra-embryonic membranes with no traces of embryonic tissues but displaying massive generalized swelling of chorionic villi and gross trophoblastic hyperplasia). The chromosomes in most of these moles have been found to be exclusively of paternal origin (Ohama *et al*. 1981). Biochemical and cytogenetical analysis demonstrated that the moles arose from the fusion of one or two sperm with a mature but anucleate ovum (Wallace, Surti, Adams, and Szulman 1982). Human moles may thus be considered as androgenetic parthenogenones. Biological models of these parthenogenones can be produced experimentally by microsurgery (see Chapter 1). Such biological models may be very useful tools in studying the origin of and the mechanisms leading to the malignant transformation of these androgenetic conceptuses.

11.8 Conclusions

This chapter deals with species-specific features as well as with features in common between humans and murine rodents. The common features serve as a logical background for the creation of experimental tools suitable for the study of different problems in medical genetics. Species-specific factors encompass differences in the quantity and quality of spontaneous chromosomal aberrations; time differences in the survival capacity of aneuploid embryos; the absence of trisomies compatible with prolonged post-natal survival in mice; the prolonged survival of triploid embryos in humans and vice versa; the better viability of XO monosomics in mice, etc. On the other hand strong experimental evidence exists for basic similarities of chromosomal aneuploidy in humans and mice. These include syndromes of non-specific changes (e.g. runting syndrome) in autosomal trisomies, specific and semi-specific malformations induced by certain types of trisomies, increased levels of chromosomal non-disjunction in translocation carriers, and similarity of factors causing sterility and potentiating the high incidence of aneuploidy in the progeny of heterozygotes. The existence of features in common witnesses to a basic similarity in the intrinsic mechanisms underlying meiotic non-disjunction and pathological changes induced by chromosomal disorders, and thus provides an experimental basis for the studying of congenital malformations, reproductive failure, and embryonic wastage induced by chromosomal aberrations.

12

Influence of genes on early mammalian development

The genetic control of early mammalian development has been thoroughly surveyed in a number of extensive reviews (Wolf and Engel 1972; Church and Schultz 1964; Engel and Franke 1976; Davidson 1976; McLaren 1976a,b, 1979; Pedersen and Spindle 1976; Biggers and Borland 1976; Epstein 1975, 1976; Adamson and Gardner 1979; Sherman 1979; Magnuson and Epstein 1981; Johnson 1981; Epstein and Magnuson 1982; Pratt, Bolton, and Gudgeon 1983).

The present chapter deals mainly with basic biochemical and genetic information related to gene expression and macromolecular synthesis during pre-implantation development in mammals, and attempts to compare these data with the results of cytogenetic studies of aneuploid mammalian embryos outlined in previous chapters. The results argue in favour of very early activation of chromosomal loci in mammalian embryogenesis.

Since the mouse has been the principal species studied biochemically, genetically, and cytogenetically during the initial stages of embryonic development, we shall consider chiefly the data obtained on mice, making only brief references to relevant data in other mammals if available.

Basic experimental approaches used in mammalian embryology for illuminating the problem of gene activation and expression in embryogenesis include (a) studies of cell cycles, (b) direct and indirect measurements of RNA transcription and translation, (c) analysis of protein synthesis, (d) expression of paternal alleles, and (e) expression of early lethal mutations.

12.1 Cell cycles in early embryos

According to our present knowledge on eukaryotic cell cycles, most genetic information is transcribed during the G_1 or partly during the G_2 phase. Consequently, studies on the cell cycle might illuminate the functional state of the genome. The absence of G_1 or G_2 phases would be evidence in favour of a dormant (inactive) state of the embryonic genome, while their presence may be taken as indirect proof of genome activation and its participation in transcription.

Cell cycles in early mammalian development have been intensively studied in a number of original works and also repeatedly reviewed (Alfert 1950; Dalcq and Pasteels 1955; Sirlin and Edwards 1959; Mintz 1962a, 1964a,c; Oprescu and Thibault 1965; Szollosi 1966; Samoshkina 1965, 1968; Zavarzin Samoshkina, and Dondua 1966; Barlow, Owen, and Graham 1972; Graham 1973a,b). According to some early studies the cell cycle of cleaving embryos is represented exclusively by S-phase (Dalcq and Pasteels 1955; Mintz 1967; Izquierdo and Roblero 1965) or by S and G_2 phases only (Gamow and Prescott 1970), while clear-cut evidence for G_1 phase was only obtained from the early blastocyst stage onwards (Mukherjee 1976). In contrast the presence of a G_1 phase from the second cell cycle and during the whole of cleavage has been experimentally established by Samoshkina (1968, 1970, 1972) and repeatedly confirmed in many later experiments (Barlow *et al.* 1972; Graham 1973a,b; Kato, Yamazaki, Kimura, Hayasaka, and Kando 1982; Flach, Johnson, Braude, Taylor, and Bolton 1982; Bolton, Cades, and Johnson 1984). Some methodological problems concerning the unscheduled re-utilization of labelled ^{3}H-thymidine stored in zona pellucida polysaccharides (Piko 1970) or accumulated in the perivitelline space are the most plausible sources of these discrepancies. The duration of the G_1 phase in cleaving embryos may vary significantly. In recent studies, no G_1 phase could be determined in the first cleavage division, that is in the zygote (Kato *et al.* 1982), while the presence of a very short G_2 phase (about 1 h) was shown in the second cell cycle (Flach *et al.* 1982). (See also Section 13.1.3.1.)

The presence of a G_1 phase from early cleavage onwards may be considered as already well-proven, not only in mice but probably in other mammals also. In rat embryos subjected to comparative autoradiographic and cytofluorometric analysis on air-dried preparations (Dyban *et al.* 1976), comparative measurements of DNA content and ^{3}H-thymidine uptake in the same nuclei enabled cell cycle stages to be much more precisely determined. This work unequivocally confirmed the existence of a G_1 phase in rat and mice embryos during cleavage. The duration of this phase in cleaving rat embryos was approximately three times greater than in mice. This difference is in good agreement with the overall length of cell generation time during cleavage in rats.

The early development of the frog or sea urchin is distinguished by very short cell generation times, as G_1 and G_2 phases are lacking and the DNA synthesis phase is also very short. In contrast, cleaving mammalian embryos possess a very long cell cycle time, due to the long DNA synthesis phase and to the presence of G_1 and G_2 phases from the very start of embryogenesis.

Thus the results of cell cycle studies are indicative of genome activation from early cleavage. More reliable information on this point has come from direct biochemical studies on RNA and protein synthesis in mammalian embryos.

12.2 Synthesis of various classes of RNA

The synthesis of RNA is the most immediate consequence of genetic activity. We will consider some basic autoradiographic and biochemical data concerning RNA synthesis in early mammalian embryos: for more details, the reader is referred to some extensive recent reviews (Epstein and Magnuson 1982; Magnuson and Epstein 1981; Johnson 1981; Pratt *et al.* 1983; Schindler and Sherman 1981).

12.2.1 Histo-autoradiographic data

This method has been almost exclusively applied for studying RNA synthesis in mice and rabbit embryos. Relevant information on other mammalian species including laboratory rats is still rather fragmentary.

RNA synthesis begins soon after fertilization in mouse embryos (Mintz 1964c) and proceeds at an extremely low level until the two-cell stage (Bernstein and Mukherjee 1972). The labelled ^{3}H-uridine does not leave the pronuclei of the zygote or the nuclei of early two-cell embryos (Samoshkina and Khozhai 1975). In probably more advanced two-cell embryos, the radioactive label could be traced to the cytoplasm (Mintz 1964c, 1967). The pattern of nuclear labelling at the two-cell stage probably depends on some unknown cytoplasmic factor. In somatic hybrids between two cell embryos and cells of adult tissues cultured *in vitro*, the incorporation of RNA precursors into somatic nuclei drastically declined after cell fusion, approaching the level of embryonic cells (Bernstein and Mukherjie 1972, 1973).

At the four-cell stage the rate of RNA synthesis increases considerably, with the level of nuclear, nucleolar, and cytoplasmic labelling rising correspondingly (Mintz 1964c; Hillman and Tasca 1969). Actively labelled 'true' (instead of non-labelled 'primary') nucleoli are first observed at this stage.

At subsequent stages a gradual increase of nucleolar and cytoplasmic labelling is seen, levelling off at the blastocyst stage (Monesi and Salfri 1967; Hillman and Tasca 1969).

Proper quantitative interpretation of these data is complicated by the lack of information concerning the state of RNA precursors in the intracellular pool, including their concentration, degree of compartmentalization, their proportionate contribution to the different species of RNA, and the extent to which labelled nucleosides are modified prior to incorporation (Epstein 1975; Johnson 1981). Such information is now available from direct biochemical analysis of pre-implantation mammalian embryos.

12.2.2 Biochemical data

The ovulated mouse egg inherits about 0.35 ng maternal RNA (Piko and

Clegg 1982), that is RNA synthesized on the chromosomes of the oocyte predominantly during its growth phase (Bachvarova 1974, 1981; Bachvarova, Burns, Spiegelman, Choy, and Chaganti 1982; Fourcroy 1982).

Almost 40 per cent of maternal RNA is degraded by the two-cell stage, within 24 h of fertilization (Bachvarova and De Leon 1980; Piko and Clegg 1982), but total RNA content steadily increases during cleavage suggesting nearly complete replacement of maternal with embryo-derived RNA by the early blastocyst stage (Piko and Clegg 1982).

The initial quantitative studies used a concentration of ^{3}H-uridine that produced maximal precursor uptake. They demonstrated only a small increase in incorporation of ^{3}H-uridine into RNA between the two-cell and 8–16 cell stages and a very large (about 15 times) increase between 8–16 cell and early blastocyst stages (Daentl and Epstein 1971). Biochemical analysis of the intracellular pool of nucleotides and measurements of their intracellular transport indicated a large endogenous pool of nucleotides during early cleavage, with a very abrupt decline thereafter. Keeping in mind these limitations as well as the progressive increase of cell number, the actual rate of RNA synthesis per genome during cleavage was found to show little or no increase (Epstein 1975; Biggers and Borland 1976). In an even more precise study by Clegg and Pico (1983), the endogenous pools of ATP and UTP precursors as well as their specific activities were measured in mouse embryos labelled with ^{3}H-uridine and ^{3}H-adenine. Significant differences in the specific activities of endogenous pools of these nucleotides were found, and it was therefore inferred that the overall rate of RNA synthesis increases approximately 30-fold during the pre-implantation period, i.e. about four-fold when measured on a per cell basis.

The rates of overall RNA synthesis in early blastocysts estimated by Epstein were 50 pg RNA h^{-1} embryo^{-1} or 1.4 pg RNA h^{-1} cell^{-1} (taking an early blastocyst as 32 cells). Consequently approximately 750 pg of RNA would be synthesized during the transition between eight-cell and early blastocyst stages (Epstein 1975). These calculated values are in good agreement with the results of fluorometric measurements of embryonic RNA content of about 912 pg between the eight and 34-cell stages (Olds, Stern, and Biggers 1973; Biggers and Borland 1976). Total RNA content is about 0.50 ng in the zygote, 0.40 ng in the two-cell embryo, 0.46 ng in the eight-cell embryo, and 1.37 ng in the morula–early blastocyst (Olds et al. 1973).

Similarly in rabbits, incorporation of ^{3}H-uridine into RNA increases in parallel with cell number up to the morula stage, with the greatest rate of synthesis per cell being observed at the 28 stage (morula to blastocyst transition) (Manes 1969).

Thus, both mice and rabbit embryos reveal a rather low rate of RNA synthesis during cleavage, with a sharp rise in net synthesis of RNA occurring at the early morula stage (Biggers and Borland 1976).

Biochemical analysis, especially column affinity chromatography, sucrose density sedimentation, and polyacrylamide gel electrophoresis, has been found indispensable in the qualitative analysis of RNA synthesis during early development. Changes in pattern of precursor uptake into qualitatively distinct species of RNA have been studied from the zygote onwards.

At the one-cell stage, most authors report little if any incorporation of ^{3}H-uridine into acid-precipitable RNAase-sensitive material, and this small amount was attributed to the synthetic activity of the polar body (Young, Sweeney, and Bedford 1978). However, more recent experiments with ^{3}H-adenosine as a labelled precursor favour RNA transcription in the pronuclei of the mouse zygote (Clegg and Piko 1983). Large heterodisperse poly (A)$^-$ RNA was identified as the major product while tRNA and poly(A)$^+$ RNA contributed to the minor fractions at the one-cell stage. The data suggested a rate of tRNA synthesis of about 0.2 pg embryo^{-1} h^{-1}, while for a stable RNA fraction which may consist of poly(A)$^-$ mRNA (histone mRNA), and a small amount of stable poly(A)$^+$ mRNA, the estimated rates were about 0.3 pg embryo^{-1} h^{-1} and 0.045 pg embryo^{-1} h^{-1}, respectively. The major heterodisperse RNA fraction assumed to be heterogeneous nuclear RNA accumulated at a rate of 1.5 pg h^{-1} embryo^{-1} (Clegg and Piko 1983). More experimental studies with other RNA precursors are needed to prove unequivocally the activation of the mammalian genome at the pronuclear stage.

At the two-cell stage mouse embryos showed clear-cut incorporation of ^{3}H-uridine into acid-precipitable RNAase-sensitive material that could be separated by Sephadex gel filtration and acrylamide electrophoresis as high molecular weight heterogeneous RNA, low molecular weight (4S) RNA (Woodland and Graham 1969; Knowland and Graham 1972), and uridine-labelled polyadenylated RNA (Levey, Stull, and Brinster 1978). Including more recent data the synthesis of all major classes of RNA, including ribosomal RNA, was evident at this stage (Clegg and Piko 1983). The time-course of accumulation of adenosine in poly(A)$^-$ RNA indicated the synthesis of stable and unstable RNA components. The former was represented by two fractions—mature rbRNA (synthesis rate about 0.4 pg h^{-1} embryo^{-1}) and stable heterogeneous poly(A)$^{-1}$ RNA (histone mRNA, 0.3 pg h^{-1} embryo^{-1}). The unstable component consisted almost entirely of heterogeneous nuclear RNA (synthesis rate about 3.0 pg embryo^{-1} or 1.5 pg cell^{-1}, Clegg and Piko 1983). These recent findings seem to contradict the earlier biochemical data (Magnuson and Epstein 1981; Johnson 1981) as well as the results of selective silver staining for detecting active nucleolar-organizing regions (NOR) in metaphase chromosomes. No active NORs were detected on the metaphase chromosomes of the first cleavage division or in the nucleoli of the early two-cell embryo (Patkin and Sorokin 1983). The NOR-positive reaction becomes evident in the metaphase chromosomes of the second mitotic division, indicating active transcription of ribosomal cistrons from the mid two-cell stage. More studies are needed to prove the synthesis of rbRNA from the early two-cell stage, but more important for our further discussion is the synthesis of heterogeneous, high molecular weight mRNA at this early stage.

Four-cell stage embryos incorporate label into both ribosomal and low molecular weight (4S) RNA as well as into high molecular weight hetero-geneous RNA (Woodland and Graham, 1969; Knowland and Graham 1972).

From the eight-cell stage onwards the synthesis of 4S,5S,18S,28S,32S, and 45S RNA has been shown (Ellem and Gwatkin 1968; Woodland and Graham 1969; Piko 1970; Church and Schultz 1964). Distribution of low molecular weight RNA (4-5S), rbRNA, and high molecular weight heterogeneous RNA according to labelled uridine incorporation was 6–10 per cent, 65–80 per cent, and 8–16 per cent respectively, at all stages studied from eight-cell to blastocyst (Ellem and Gwatkin 1968; Piko 1970).

It is known that most messenger RNA contains poly(A) residues at its 5′ ends and this trait has been used to estimate the actual proportion of newly synthesized mRNA during cleavage (Levey et $al.$ 1978). Polyadenylation of RNA transcribed from the embryonic genome occurs as early as the two-cell stage (Levey et $al.$ 1978), and there is some recent indication that this mechanism is operative from the zygote stage onwards, though the actual amount of mRNA synthesized at these early stages is still very low (Clegg and Piko 1983). The proportion of poly(A)$^+$ RNA is between 2.2 per cent and 3.5 per cent of all newly synthesized mRNA at the eight-cell to blastocyst stage.

One more feature of embryonic derived mRNA is the change in its stability. The average half life of mRNA synthesized at the morula stage was estimated at about 9.5 $\pm$ 0.3 h. The decay curve of ^{3}H-uridine labelled mRNA at the early blastocyst stage appeared to be biphasic, consisting of short-lived (less than 6 h) and long-lived (over 30–40 h) components (Kidder and Pedersen 1982). Thus, according to these data, the stability of embryonic mRNA increases by at least twofold during the morula to blastocyst stage. We shall come back to these findings in our discussion of the results of transcription inhibitor experiments (see next section).

In summary, all available experimental data are consistent with very early transcription of the embryonic genome in mice, probably already from the two-cell stage. Heterogeneous poly(A)$^+$ and poly(A)$^-$ RNA as well as low molecular weight RNAs comprise the bulk of newly synthesized RNA at this stage. There are some preliminary indications of heterogeneous RNA synthesis even from the one-cell stage, though more studies are needed to prove these findings. From the late two-cell stage and more regularly from the four-cell stage onwards all classes of RNA are transcribed, with rRNA being the major RNA product up to the blastocyst stage. The major increase in the net synthesis of RNA occurs from the eight-cell stage onwards, rising approximately 30-fold during the pre-implantation period which amounts to a fourfold increase per cell. Preliminary results suggest an active polyadenylation of newly synthesized heterogeneous high molecular weight RNA and its association with functional polyribosomes, indicating that translation of actively transcribed embryonic mRNA is occurring. A transition from short-lived mRNA synthesized at the morula stage to long-lived mRNA at the blastocyst stage has been postulated.

In spite of some specific temporal shifts, the overall pattern of different RNA classes synthesized in rabbit embryos is similar to that in mice. It has been shown by polyacrylamide gel electrophoresis that heterogeneous and low molecular weight transfer RNA are synthesized from the two-cell stage, while the synthesis of rRNA is significantly delayed compared to mouse embryos, being detected from the 16-cell stage onwards (Manes 1969). About 20 per cent of total heterogeneous RNA and about 70 per cent of the polysomal RNA are polyadenylated from the 16-cell stage (Schultz, Manes, and Hahn 1973a; Manes 1975). Hybridization of non-ribosomal and non-transfer RNA sequences with total DNA has shown that most of the heterogeneous RNA in early rabbit embryos (70 per cent) comes from unique sequences of DNA (Schultz, Manes, and Hahn 1973b).

Thus in early rabbit embryos both mRNA and tRNA are synthesized very early in development, while the synthesis of rRNA is postponed to the morula stage.

Active transcription of all classes of embryonic-derived RNAs soon after fertilization is suggestive but not conclusive evidence in favour of translation of these newly synthesized RNAs. There is some doubt as to the participation of their protein products in early development. Most evidence consistent with the participation of the embryonic genome in controlling morphogenetic processes from the earliest stages of embryonic development has been obtained in experiments with specific RNA inhibitors, outlined in the next section.

12.2.3 Inhibitors of RNA synthesis

Specific inhibitors of RNA synthesis have been used in numerous studies as valuable experimental tools to assess embryonic gene expression (Epstein 1975; Manes 1975; Sherman 1979; Magnuson and Epstein 1981; Johnson 1981). Only basic data from these studies will be given in this section.

Actinomycin D (AD) was one of the first RNA-specific inhibitors used extensively for these studies. At low concentrations it specifically inhibits the synthesis of rRNA while in higher concentrations it inhibits the synthesis of all types of RNA.

Since the early studies of Mintz (1964c) it has repeatedly been shown that AD at a dose level of 1 mmol ml^{-1} of culture medium completely inhibits the cleavage of two-cell embryos in mice (Thomson and Biggers 1966; Ellem and

Gwatkin 1968; Skalko and Morse 1969; Piko 1970; Molinaro, Siracusa, and Monesi 1972), and may even prevent the first cleavage (Golbus, Calarco, and Epstein 1973). The cleavage block was especially efficient if the drug was applied during the initial stages of the cell cycle (Molinaro *et al.* 1982), suggesting the synthesis of AD-susceptible RNA during the G_1 phase. In spite of possible side effects of AD on cell respiration and ATP synthesis (Manes 1969), these experiments showed that cleavage-stage embryos blocked with AD do not die within a 24-h period, and that AD does not block compaction if applied at or later than the eight-cell uncompacted stage (Epstein and Smith 1978).

This has been amply confirmed in experiments with α-amanitin— a highly specific inhibitor of type-II RNA polymerase, responsible for the synthesis of heterogeneous messenger RNA. At low concentrations, which selectively inhibit only type II RNA polymerase, α-amanitin was shown to block compaction and blastulation after prolonged or transient treatment at the early eight-cell stage or at the morula stage, respectively (Golbus *et al* 1973; Warner and Vestergh 1974; Epstein 1975). Direct biochemical studies of affected embryos revealed a substantial fall in the polyadelated RNA concentrations and thus indicated that both compaction and blastocyst formation are transcriptionally dependent. Two-dimensional SDS poly-acrylamide electrophoresis of ^{35}S-methionine labelled embryos previously treated with α-amanitin before 80 h post-HCG (human chorionic gonado-tropin) injection revealed the absence of certain polypeptides normally associated with blastocyst formation. On the other hand, qualitative and quantitative polypeptide patterns of mouse embryos treated with α-amanitin at more advanced stages (80 h after HCG injection) resembled the controls (Braude 1979). The critical transcriptional events concerned with blastocyst formation therefore probably occur around 80 h after HCG injec-tion and may be associated with the completion of the fifth cleavage division. It is still unknown whether mRNA synthesized during this crucial stage results in a specific protein that participates in blastocyst formation, or is necessary as a regulator for activating the translation of already existing mRNA (maternal RNA) (Braude 1979).

At a concentration of $1\mu g\ ml^{-1}$, α-amanitin does not interfere with embryonic survival, although the synthesis of poly(A)$^+$ RNA as well as the incorporation of ^{35}S-methionine into embryonic polypeptides are signifi-cantly reduced (Schindler and Sherman 1981). Not all embryonic proteins are equally susceptible to α-amanitin. Treatment of cultured mouse embryos with α-amanitin for 22 h did not affect the levels of β-glucuronidase or plas-minogen activator, while the activity of 3β-hydroxysteroid dehydrogenase was significantly reduced. These findings suggest that α-amanitin resistant proteins were probably synthesized on stable mRNA templates produced during oogenesis or soon after fertilization (Schindler and Sherman 1981).

The production of such long-lived embryonic mRNA is not consistent with recent findings (see above), indicating the synthesis of predominantly short-lived mRNA at the morula stage (Kidder and Pedersen 1982). However, some mRNA templates of maternal origin may survive up to the blastocyst stage. Forty per cent of total maternal mRNA decays during the first 24 h of development, no further loss occurs from the two-cell up to the eight-cell stage, and about 30 per cent cf all maternal mRNA is destroyed during the third day (late morula–early blastocyst) (Bachvarova and De Leon 1980). The existence of long-lived maternal mRNA could therefore be responsible for the relative resistance of protein synthesis in cleaving embryos to α-amanitin.

α-Amanitin application to the zygote stage does not appear to affect the normal sequence of morphological and molecular events that take place during development to the early two-cell stage (Braude *et al.* 1979). This is quite consistent with the present notion that the genetic information necessary for at least the first 24 h survival of mouse embryos is provided by maternal mRNA (Johnson 1981). Evidence in favour of embryonic mRNA synthesis and translation in two-cell stage mouse embryos was obtained in recent α-amanitin studies carried out on a highly synchronized population of two-cell embryos (Bolton *et al.* 1984). These revealed two bursts of α-amanitin sensitive activity (mRNA synthesis), one before (G_1-phase) and the other (G_2-phase) after DNA replication. The translation products were found in each case within 3–4 h of the bursts. One of the first embryonic mRNA translation products has been identified as an embryo-specific polypeptide of molecular weight 67 000. Prevention of cytokinesis or impairment of the second round of DNA replication does not affect the synthesis of this α-amanitin sensitive polypeptide. If applied during the second sensitive stage (G_2), α-amanitin inhibits the production of numerous mRNAs responsible for the synthesis of proteins participating in cleavage, and division to the four-cell stage may be completely blocked. However, if α-amanitin is given at 30 h after fertilization, the new proteins are already synthesized and development to the four-cell stage proceeds. Thus new mRNA templates produced at these early stages are translated within a few hours of transcription (Flach, Johnson, Braude, Taylor, and Bolton 1982).

Experiments with rabbit embryos treated with α-amanitin demonstrated that mRNA synthesis was completely inhibited in two, four, and 16-cell stage embryos, though the embryos survived for at least a couple of days. Development during this lag period may have proceeded at the expense of mRNA templates stored in the ovulated egg. Blastocyst formation in the rabbit was especially sensitive to α-amanitin treatment.

The experiments with specific RNA inhibitors suggest that early mouse and rabbit embryos possess a bulk of inherited RNA species that may support embryonic survival for at least the first 24 h in mice and somewhat longer in rabbits. At the early and especially at the late two-cell stage a large number of new species of α-amanitin sensitive polypeptides appear, and their appearance is critical for cleavage from two to four cells. Compaction and blastulation depend at least partly on transcriptional and translational activities of the embryonic genome, as both of these processes can be easily blocked by these specific RNA inhibitors applied at the proper stage of development. The critical transcriptional events concerned with compaction occur at the early eight-cell stage and with blastocyst formation, just after completion of the fifth cleavage division.

12.3 Protein synthesis

Detailed studies of protein profiles, both quantitative and qualitative, at different stages of embryonic development is the next logical step to elucidate the mechanisms of genome action in embryogenesis. Most of these studies are based on high resolution one-dimensional or two-dimensional polyacrylamide gel electrophoresis of embryonic polypeptides previously labelled with radioactive amino acids.

The total protein content of newly ovulated eggs in mice is about 28 ng and it remains almost constant throughout cleavage, thus indicating active protein turnover during pre-implantation development (Biggers and Borland 1976). Since the early autoradiographic studies initiated by Mintz (1964c) it has been repeatedly demonstrated that cleaving embryos of mice and other mammals readily incorporate labelled amino acids. This incorporation becomes especially active between the eight-cell and blastocyst stages (Weitlauf and Greenwald 1967; Monesi and Salfri 1967; Tasca and Hillman 1970). It has been shown with more refined quantitative techniques that in mice there is an eightfold increase in the total rate of protein synthesis from the eight-cell stage onwards (Brinster 1967). This corresponds to a synthetic rate of 1.34 ng protein per day for 8–16 cell embryos and 4.75 ng protein per day for the blastocyst (Epstein and Smith 1973; Epstein 1975). Thus, balanced synthesis and degradation of proteins occur at all stages of embryonic development in mammals.

The same correlation has been found in rabbit embryos. Incorporation of labelled precursors remained constant throughout cleavage and increased 8–10 times during blastulation (Karp, Manes, and Huahn 1974), while the relative rate of protein synthesis per cell remained constant throughout the whole of pre-implantation development (Biggers and Borland 1976).

Quantitative studies of protein synthesis have recently been extended from overall protein content to the study of particular protein groups or even individual proteins. In one extensive study, a double-labelling technique was applied to compare the relative synthesis of 95 proteins in mature ova and in fertilized eggs. The synthesis of 78 proteins remained constant, it somewhat increased for 6 proteins and decreased for the remaining 11 proteins

(Chen, Brinster, and Merz 1980). Quantitative studies of individual proteins have been predominantly concentrated on actin, tubulin, and lactate dehydrogenase (see Magnuson and Epstein 1981 for review). The rate of synthesis of LDH and its numerous isozymes has been thoroughly studied both before fertilization and during pre-implantation development in mice (Mangia, Erickson, and Epstein 1976; Kolombet 1977; Lubimova-Kerkis, Kuzin, and Korochkin 1981) and in rats (Poznakhirkina, Serov, and Korochkin 1975). Each of these proteins has revealed a quite specific developmental profile. The rate of synthesis of many other individual proteins as well as their participation in developmental processes during early embryogenesis remains unknown.

In contrast to the rather monotonous quantitative turnover of overall proteins, qualitative changes in protein synthesis in pre-implantation mouse embryos are very marked.

The overall number of proteins synthesized by pre-implantation embryos in mice, rats, and rabbits approximates to 100 or so (Van Blerkom and Brockway 1975a, b; Van Blerkom and Manes 1974; Van Blerkom 1975, 1981). Some are synthesized only during a short period of development (e.g. unfertilized eggs and first cleavage divisions), others appear at fertilization and their activity gradually increases. Synthesis of some proteins does not begin until the eight-cell stage (Biggers and Stern 1973; Biggers and Borland 1976; Van Blerkom 1981).

High-resolution two-dimensional gel electrophoresis demonstrated many similarities between protein patterns in ovulated eggs and those in one-cell mouse embryos, though a few characteristic differences can already be seen. A singular change between fertilized and non-fertilized eggs involves six polypeptides designated 'fertilization proteins' (FP-1 to FP-6), the synthesis of which proceeds for approximately 36 h and terminates by the eight-cell stage (Cascio and Wasserman 1982). The changes in protein pattern progressively increase, though no major changes are evident until two to two and a half days into embryonic development. However, between the two and eight-cell stages a very different pattern of proteins is synthesized. The qualitative reprogramming of protein pattern is especially pronounced by the eight-cell stage, whereas no major changes in protein synthesis could be detected between the eight-cell and blastocyst stages (Epstein and Smith 1974; Epstein 1975). The qualitative protein pattern seems to be quite normal in haploid parthenogenetic mouse embryos (Van Blerkom and Brockway 1975a, b) as well as in microsurgically produced androgenetic and gynogenetic mouse embryos (Petzoldt, Illmensee, Burki, Hoppe, and Illmensee 1981; see also Chapter 1).

Progressive changes in the qualitative protein pattern in developing rabbit embryos are much more gradual and proceed up to blastocyst stage, when the spectrum of proteins synthesized becomes somewhat stabilized (Van

Blerkom and Manes 1974). The protein pattern turns out to be identical or very similar in rabbit embryos grown *in vivo* or *in vitro* (Van Blerkom and Manes 1974).

Immunoradiochemical studies have also been carried out to detect the appearance of individual proteins (some have already been mentioned in previous sections). These are tubulin, actin, LDH (see above), specific trophectodermal proteins, and cell membrane antigens that probably play a key role in cell allocation at compartmentalization and thus are very important for the primary embryonic differentiation into trophectoderm and inner cell mass (Van Blerkom 1981; Magnuson and Epstein 1981).

Of special interest are the earliest embryo-derived proteins recently discovered at the two-cell stage. Using both one- and two-dimensional electrophoresis, the appearance of α-amanitin-sensitive new polypeptides of molecular weight about 67–70 000 dalton has been recently shown at the early two-cell stage (Flach, Johnson, Braude, Taylor, and Bolton 1982). The synthesis of these proteins has been confirmed in independent studies by two-dimensional gel electrophoresis analysis of ^{35}S-methionine-labelled proteins (Bensaude, Babinet, Morange, and Jacob 1983). Two of these polypeptides were identified as heat shock proteins—HSP-68 and HSP-70—synthesized in F9 cells after exposure to sodium arsenite. None of these proteins is contained in mature eggs and they do not appear after transient treatment of the fertilized oocyte with α-amanitin. The synthesis of these heat shock proteins was suggested to be related to the onset of embryonic genome expression at the early two-cell stage (Bensaude *et al*. 1983). The same authors have pointed out that all experimental conditions that induce heat shock protein synthesis may also promote parthenogenetic activation of mouse oocytes (hyperthermia, ethanol, etc.—see Chapter 1).

The exact origin of the overwhelming majority of polypeptides synthesized by the cleaving eggs of mice remains unknown, so qualitative and quantitative studies of embryonic proteins *per se* do not make a great contribution to our understanding of genome function. There are, however, at least three main approaches that may help to discriminate between the proteins synthesized on maternal (ovum-stored) mRNA templates and on the newly synthesized (embryonic genome derived) mRNAs. These are (a) detection of paternal allele products, (b) protein analysis in α-amanitin treated embryos, and (c) comparative analysis of microsurgically operated eggs.

We shall deal with paternal allele expression in the next section and we have already partly discussed the susceptibility of embryonic protein synthesis to α-amanitin. It should be remembered, however, that at no stage of pre-implantation development is α-amanitin capable of preventing the synthesis of all proteins. Protein synthesis shows especially high resistance to α-amanitin treatment during the first 24–27 h after fertilization. Thereafter, the number of α-amanitin-sensitive proteins gradually increases, especially between the two- and eight-cell stages, and they reach about 50 per cent of the overall protein content on the fourth day of development (Schindler and Sherman 1981).

Thus the actual number of proteins synthesized from embryo-derived mRNA, as inferred from α-amanitin studies, is still very low at the two-cell stage, dramatically increases between the two- and eight-cell stages, and gradually reaches a plateau at the blastocyst stage.

These findings are in good agreement with biochemical studies of cytoplasts produced by the microsurgical withdrawal of both pronuclei from recently fertilized eggs (see Petzoldt, Hoppe, and Illmensee 1980). Two-dimensional polyacrylamide gel electrophoresis of fertilized and unfertilized C57BL/6J eggs after operation revealed that in the cytoplasts of both, protein synthesis continued for three days. After two days in culture, the pattern of 250–300 polypeptides typical for control eggs was reduced to less than 50 spots in enucleated fertilized and unfertilized cytoplasts. Nonetheless, some new proteins specific for normal two-cell embryos appeared at this stage in cytoplasts from fertilized but not from unfertilized eggs. The microsurgical approach supports the notion that maternal mRNA stored during oogenesis is utilized for at least two days after fertilization.

Thus quantitative and qualitative analysis of embryonic proteins indicates balanced synthesis and degradation of proteins during pre-implantation development in mammals. The rate of overall protein synthesis per cell remains constant throughout pre-implantation development, while each individual protein studied so far reveals a unique developmental profile. Some proteins are synthesized only during a rather short period of development and these are often identical to the proteins of the mature eggs, while other proteins appear at successive stages of development, beginning at the two-cell stage. The earliest embryo-derived proteins were identified at the early two-cell stage, and two were identified as known heat-shock proteins. Drastic qualitative changes in protein pattern occur between the two and eight-cell stages in mice, and seem to be more gradual in rabbit embryos. Studies in α-amanitin treated embryos and in cytoplasts after withdrawal of both pronuclei prove that the major changes in protein pattern between the two and eight-cell stages reflect a cessation of translation of the bulk of maternal mRNAs and the commencement of translation of embryo-derived mRNA. Quantitative and qualitative studies of several individual proteins in mouse and rabbit embryos are in progress. Of special embryological interest are the proteins synthesized during the two-cell stage, as well as the proteins participating in cell membranes and providing efficient cell interactions.

12.4 Expression of paternal alleles in early development

One of the most decisive lines of evidence in favour of very early expression of the embryonic genome in mammals comes from paternal gene products in developing embryos. These investigations have recently been reviewed by Magnuson and Epstein (1981) and by Johnson (1981).

The indispensable prerequisite for such studies is the existence of genetically determined protein variants (isoenzymes) that may be encoded by autosomal or gonosomal genes that can be recognized by their unique electrophoretic, physical, and immunological properties (Wolf and Engel 1972; Brinster 1973; Epstein 1975; Korochkin 1981). Since spermatozoa do not appear to carry paternally derived proteins or mRNA into the egg, the appearance of paternal gene products may be taken as a proof of the expression of the embryonic genome. A number of experimental studies have been undertaken to look for paternally derived proteins in the pre-implantation embryos of mice (see Magnuson and Epstein 1981; Johnson 1981 for reviews) and rats (Serov and Manchenko 1974; Khlebodarova, Serov, and Korochkin 1976). The results of such studies in mice are summarized in Table 38. Two-cell embryos are the earliest stage at which the expression of paternal genes can be detected at present. Two lines of evidence substantiate this conclusion.

The first is the expression of β_2 microglobulin, a small polypeptide that is bound to the extracellular portion of *H-2* gene products and may also be coupled to minor non-H-2 alloantigens and perhaps to HY-antigen as well (Magnuson and Epstein 1981). The existence of an electrophoretic variant of

Table 38. Expression of paternal alleles in early mouse embryos

Embryonic stage	Enzymes or protein markers	Chromosome No.	Reference
One-cell	—	—	—
Two-cell	β_2-microglobulin	2	Sawicki *et al.* 1981
	Hypoxanthine phosphoribosyl transferase (HPRT)	X	Kratzer 1983
Four-cell	β-glucuronidase	7	Wudl and Chapman 1976
Six- to eight-cell	Non-H-2 alloantigens	2	Muggleton-Harris and Johnson 1976
Eight-cell	Glucosephosphate isomerase (GPI)	8	Brinster 1973
	HY-antigen	Y	Epstein, Smith, and Travis 1980 Shelton and Goldberg 1984
	H-2-antigens	17	Cozad and Warner 1982

β_2 microglobulin in inbred strains of mice established the genetic basis for these studies (Sawicki, Magnuson, and Epstein 1981). With radioactive labelling and immunofluorescence methods it has been unequivocally established that the synthesis of paternally derived β_2 microglobulin first becomes detectable at the two-cell stage. This stage of paternal gene expression is the earliest one found so far for any specific locus.

The second, less direct indication that the paternal genome is functional at the two-cell stage (see Table 38) comes from recent studies of the X-linked *HPRT* gene activity profiles in embryos from In(X)/X mice (Kratzer 1983). Beginning from the two-cell stage, the distribution of *HPRT* activity was bimodal, which is evidence for embryonic synthesis of this enzyme. Whether the activity derived from the maternal or paternal X chromosome or both remains obscure. More direct observations suggest that both paternal and maternal X chromosomes are active by the eight-cell stage (Burgoyne and Biggers 1976; McLaren 1979).

The synthesis of paternally derived β-glucuronidase is detected from the four-cell stage onwards (Wudl and Chapman 1976). Several other paternal gene products (non-H-2 alloantigens, GPI-1, HY-antigen) are expressed at the eight-cell stage (see Table 38 for references; also Magnuson and Epstein 1981; Johnson 1981 for reviews).

Much controversy exists as to the exact time of expression of the major (H-2) histocompatibility antigens (see Johnson and Calarco 1980 for review). Most investigators (e.g. Magnuson and Epstein 1981) were unable to detect expression of H-2 antigens on early mouse embryos. On the other hand, there is some evidence in favour of both paternal and maternal expression of *H-2* genes from the eight-cell stage onwards (Searle *et al.* 1976; Krco and Goldberg 1977), though at a very low level. In the most recent complement-dependent cytotoxicity assay in mouse embryos of different haplotypes, it has been demonstrated that eight-cell embryos of b, k, and d haplotype express H-2 antigens, but no H-2 antigens were detectable on eight-cell embryos of the a-haplotype (Cozad and Warner 1981). Thus the use of mice of different haplotypes in earlier work might at least partly explain the conflicting data on H-2 antigen expression.

More studies are needed on the expression of *H-2* genes at early stages of embryonic development in mammals, especially in view of recent experimental studies in mice, postulating the existence of a major gene(s) located inside the *H-2* complex, which influences early development in mice (Goldbard *et al.* 1982a,b). Analysis of cell number in early mouse embryos of different *H-2* haplotype showed that embryos of *H-2^k* haplotype had fewer cells per embryo than those of *H-2^b* haplotype. Further genetic analysis of reciprocal congenic pairs demonstrated unequivocally that developmental rates correlate with *H-2* haplotype. A slow rate is linked with *H-2^k* and a fast one with *H-2^b*. Differences in cell number of cleaving eggs from mice of different strains has been known for a long time (Gates, Doyle, and Noyes 1961; Dickson 1967; Bowman and McLaren 1970a,b; Graham 1973a,b), and effects of the paternal genome on cleavage rate have been suggested (Whitten and Dagg 1962). Embryological studies of this phenomenon revealed that it was a difference in the onset of the first cleavage division rather than in the cleavage rate *per se* that is responsible for 'slow' and 'fast' developing strains (McLaren and Bowman 1973). The recent studies of congenic strains support the notion that gene(s) in *H-2* complex (termed *Ped* genes—pre-implantation embryonic development) may influence the timing of pre-implantation development both by delaying the onset of the first division and also affecting the cleavage rate (but see Section 13.1.3.2). What *Ped* genes actually are and what their gene products are is open to speculation. Are they discrete genes of special function, or are they indeed *H-2* genes with pleiotropic functions? The latter assumption seems unlikely because of the absence of detectable H-2 antigens during early cleavage (see above).

Studies on *Ped* genes should still be considered as preliminary as they have not yet been confirmed in other laboratories. More studies are needed before the existence of these very interesting genes can be confirmed or refuted.

Evidence in favour of very early expression of the paternal genome in mice also comes from biochemical studies of androgenetic parthenogenones produced by removal of the female pronucleus and subsequent doubling of the male genome by cytochalasin B treatment (see Chapter 1). Protein synthesis in androgenetic and gynogenetic parthenogenones has been examined by two-dimensional polyacrylamide gel electrophoresis. Both types of uniparental embryo synthesized a similar set of proteins, almost identical to that in non-operated diploid controls at different stages of pre-implantation development (Petzoldt *et al.* 1981). These results suggest that the diploidized paternal pronucleus is indeed quite active during pre-implantation development and can support developmental processes up to the blastocyst stage.

Paternal gene expression in the embryos of other mammalian species has been studied only in laboratory rats. Two alleles, Pgd^b and Pgd^a, are known to control the synthesis of two isoenzyme forms of 6-phosphogluconate dehydrogenase in wild grey rats and in laboratory Wistar strain rats, respectively (Serov and Manchenko 1974). Immunochemical analysis of early embryos from crosses of wild Pgd^b/Pgd^b rats to Pgd^a/Pgd^a rats of laboratory strains have revealed the paternal form of *Pgd* from the eight-cell stage onwards (Khlebodarova *et al.* 1976).

Thus all available data indicate very early expression of paternal genes during embryonic development in mice. There is evidence in favour of the appearance of paternally controlled enzymes from the two-cell stage onwards, and the expression of paternal genes at four- to eight-cell stages has already been shown for a number of enzymes as well as for some cell surface markers (HY antigen, non-H-2 alloantigens). The existence of special *Ped* genes responsible for the timing of the first cleavage division as well as for cleavage rate has been suggested but not yet definitely proved.

12.5 Expression of early lethal mutations

The synthesis of embryo-derived mRNAs (see Section 12.2), as well as their translation in the form of different enzymes or cell-membrane polypeptides (see Sections 11.3, 11.4, and 11.5), still do not provide much information either on the exact embryonic stage when these genetic products are needed or on their exact functions. Much of this evidence is provided by the use of specific inhibitors (see Section 12.3). Unfortunately, even the most specific of them, α-amanitin, may possess some secondary or 'non-specific' effects, hindering interpretation of the experimental results (Braude 1979). Lethal or semilethal mutations affecting pre-implantation development are more convenient and reliable tools for such studies (see McLaren 1974, 1976a,b for reviews), but even this traditional and fruitful approach to the developmental biology of mammals is not free of limitations. First, these 'early' mutations are very few in number. Second, they are difficult to identify before advanced pathological changes occur.

Of more than 500 mutant genes known in the laboratory mouse (Hadorn 1961; Kalter 1968; Gluecksohn-Waelsch 1953, 1964; Green 1966, 1975), only a few are expressed during embryogenesis (Konyukhov 1969, 1975, 1980) and still fewer before implantation (McLaren 1976b).

The second limitation to these studies is even more important. A gap of unknown length exists between the action of a mutant gene (or genes) and its morphological manifestations. So mutant embryos with gross abnormalities already visible are used for detailed morphological or biochemical studies. The absence of proper genetic and biochemical markers causes serious problems for a step-by-step analysis of the affected embryos.

We are still almost completely ignorant of the primary biochemical effects induced by any of the early lethal mutations, as well as of the exact succession of events underlying the early death of affected embryos. Thus data on developmental effects of early lethal mutations have their own serious limitations and provide us with only a pilot look at gene expression in mammalian embryogenesis.

Keeping in mind these limitations, we will describe the developmental abnormalities caused by each of six early lethal mutations in the mouse, all expressed in homozygous condition before implantation and all lethal at different stages of pre-implantation development.

12.5.1 T/t complex

The *T/t* complex (or *T-t* locus) encompasses approximately 15cM of chromosome 17 in the mouse, beginning at the T mutation and extending distally to the *H-2* complex (Bennett 1975, 1981; Lyon *et al.* 1979; Artzt, Shin, and Bennett 1982). Crossover suppression within this 15cM region has led to the speculation that *T*-locus mutants maintain this substantial chromosome region as a 'supergene' complex (Snell 1968). The existence of a similar gene complex in close proximity to the major histocompatibility complex has recently been suggested in rats and also suspected in humans (Gill, Siew, and Kwe 1983). Genetic, immunogenetic, and biochemical data have prompted the view that the *T*-locus region might be the evolutionary precursor of *H-2* (Bennett, Goldberg, Dunn, and Boyse 1972). This interesting speculation is at present under intensive experimental study by direct DNA analysis of corresponding regions using specific DNA probes (Silver 1982; Silver and White 1982; Shin, Stavnezer, Artzt, and Bennett 1982).

Several dominant (*T*) and recessive (*t*) mutations have been identified in the *T/t* complex. Some of these mutations affect embryonic development, sperm production, tail elongation, and genetic recombination within this chromosome region (Bennett 1975, 1981; Glueksohn-Waelsch 1953; Glueksohn-Waelsch and Erickson 1970).

A very rare recombination event has been detected near the *Tcp* locus (*T*-complex protein-1), dividing the complex *T/t* into proximal and distal regions (Artzt *et al.* 1982). The proximal part (about 4 cM) contains the *t* tail-interaction factor (t^T), responsible for the tailless phenotype of T/t^x mice. The distal region extends about 10cM, including the recessive *tf* mutation, and contains *t*-lethal (t^l) factors, factors affecting male fertility, and the segregation distortion factor. Eight different complementation groups of t^l have been identified in the distal region, each of which blocks embryonic development at a specific stage (Bennett 1975). Detailed morphological studies have been carried out on embryos homozygous for each of the well-known lethal *t*-haplotypes, and many comprehensive summaries of these studies have been published (Glueksohn-Waelsch and Erickson 1970; Bennett 1975, 1981; Sherman and Wudl 1977; Magnuson and Epstein 1981). Only a few *T/t* complex mutations are lethal before implantation, including the t^{12} recessive lethal and the deletion mutations T^{hp} and T^{Orl}.

12.5.1.1 t^{12} and t^{w32} alleles. The t^{12} allele was one of the first recessive lethals studied in the *T/t* complex. Homozygotes develop on normal schedule until the morula stage (Smith 1956) but fail to form blastocysts and die at about the 30-cell stage (Mintz 1964b). The t^{12} allele behaves as a true cell lethal as no viable $t^{12}t^{12}$ cells could be rescued in aggregation chimaeras between $t^{12}t^{12}$ and + + embryos (Mintz 1964a). At the light microscope level t^{12} homozygotes are very similar to embryos homozygous for the t^{w32} allele—a second recessive allele assigned to the t^{12} complementation group. More detailed analysis of both mutations, however, has shown that t^{w32} homozygotes usually die somewhat earlier than t^{12} ones and may also be distinguished by some minor ultrastructural differences (Hillman 1975).

Three basic routes leading to the lethal manifestation of t^{12} (and most probably of t^{w32}) could be inferred from the various experimental studies (Magnuson and Epstein 1981). These are (1) defects of RNA and protein synthesis, (2) energy failure, and (3) impairments of membrane antigens.

1. Earlier morphological studies (both histochemical and autoradiographic) suggested that the lethality of $t^{12}t^{12}$ embryos was related to abnormal nucleolar development and reduced RNA synthesis at the late morula stage (Smith 1956; Mintz 1964c). This suggestion however was refuted by later studies, which clearly demonstrated that $t^{12}t^{12}$ embryos do not differ from their littermates either in total RNA content or in nucleolar ultrastructure prior to advanced degenerative changes (Calarco and Brown 1968). Moreover, autoradiographic studies of rRNA and other RNA species in $t^{12}t^{12}$ embryos indicated that RNA synthesis is normal until the time of death (Hillman et al. 1970; Hillman 1972; Hillman and Tasca 1973; Erickson et al. 1974a; Epstein 1975). No specific abnormalities of protein synthesis have been reported for t^{12} homozygotes (Erickson, Siekevitz, Jacobs, and Gluecksohn-Waelsch 1974c; Hillman 1975). The activity of a single enzyme measured in individual embryos from $t^{12}/+ \times t^{12}/+$ crosses remained within the normal range up to the morula stage and dropped abruptly at death (Wudl and Sherman 1976). Thus disturbances of nucleolar functions and RNA synthesis can be considered as secondary but not as primary effects of the t^{12} allele.

2. It has been also shown that $t^{12}t^{12}$ embryos can be distinguished at the ultrastructural level as early as the two-cell stage by the presence of nuclear lipid droplets, nuclear fibrillo-granular bodies, and excessive cytoplasmic lipids, and by the appearance of binucleate cells at later cleavage stages (Hillman 1975). Moreover, some direct biochemical measurements indicated an abnormally high level of ATP as well as an unusually low ATP:ADP ratio in cleaving t^{12} or t^{w32} homozygotes (Ginsberg and Hillman 1973). All these pathological changes and biochemical findings could have as a common cause some intrinsic defect of energy metabolism in both types of t^{12} homozygotes.

3. One further suggestion that is not completely incompatible with mechanisms 1 or 2 correlates pathological changes of t^{12} homozygotes with membrane abnormalities associated with this mutation. Numerous experimental data suggest that t^{12} and t^{w32} homozygotes fail to compact properly, the intercellular junctions between their cells at the morula stage are less numerous and much less stable than in normal embryos (Sherman and Wudl 1977; Bennett 1981). Some attempts have been made to relate the expression

of *t*-alleles with the specific cell surface protein p63/6,9a (Silver and White 1982) or with a mutated form of the normal F-9 antigen (Kemler, Babinet, Condamine, Gachelin, Guenet, and Jacob 1976). The latter is thought to play a major role in compaction and its alteration or suppression may mimic the effects of t^{12} or t^{w32} alleles. No conclusive immunological data in favour of this suggestion have been obtained so far, but two-dimensional gel electrophoresis of $t^{12}t^{12}$ embryos has failed to reveal four proteins normally synthesized at the morula stage. The appearance of these four proteins was also completely suppressed by tunicamycin, a specific inhibitor of F-9 antigen (Iwakawa, Nozaki, and Matsushiro 1982). Embryonic death of t^{12} mutants could be caused by the lack of these four stage-specific proteins necessary for the blastocyst stage. More recent biochemical studies of *T-t* antigens have provided evidence that purified t^{12} antigen is an immuno-precipitating glycoprotein (M.W. 87 000) containing galactose (Artzt and Cheng 1982). Furthur studies of this particular antigen might make a great contribution to our understanding of the primary effect of $t^{12}(t^{w32})$ mutations in mice.

One of the most serious complications of all these studies, especially the biochemical ones, comes from the need to work with a mixed population of embryos ($t^{12}t^{12}$; $t^{12}+$; $++$). This disadvantage might be successfully overcome by the use of an appropriate chromosome marker for chromosome 17, such as Rb(16.17)7 Bnr or Rb(8.17)1 Iem (Baranov 1979; Magnuson and Epstein 1981). After crosses of appropriate heterozygotes (Rb7 $+/+t^{12}$ or Rb1 $+/+t^{12}$), the embryos may be divided at the two-blastomere stage (Epstein *et al.* 1978) and each part may be grown separately ('twin embryo technique'). At a later stage one of the twin embryos, cultured *in vitro*, is used for karyotyping while its counterpart remains intact and ready for morphological, biochemical, or other investigations. This interesting experimental approach will help to gain more precise information on the earliest manifestation of the t^{12} lethal allele in mouse embryos.

12.5.1.2 T^{Orl} and T^{hp} mutations. Besides the famous *T*-mutation (Brachyury), the first dominant allele to identify the region of the *T-t* complex, four other dominant mutations are now known in this region. These are T^{C} (curtailed), T^{hp} (hair-pin), T^{Orl} (Orleans), and T^{Or} (Oak Ridge). Two of these are spontaneous in origin (T^{hp} and T^{Orl}) and the other two (T^{C} and T^{Or}) are radiation-induced (Bennett 1975). Three (T^{Or}, T^{Orl}, and T^{hp}) are genetically proved to be deletions (Erickson, Lewis, and Slusser 1978). A morphological study of mouse embryos homozygous for each of these three deletions has recently been completed (Babiarz 1983). The results show that each deletion produces a unique homozygous embryonic lethal phenotype, which correlated well with the length of each deletion. The T^{hp} deletion is the largest and covers a significant part of the *T-t* complex (from the *T*-locus to the Low mutation—4–5 cM in length). In the homozygous condition it inhibits embryonic development beyond the morula stage. Thus nullisomy for about 4–5 cM of the T-t complex region has a lethal phenotype very similar to the recessive lethal haplotype t^{12} (see above). Intermediate in length, the T^{Orl} deletion (from the T-locus to the Tsp-1 gene—a distance of about 3–4 cM) is compatible with successful development to an expanded blastocyst by 4.5 days, but soon thereafter all homozygotes display serious abnormalities of embryonic ectoderm and are eliminated. (The lethal phenotype of these nullisomics is very similar to that of the recessive lethal haplotype t^{o}.) T^{Or} deletion, the smallest of the three, extends from the T-locus to the qk-mutation, a distance of about 2–3 cM. Homozygotes develop normally up to the egg cylinder stage (5.5 days of development), but it is obviously detrimental thereafter, causing abnormal development of the egg cylinder which looks very stunted by the tenth day and contains no mesodermal derivatives. The lethal phenotype is similar to the recessive lethal t^{w5}.

Embryological studies of T-deletions thus allow one to map the regions of chromosome 17 critical for early development. The region including Tcp-1 and close to Low must be responsible for normal development up to the morula stage, the region between Tcp-1 and qk (see Fig. 36) up to the blastocyst stage, and the region between T and qk up to the egg-cylinder stage (Babiarz 1983). The relation between T-deletions and the respective recessive lethals (t^{x}) remains obscure. Unexpectedly, T^{hp}/t^{12}, T^{Orl}/t^{o}, and T^{Or}/t^{w5} embryos were viable, proving that the t-recessive alleles are not themselves deletions of the T-t region. Each of the dominant deletions and the corresponding genetic defects must do something to compensate for the deficiencies of its abnormal counterpart and thus permit normal development. These studies may therefore have serious implications for understanding the nature of recessive t-alleles as well as the mechanisms of their action in early mouse embryos.

12.5.2   Yellow (A^{y}/A^{y})

This allele at the agouti locus of chromosome 2 was the first recessive lethal described in mice (see McLaren 1976a for review). Heterozygous A^{y}/a mice show numerous minor abnormalities including obesity, susceptibility to certain cancers, mild glucose intolerance, hyperinsulinaemia, reproductive insufficiency in females, etc. According to DNA restriction analysis, A^{y} mutation has been suspected of resulting from an ectopic virus integration inside the agouti locus (Copeland, Jenkins, and Lee 1983).

In the homozygous state the A^{y} allele is lethal during early development, usually at the time of implantation (Robertson 1942; Eaton and Green 1963). The presumptive $A^{y}A^{y}$ embryos, which should comprise about 25 per cent of all littermates in crosses $A^{y}/a \times A^{y}/a$, are able to form blastocysts but fail to implant, and die on the fifth day of development. A primary pathological effect of $A^y A^y$ alleles on trophoblast giant cells has been suspected (Eaton and Green 1963). More detailed analysis of presumptive A^y homozygotes developing *in vitro* revealed conspicuous abnormalities in 24 per cent of embryos from $A^y/a \times A^y/a$ matings at the morula and blastocyst stage. The abnormalities consisted primarily of arrest of blastomere survival and exclusion of affected blastomeres from the embryo. At somewhat later stages they included failure of hatching and inner cell mass development (Pedersen 1974; Johnson and Granholm 1978; Granholm and Johnson 1978). Mitotic arrest or extensive disorganization of a single blastomere with its subsequent fragmentation and disintegration has also been detected histologically and by means of time-lapse cinematography in four- to eight-cell $A^y A^y$ embryos (Pedersen and Spindle 1978). Developmental arrest during early cleavage provides an opportunity for identification of $A^y A^y$ embryos from the four-cell stage onwards (Granholm and Johnson 1978). However, no specific ultrastructural abnormalities (Calarco and Pedersen 1976) or changes in DNA synthesis (Pedersen and Spindle 1976) have been found in $A^y A^y$ blastomeres during cleavage. The phenocritical period of the A^y mutation seems to be very extended, starting soon after the second cleavage division and proceeding up to the blastocyst stage.

The time of A^y allele expression in homozygotes varies considerably with genetic background. No evidence of pathological changes prior to implantation has been found in A^y homozygotes on AG/Can background, but in C57BL/6J $A^y A^y$ embryos impairment of pre-implantation development becomes evident both *in vivo* and *in vitro* (Pedersen 1974; Pedersen and Spindle 1976; Gardner 1976).

An attempt to pinpoint the primary cell type affected by the A^y allele has recently been undertaken by making injection chimaeras between putative $A^y A^y$ and a/a embryos (Papaioannou and Gardner 1979). When a presumptive $A^y A^y$ inner cell mass was injected into normal embryos no decrease in implantation or frequency was noticed at 10.5 days. However, when a normal ICM was injected into a presumptive $A^y A^y$ blastocyst, no improvement of embryonic survival was evident. The A^y allele is therefore probably not a cell lethal up to 10.5 days, and the trophectodermal cells are the most plausible target for the primary action of this mutation.

In summary, the A^y allele in the homozygous condition reveals at least two different phenocritical stages. The first coincides in time with the second cleavage, while the second corresponds to blastocyst expansion and hatching. Thus the A^y mutation probably affects some crucial cell functions important for normal cleavage but the exact mechanism of its action is still unknown. Expression of A^y depends on genetic background. More studies are needed on known $A^y A^y$ embryos, not only on presumptive ones, to find out the primary effect of A^y in mouse embryos.

12.5.3 Oligosyndactylism (Os)

In adult heterozygotes the Os mutation, mapped to chromosome 8, produces syndactyly of all four feet. About 25 per cent of late blastocysts (64-cell stage) from Os/+ × Os/+ mating showed abnormal cells with fragmented pyknotic chromatin, pale cytoplasm, and absent nucleoli (Van Valen 1966). By the following day the embryos were degenerating. Histological investigations of these presumed Os/Os embryos confirmed the previous report and showed that progressively more cells in these specimens are arrested in mitosis. In air-dried preparations this looks similar to the typical mitosis induced by colcemid treatment (Paterson 1979). Os does not seem to affect trophoblast cells undergoing endoreduplication, but predominantly affects mitotically dividing cells of the inner cell mass. The nature of this abnormality suggests that in the homozygous condition Os may affect the synthesis or structure of tubulin or some mitotic spindle-related protein. (See Section 13.1.4.)

12.5.4 Ovum mutant (om)

A mutation has been detected in mice of the DDK strain which seems primarily to affect some unknown cytoplasmic component of the ovum (Wakasugi, Tomita, and Kondo 1967). Females of the DDK strain are fully fertile in intrastrain matings but showed reduced litter size if crossed to males of other strains. Lowered fertility of outcrossed DDK females was found to be due to the death of F_1 embryos during trophoblast differentiation at 3–4 days post coitum (Wakasugi 1973). The detrimental effect of the om mutation could not be overcome by transplantation of ovaries from DDK females to females of other strains (Wakasugi 1974). F_1 embryos from the semi-sterile cross (DDK females to BS males) showed a very low mean cell number and a reduced mitotic index compared with DDK morulae (Wakasugi and Morita 1977). It has been suggested that the abnormalities and arrest of development in om/om embryos are due to some incompatibility between DDK cytoplasm and alien spermatozoa, resulting in a detrimental effect on development during cleavage. It has been postulated also that the normal allele of om codes for a factor in the egg that interacts specifically with or maybe activates some gene of sperm origin, to synthesize some substance necessary for the formation of trophectoderm. More experimental studies are needed to confirm the proposed mechanism.

12.5.5 Albino deletion (c^{25H})

A series of radiation-induced mutations at the albino locus (chromosome 7) in the mouse have been found to be homozygous lethals or semi-lethals associated with a number of biochemical, morphological, and ultrastructural defects at various stages of development (see Glueckoosohn-Waelsch 1979 for review). The longest of these deletions, c^{25H}, has been identified both genetically and cytologically, with the size of the deleted segment approaching 7 per cent of the total length of chromosome 7 (Miller and Miller 1975). In adult heterozygotes c^{25H} is associated with ultrastructural abnormalities of liver cells and changes in serum proteins (Erickson *et al.* 1974c; Gluecksohn-Waelsch 1979). The embryos homozygous for c^{25H} were found to be developmentally arrested at the two- to six-cell cleavage stages but nonetheless remained alive until 3.5 days of gestation. Their lethal period thus lasted for almost two days after cessation of cleavage (Lewis 1978). At the light microscope level c^{25H} homozygotes revealed the presence of binucleate cells and disturbed mitosis (Lewis 1978). Electron microscope studies identified highly contorted and convoluted nuclei, whereas other subcellular organelles remain normal (Nadijcka, Hillman, and Gluecksohn-Waelsch 1979). All mutant embryos cease development before they can be distinguished ultrastructurally and thus the primary lesions caused by this mutation remain completely unknown.

The c^{25H} deletion induces developmental arrest of all homozygotes prior to the second or during the third cleavage division and thus may be considered as the earliest acting lethal mutation reported in the mouse so far.

12.5.6 Tail short (Ts)

According to Paterson (1980), the *Ts* mutation in homozygous condition substantially reduces cleavage rate though it does not interfere with blastocyst formation. Most *TsTs* embryos die as small abnormal-looking blastocysts with a mean cell number below 30. The abnormal blastocysts fail to induce a proper decidual reaction and give poor growth if cultured *in vitro*.

The studies of early lethal mutations in mice demonstrate that most of these mutations in the homozygous condition do not allow embryonic survival beyond even the early blastocyst stage. Only some (A^y and t^{12}) become evident during cleavage and only one mutation is known at present (c^{25H}) that induces developmental arrest between the two- and six-cell stages. Analysis of these early lethal mutations is quite consistent with very early activation of the embryonic genome in mammals.

12.6 Conclusions

The experimental evidence concerning the activity of the embryonic genome during early development in mice may be summarized in the following general conclusions (see also Johnson 1981; Magnuson and Epstein 1981).

1. Development of mouse embryos during the first 24–27 h after fertilization is entirely dependent on inherited maternal mRNAs and stored egg proteins.

This is supported by (a) the relative resistance of mouse zygotes to the action of specific inhibitors of RNA synthesis, (b) the absence of reliable evidence for RNA or protein synthesis at this early stage, and (c) the absence of any lethal mutation able to block zygote development to the two-cell stage.

2. Most but probably not all maternal mRNA templates cease to be active after the two-cell stage.

This is supported by the drastic qualitative changes in protein pattern between the two-cell and eight-cell stages, and by the fact that if these are blocked by α-amanitin or by previous enucleation, synthesis of proteins in maternal RNA templates declines rapidly with a concomitant sharp fall in total mRNA content.

3. The embryonic genome is already switched on at the early two-cell stage and at least some of the newly synthesized embryonic mRNAs are translated within a few hours of transcription.

This conclusion may be inferred from (a) the appearance of G_1 and G_2 phases in the cell cycle of two-cell embryos, (b) the active transcription at this stage of at least some classes of RNA, including poly(A)$^+$ heterogeneous high molecular weight RNA, (c) the appearance of the first α-amanitin-sensitive polypeptides (heat shock proteins), and (d) the first clear-cut evidence of paternal gene expression.

4. From the late two-cell stage (27–30 h after fertilization) up to the eight-cell stage a good many embryonic genes are already switched on and most of their transcripts are readily translated, resulting in proteins essential for immediate morphogenetic processes.

This is substantiated by (a) the synthesis of embryo-derived RNA products of different types, (b) drastic qualitative changes in protein pattern, (c) high sensitivity of at least some newly synthesized proteins to α-amanitin treatment, (d) expression of paternal alleles (β-glucuronidase, non-H2 allo-antigens, glucosephosphate isomerase, HY-antigen, H-2 antigen), and (e) expression of some early embryonic lethals (c^{25H}, A^y, and t^{12}).

13

Functional activity of chromosomes and control mechanisms of early embryonic development

The data presented in the previous chapter leave no doubt that paternal gene products, both transcriptional and translational, are already found in mammalian embryos at the earliest developmental stages. Such an early expression of the embryonic genome seems to be a specific feature of mammalian species, distinguishing them from other members of the animal kingdom such as fishes, amphibians, etc., whose genome remains dormant during cleavage (Neifakh and Timofeeva 1977; Kafiani and Kostomarova 1978; Ignatieva 1980; Davidson 1976).

However, the early expression of the mammalian genome can hardly be accepted as unequivocal evidence of direct control of early embryogenesis by the embryonic genome until one can discover when these newly formed gene products are used and what particular morphogenetic processes they participate in.

Which of the early morphogenetic events are governed by genetic information stored in the mature egg (maternal gene expression products)? Which are primarily controlled by the embryonic genome itself? What morphogenetic processes depend on genetic information from both sources? These and many other related questions are still completely unanswered, and their solution is a major task of modern developmental biology.

In more general terms all these questions may be considered as part of a more fundamental biological problem concerned with nucleo-cytoplasmic interrelationships in early development.

Various experimental approaches applied in this field are outside the scope of this chapter. Experiments with X-rays or chemically induced nuclear inactivation yielded much useful information regarding gene action in the early development of invertebrate and lower vertebrate embryos. This 'inactivation' approach has been especially useful for precise determination of the embryonic stages controlled by the embryonic genome. Theoretical considerations of these experimental data have initiated the fruitful hypothesis of morphogenetic functions of the embryonic genome, i.e. correlation between morphogenetic events and the output of genetic information from the embryonic nuclei (Neifakh 1962, 1965a,b, 1971, 1974; see also Neifakh and Timofeeva 1978; Ignatieva 1980). This experimental approach is hardly applicable to studies on gene action in mammalian embryos. The expression of embryonic genes and the developmental role of genetic information stored in the egg are much more readily disclosed by a comparative analysis of gene expression from the zygote stage onwards on the one hand, with developmental manifestations of gene mutations, structural chromosomal aberrations (duplications, deficiencies, inversions), aneuploids (nullisomy, monosomy, trisomy), and genome imbalance (haploidy, triploidy, tetraploidy) on the other.

A brief comparative survey of this broad issue constitutes the major goal of this chapter. Maternal effects in this context will be primarily attributed to the cytoplasmic ovum factors stored during egg maturation, though these do not exhaust all aspects of the maternal influence, in mammalian development as a whole (McLaren 1979, 1981b).

Early development in mice in the present context means the first 7 or 8 days of life, and thus it encompasses all morphogenetic events occurring before implantation as well as during the peri-implantation period, including primitive streak formation, neurulation, and the onset of major organogenesis (see Chapter 11).

Such subdivision is justified by the schedule of embryolethal effects due to genome (haploidy, triploidy, tetraploidy, androgenesis, gynogenesis, parthenogenesis) or chromosome abnormalities (aneuploidy, structural aberrations).

13.1 Cytoplasmic factors, gene action, and chromosomal activity during pre-implantation development in mice

This section deals mainly with morphogenetic processes occurring during fertilization, cleavage, compaction of the morula, cavitation [with the formation of a blastocyst composed of trophectoderm (TE) and inner cell mass (ICM)], commitment of both cell populations, and the onset of cell differentiation. Participation of both cytoplasmic factors (stored maternal gene products) and newly formed products of the embryonic genome in the effective control of all these morphogenetic processes is discussed.

13.1.1 Cytoplasmic factors and chromosomal behaviour in one-cell embryos

The following morphogenetic processes are confined to the one-cell stage: egg penetration and activation by the sperm; completion of the second meiotic division: chromosome disjunction and second polar body extrusion;

transformation of the sperm head and the female haploid chromosome set into male and female pronuclei respectively; fusion of both pronuclei on the first metaphase plate with subsequent splitting of each chromosome into chromatids and incorporation of these into the nuclei of the first two blastomeres. All of these events are determined almost exclusively by the gene products accumulated in the pre-ovulatory oocytes, that is by stored cytoplasmic factors.

13.1.1.1 Egg maturation and cytoplasmic factors responsible for chromosome behaviour. All cytoplasmic factors arise in the course of transcription and translation of oocyte genes, as well as by post-translational modification of their protein products. Ooplasmic maturation in the eggs of mammals and other animals is determined by the transcriptional activity of lamp-brush chromosomes in the oocytes of growing follicles. Apart from numerous ribosomal cistrons, we are still almost completely ignorant of what other genes are transcribed during these stages and what proteins they encode. Numerous meiotic mutations ('mei'-mutations in plant genetics; Golubovskaya 1979) are known to affect meiotic chromosome behaviour in plants and in the animal kingdom, including in mammals. Some of these mutations bring about egg elimination at various stages of oogenesis, or drastically reduce the fertilizing capacity of the eggs (gamete lethals) or cause embryonic death (zygotic lethals). Reduced fertility and complete sterility in female mice and other mammals may well be produced by gamete lethals, but mutant genes affecting oogenesis in mammals are still poorly studied.

The processes concerned with egg maturation shortly before ovulation are somewhat better known. This stage coincides with the resumption of meiosis and the maturation of egg cytoplasm, which at this time acquires factors indispensible for early embryonic development. These events seem to be common to all mammals including mice. Six to eight hours before ovulation the extended oocyte nucleus loses its envelope, pouring its contents directly into the ooplasm.

The factor initiating the breakdown of the germinal vesicle membrane is not species-specific and closely resembles or is identical to the cytoplasmic factor responsible for mitotic induction in somatic cells (Tarkowski 1982). The meiotic trigger substance appears in the ooplasm at the appropriate time, in the presence or absence of the germinal vesicle (Balakier and Tarkowski 1978). Germinal vesicle breakdown results in dramatic changes in the properties of the ooplasm. These changes are probably due to substances stored in the germinal vesicle and released after membrane dissolution. The leakage of germinal vesicle stored proteins prevents the synthesis of some proteins, including phosphoproteins (M.W. 28×10^3 dalton), the concentration of which in the germinal vesicle is 1000 times higher than in the ooplasm (Wassarman, Schultz, and Letourneau 1978). On the other hand it initiates a cascade of molecular reactions (translation of stored transcripts, post-translational maturation of some proteins, etc.) and morphological changes both in the centre of the egg and in its cortical part (Van Blerkom and Runner 1984). Germinal vesicle breakdown coincides with meiotic resumption. After completion of the first meiotic division and extrusion of the first polar body, the secondary oocyte is induced to embark on the second meiotic division by the action of special cytoplasmic factors (proteins). Fertilization or artificial activation of the egg results in specific ooplasm changes and reverses the action of the meiosis-blocking factor. The activated egg undergoes sequential changes starting from Ca^{2+} extrusion from the plasma membrane, increasing Ca^{2+} concentration in the ooplasm, and termination of meiosis-preventing protein synthesis (Whittingham 1980). Concomitantly, the metaphase chromosomes of the egg resume the second meiotic division, with one group of chromosomes being incorporated in the second polar body and the remaining one giving rise to the female pronucleus.

At this stage some cytoplasmic factors of the egg govern transformation of the sperm head into the male pronucleus. Whether both male and female pronuclear transformation are under the control of the same cytoplasmic factor or whether these factors are sex-specific remains unknown. The factors are probably formed long before fertilization, i.e. at some time during egg maturation, since for example mouse oocytes cultured *in vitro* without surrounding follicle cells can readily be fertilized, but the penetrating sperm head fails to form a normal male pronucleus and development cannot proceed beyond one or two cleavages (Pratt *et al.* 1983). Transformation and maturation of the male pronucleus was found to be much more vulnerable to the amount and quality of cytoplasmic factors than was the female pronucleus (Thibault 1977).

The amount of the male pronucleus factor seems to be very limited, so that it is easily exhausted after fertilization, especially with polyspermy. The head of the first sperm penetrating the egg usually transforms into a somewhat larger pronucleus than the rest of the sperm reaching the ooplasm (Witkowska 1981). Differences in the amount of pronucleus factor might explain differences in the size of the male and female pronuclei, as well as the well-known phenomenon of heterocycling (allocycling) of the parental chromosomes (Prokofieva-Belgovskaya 1946, 1947). Our studies of diploid parthenogenones (Dyban and Noniashvili 1985a,b) indicate the marked asynchrony of the two pronuclei, in spite of their origin from the same chromosomes of the same metaphase plate. These data favour the existence of a single cytoplasmic factor for both male and female pronuclear transformation. The effect of this factor would depend on its concentration and on the position of the pronuclei in the ooplasm. The hypothesis of compartmentalization suggests the existence of a concentration gradient specific for the cytoplasmic factor responsible for pronuclear transformation and chromosome behaviour (Dyban 1984). This hypothesis is in line with other experimental observations on nucleo-cytoplasmic interrelationships (Tarkowski 1982).

Thus parental genomes specifically respond to cytoplasmic factors, though embryonic genes are probably still dormant at the one-cell stage. Both sperm and egg may bear various chromosomal aberrations (hypo- and hyperhaploidy, structural rearrangements, etc.), but these do not affect pronuclear formation or the onset of the first cleavage division. This may occur even in the case of heterologous fertilization (for instance, hamster eggs fertilized by human sperm or rat eggs fertilized by mouse sperm).

Metabolic processes in one-cell embryos are not determined by the genetic products of the embryonic genome, so enucleation of fertilized or unfertilized mouse eggs does not prevent protein synthesis on stored maternal mRNA (Petzoldt *et al*. 1980). The resulting cytoplast may undergo cleavage divisions, but these proceed abnormally.

Normal division requires a proper nucleo-cytoplasmic interrelationship. The nucleus can somewhat mitigate an excessive cortical reaction, and can also provide a proper and more regular organization of the subcortical microfilaments and thus regulate the localization of the cleavage furrow (Waksmundska *et al*. 1984).

Thus normal regular cleavage of mammalian eggs from the very beginning of embryonic development depends on the proper relationship between the nucleus and the ooplasm.

13.1.1.2. Fusion of parental genomes into the integrated cell nucleus, and chromatin reprogramming. The pronuclei of the fertilized egg do not actually fuse to form a single zygote nucleus in those mammalian species studied so far. Instead, the chromosomes in each pronucleus condense, and after concomitant breakdown of the pronuclear envelopes they come together and intermix, forming the common metaphase plate of the first cleavage division.

The one-cell mammalian embryo contains two pronuclei with asynchronously condensing chromosomes. This unusual situation of a physiological dikaryon poses the question whether the two pronuclei lying apart exchange any genetic information, or whether their genomes cannot communicate with each other until they come together to form a single integrated cell nucleus. It has been suggested that once the parental chromosomes integrate into a single functional system, they acquire a specific arrangement that remains constant throughout life, though it may undergo some changes in meiosis (Ashley and Pocock 1981).

Non-random arrangements of chromosomes in an interphase nucleus may be traced to some degree to their arrangement on the metaphase plate. Regular arrangements of parental chromosomes on the metaphase plate of the first cleavage division have occasionally been found both in whole-mount preparations under phase-contrast microscopy (Khozhai 1981) and also after fluorochrome staining of air-dried preparations (Patkin 1983). These rare findings suggest the existence of intimate contact between the chromosomes of the two parents at the first cleavage division.

However, detailed cytogenetical analysis of banded chromosomes at the first cleavage division lend little support to the view that the homologous chromosomes are in contact. The two parental chromosome sets were found to lie apart both in prometaphase and metaphase of the first cleavage division (Dyban and Sorokin 1983). In autoradiographic studies of interphase nuclei at the two- or even four-blastomere stage, the chromosomes of the two parents still remain apart and do not intermix (Keneklis and Odartchenko 1974). But a proportion of metaphases at the second cleavage division sometimes reveal nearby localization of the two parental homologous chromosomes (Sorokin 1984).

Thus the formation of a common metaphase plate at the first cleavage cannot be considered to be a crucial event either for chromatin reprogramming and interphase nucleus reorganization, or for subsequent genetic control of the developmental programme. A few examples can illustrate this statement. The blastocyst stage can be successfully reached by fertilized eggs with maternal or paternal pronuclei microsurgically withdrawn (gynogenetic and androgenetic haploids), and by haploid parthenogenones with an initially single maternal pronucleus (see Chapter 1). Diploidization of the paternal or maternal genomes substantially improves development. Biochemical analysis of both androgenetic and gynogenetic embryos during cleavage has revealed all stage-specific proteins typical for normal diploids of equivalent stages (Petzoldt *et al*. 1981). Thus early development does not seem to require integration of parental chromosomes into a single nucleus.

Cytochalasin B treatment of mouse zygotes prevents both cytotomy and karyokinesis. These embryos contain two separate nuclei, corresponding to the maternal and paternal pronuclei, after two additional DNA replication rounds. However, none of these changes has much influence on the general developmental programme. The appropriate genes become functional at the proper time, though the number of blastomeres is reduced compared to the controls (Petzoldt, Bürki, Illmensee, and Illmensee 1983; Bolton *et al*. 1984).

Aphidicolin-induced DNA synthesis block in both pronuclei does not prevent chromosome arrangement on the metaphase plate or subsequent cytotomy. No stage-specific gene expression has been detected in these two-cell embryos, so their developmental programme has not yet been switched on (Johnson, Mc Connell, and Van Blerkom 1984).

DNA synthesis usually starts 7–8 h after fertilization, i.e. 4–5 h after pronuclear formation, and it lasts for 3.5–4 h (Luthardt and Donahue 1973; Abramczuk and Sawicki 1975; Molls *et al.* 1983) or only 1.7 h according to more recent data (Doman 1983). If the high specificity of the DNA replication inhibitory effect of aphidicolin is confirmed, it will mean that during the rather short S-phase, reprogramming of the parental chromatin takes place and these drastic changes are somehow crucial for the developmental programme to be switched on as an indispensable prerequisite for normal functioning of the embryonic genome from the two-cell stage onwards. The identity of the trigger mechanism remains totally unknown. However, a good deal of information has already accumulated on the developmental characteristics of two-cell stage embryos (see Chapter 12).

13.1.2 Nucleo-cytoplasmic relationships at the two-cell stage

At least two developmental peculiarities make the two-cell stage unique:
1. It is the earliest stage of embryonic genome activation.
2. Both maternal and embryonic mRNAs are translated.

Most maternal mRNAs, however, are inactivated at this stage, though their translational products may persist up to the blastocyst stage (Pratt *et al.* 1983; West and Green 1983) or even somewhat later (see Chapter 12, and also McLaren 1979; McLaren and Buehr 1981; Magnuson and Epstein 1981; Johnson 1981). All these molecular events concerned with gene activation and mRNA translation and degradation are highly coordinated in time and correspond to a quite specific biological age. They may be observed both in one-cell embryos of corresponding age with a block to cell division induced by cytochalasin B treatment, and also in two-cell embryos in which the DNA replication cycle is blocked and cell division has failed to occupy (Bolton *et al.* 1984; Johnson *et al.* 1984). In spite of transcriptional and translational mechanisms operating at the two-cell stage, newly formed gene products hardly participate in the control of morphogenetic processes, and probably do not determine the transition from the two- to the four-cell stage. The absence of embryonic lethals or chromosomal aberrations expressed at this stage supports this view. Direct proof of cytoplasmic control operating at this early embryonic stage has been obtained recently (Muggleton-Harris, Whittingham, and Wilson 1982), since it seems that the developmental block shown by two-cell embryos cultured *in vitro* from the zygote stage is due to cytoplasmic and not to nuclear factors. If eggs unable to pass the two-cell stage *in vitro* are injected with ooplasm of mice lacking this two-cell block, they readily overcome the block and proceed to the blastocyst stage. It might be inferred therefore that some unknown cytoplasmic factors determine the ability of the egg to cleave from two cells to four cells. Production of these cytoplasmic factors is presumably controlled by gene action during oogenesis. Strain differences for the factors depend on maternal gene expression, proper mRNA storage, enzyme characteristics, and plasma membrane permeability. The two-cell block is often encountered in artificially activated eggs, especially in haploid eggs (see Chapter 1), as well as after *in vitro* fertilization. The shortage or absence of proper cytoplasmic factors is probably responsible for this developmental failure. However, not only cytoplasmic factors but gene products of the embryonic genome itself determine the progess of cleavage.

13.1.3 Genes and the timing of cleavage

Duration of the mitotic cycle during the period of synchronic cleavage is a convenient biological unit (T_o) for the relative measure of cleavage rate in eggs with complete (holoblastic) cleavage (sea urchin, hanoid fishes, amphibians; Detlaff 1964, 1977; Detlaff and Detlaff 1961). The same unit has been successfully applied to the eggs of higher fishes with meroblastic cleavage (Ignatieva 1980).

This fruitful approach has not yet been adapted to mammalian egg. Cleavage rate in mammals is measured in ordinary time units, which are not very convenient for estimations of the biological age of cleaving eggs (Smith and McLaren 1977). It takes about seven to eight cell cycles and approximately 4–4.5 days for orre-cell mouse embryos to reach the blastocyst stage. Rabbit embryos start blastulation 3 days later and the mature blastocyst consists of several thousand cells (see also Dyban *et al.* 1975). The duration of each cleavage in mammals is much longer than in fishes or amphibians, where it lasts for 30–40 min only (Davidson 1976; Ignatieva 1980). Cell cycles of cleaving mammalian eggs somewhat resemble those of somatic tissues. The first and second cell cycles in mice last for about 18 h and subsequent cycles occupy about 12 h each (Johnson *et al.* 1984). Mouse embryos of the same chronological age may vary substantially in their blastomere number. Strains of mice with many-celled blastocysts, as well as those with blastocysts of relatively low cell count, have been known for a long time (Gates *et al.* 1961; Dagg 1960; Whitten and Dagg 1962; Dickson 1967). These data suggest that cleavage processes are correlated with some characteristics of embryonic genome. It is not cleavage rate by itself that determines these differences, but perhaps the duration of the one-cell stage may be involved. Some data in favour of this view are presented in the next section.

13.1.3.1 Characteristics of cell cycles in cleaving animal embryos
Cell cycles of animal embryos with fast cleavage rates are rather short. Each cycle includes mitosis itself, plus a short intermitotic interval used for DNA replication (Davidson 1976; Ignatieva 1980). In contrast, cell cycles of mammalian embryos consist of all the phases known to occur in the cells of somatic tissues. Detailed analysis of all the cell cycle phases of mammalian embryos is beyond the scope of this chapter (see Section 12.1), and is

thoroughly covered by numerous comprehensive reviews and original studies (Samoshkina 1965, 1968, 1972, 1978; Zavarzin *et al.* 1966; Zavarzin 1967; Barlow *et al.* 1972; Graham 1973a,b; Gamow and Prescott 1970; Luthardt and Donahue 1973, 1975; Mukherjee 1976; Sawicki, Abramczuk, and Blaton 1978; Abramczuk and Sawicki 1975; Krishna and Generoso 1977; Siracusa, Coletta, and Monesi 1975; Streffer, van Beuningenen, Molls, Zamboglon, and Schultz 1980; Molls, Zamboglon, and Streffer 1983; Domon 1983).

The existence of all cell cycle phases, including the G_1 phase, in the cleaving eggs of mice was initially discovered by Samoshkina (1968, 1978) in our laboratory. The presence of a G_1 phase was however put in doubt by later observations (Gamow and Prescott 1970), claiming that a G_1 phase cannot be detected in mouse embryos up to the 32-cell stage or even somewhat later (Mukherjee 1976). However, simultaneous use of cytofluorometry and ^{3}H-thymidine histoautoradiography on the same preparations of cleaving eggs of rats and mice ultimately proved the presence of a G_1 phase from the onset of cleavage (Dyban *et al.* 1976). Initially it is extremely short, taking only about 1 h during the second cell cycle (Pratt *et al.* 1983; Bolton *et al.* 1984). This time interval is however quite enough for the transcription of activated genes. Some of these transcripts are responsible for heat-shock-like protein synthesis (see Chapter 12). Thus the rather short G_1 phase of the early cell cycles cannot be accepted as evidence that the embryonic genome is inactive.

Most of the cell cycle (50–60 per cent) is occupied by the S phase, the duration of which is determined by embryonic genome activity at the two-cell stage (Sawicki *et al.* 1978; Streffer *et al.* 1980; Molls *et al.* 1983) and during the subsequent three to five cell cycles (Samoshkina 1978; Dyban *et al.* 1976). These data suggest that differences in cleavage duration could be responsible for the cell number variability in mature blastocysts.

However there is more evidence relating this variability to the duration of the first cleavage cycle.

13.1.3.2 Correlation of cell number in cleaving eggs with duration of the one-cell stage. A detailed experimental study in five strains of mice differing in cell number at the blastocyst stage, unequivocally indicated a maternal influence on this phenomenon. Variations in the duration of the one-cell stage were found to be responsible for the cell number differences (Bowman and McLaren 1970a,b).

Not only the female but also the male genome seems to be involved in the first division schedule (Shire and Whitten 1980a,b). Experiments on mouse eggs fertilized *in vivo* and then maintained *in vitro* support this view. Additional evidence favouring this view was obtained on *in vitro* fertilized eggs from four different inbred strains and two random-bred ones. Both the time of penetration of the eggs and the onset of the first cleavage division were determined very precisely (Niwa, Araki, and Iritani 1980). The duration of the one-cell stage in this study was found to be in line with the activating capacities of the sperm. A prolonged one-cell stage was typical for mice with moderate penetration of sperm into the ooplasm, while enhanced penetration correlated with a relatively short one-cell stage. For example, 20 h after insemination only 45 per cent of fertilized eggs from C3H mice (moderate cleavage) and 85–99 per cent of eggs of C57BL mice (fast cleavage) had completed the first cleavage. Thirty hours after *in vitro* insemination, all C3H eggs were at the two-cell stage (Niwa *et al.* 1980).

Thus not only the culture conditions but also intrinsic genetic characteristics of the eggs tested were responsible for the different timings of the first cleavage division in these mice.

The data from *in vivo* studies are more difficult to interpret. The fluctuations in fertilization time, peculiarities of ovulation and capacitation, differences in sperm passage interval along the female reproductive tract as well as variations in penetration, significantly hamper proper elucidation of the role of the embryonic genome in determining the timing of cleavage. Nonetheless, the results of these experiments are in line with those from *in vitro* studies (Krzanowska 1970; Frazer 1977; Frazer and Drury 1976). Strain differences in cell number are especially marked at the late morula–blastocyst stages (Titenko 1977; Verbanac and Warner 1981; Goldbard *et al.* 1982a,b). A correlation of cleavage rate with *H-2* haplotype has also been suggested. Mouse embryos of $H-2^k$ haplotype are distinguished by rather low cleavage rates compared to those with $H-2^a$, $H-2^b$, $H-2^d$, and $H-2^q$ haplotypes. The delay in cleavage rate of $H-2^k$ embryos has been attributed to a cluster of genes known as *Ped*-genes (pre-implantation embryonic development genes—see Chapter 12), assigned somewhere close to the *H-2* region of chromosome 17 (Goldbard *et al.* 1982a,b). The expression of *Ped*-genes from the one-cell stage and throughout cleavage has been postulated (Goldbard *et al.* 1982a,b), but this suggestion is not in line with numerous data that argue against embryonic gene expression at the one-cell stage (see Chapter 12 and also Johnson *et al.* 1984; Pratt *et al.* 1983; Bolton *et al.* 1984). Moreover, major histocompatibility antigens controlled by the *H-2* locus can be detected only from the late blastocyst stage onwards. Thus reduced cell number in $H-2^k$ embryos can hardly be correlated with *H-2* gene expression. It cannot however be excluded that some other characteristic of the egg acquired during oogenesis of $H-2^k$ haplotype mice can influence the rate of capacitation or sperm entry into the ooplasm, and thus be responsible for cleavage delay. H-2 alloantigens are readily detectable in the plasma membrane of ovulated eggs but they completely disappear by the two-cell stage (Heyner and Hunziker 1979). The presence of $H-2^k$ alloantigen in the plasma membrane of the fertilized egg may in some unknown way prolong fertilization and thus increase the duration of the one-cell stage.

More experimental studies are needed to prove or refute the existence of special cleavage timing genes in mammals.

13.1.4 Cleavage characteristics in fertilized eggs with gene, genome, or chromosome aberrations

Gene action in cleaving eggs has been thoroughly surveyed in the previous chapter. Genetic influences are obvious in studies of cleaving embryos with inherited abnormalities at the level of specific genes but also specific chromosomes or the genome as a whole.

It has been known for a long time that cell number in pre-implantation embryos is negatively correlated with nuclear ploidy, i.e. an increased genetic content corresponds to a reduced cleavage rate (Beatty and Fischberg 1951b, 1952; Fischberg and Beatty 1952; Edwards 1958b). Further cytogenetic studies fully confirmed these early findings (see Chapters 1–3).

It has recently been demonstrated that the mean cell count in triploid mouse embryos cultured *in vitro* is much less than in their euploid littermates of the same developmental age. These differences are probably due to an increase in the length of the cell cycle from the eight-cell stage onwards (McGrath and Hillman 1982).

A low cell count is very marked in tetraploid embryos. It is especially pronounced by the end of the cleavage and at the start of blastulation (see Chapter 3), suggesting that blastulation does not depend on cell number but is primarily determined by embryonic age (Smith and McLaren 1977; see Section 13.1.6).

Data on the cleavage rate of haploid embryos are less conclusive (see Chapter 1). Cleavage proceeds at a normal rate up to the eight-cell stage or somewhat later, but appears to slow down soon after (Kaufman 1983a).

Assuming that cleavage is a genetically determined process, one must distinguish the contribution of genetic information from two possible sources, namely maternally determined mRNA templates stored during oogenesis, and transcriptional–translational products of the activated embryonic genome.

It has already been mentioned that cell cycles of cleaving eggs are similar to those of other tissues. Each phase of the cell cycle is known to require special transcriptional–translational products or so-called 'proteins of mitotic cycles'. Some of these signal entry into the mitotic cycle, others control DNA replication and mitotic disjunction of chromosomes, while still others govern the schedule of mitotic phases (Epiphanova 1973; Prescott 1976; Brodsky and Urivaeva 1981). Protein blockage at a definite time of the G_1-stage prevents cell transition into the S-phase (Brooks 1977). Duration of the G_1 phase depends on the concentration of trigger proteins as well as of other proteins involved in DNA replication (Rao et al. 1978). At least nine proteins are known at present to be specific for S-phase and for cytokinesis determination (Brodsky and Urivaeva 1981).

Because there are so many different proteins and hence also genes involved in mitotic divisions, one should expect many inherited aberrations, both genetic and chromosomal in nature, to affect cell cycles in early mammalian embryos and thus prevent or significantly hamper normal cleavage. However, surprisingly enough there are as yet few mutations or chromosomal aberrations with such an early damaging effect. Presumably this type of mutation is not so rare but most undergo early elimination and are not transmitted. Only mutations which do not induce a mitotic block in heterozygous condition are compatible with post-natal survival. Most chromosomal anomalies are rather early embryo lethals in mice, but few affect cleavage. Cleavage and blastocyst formation can proceed successfully in embryos with haploidy, triploidy, and tetraploidy as well as with trisomy of any autosome, monosomy of most (except autosomes 2,5,7,17), and also with dominant lethals induced by X-rays or radiomimetic substances. These latter, in spite of severe genetic imbalance of several different autosomes of paternal or maternal origin, are capable of several cleavages at least and undergo blastulation (Dyban and Udalova 1967; Dyban 1970a,b).

At least three main developmental mechanisms may be responsible for the apparent non-susceptibility of cleavage to adverse genetic effects:

(a) cell cycles of early mammalian embryos may be under dual control, by stored maternal gene products as well as by those of early activated genes;

(b) genetic information provided by a single haploid genome may be sufficient for the accomplishment of normal cleavage and blastulation;

(c) the genes responsible for 'mitotic' proteins may not be uniformly dispersed throughout the genome but may be concentrated in specific sites on specific chromosomes (Dyban 1974, 1979, 1984). Similar clustering effects are known for genes governing meiosis in the germ cells and mitosis in somatic cells (Baker et al. 1976; Gehring 1976).

The mammalian X chromosome is a plausible candidate for 'early gene' clustering. This chromosome can be taken as a good example of precisely balanced factors of maternal and embryonic origin indispensable for the initial stages of embryogenesis in mice. Only one X chromosome is active in the definitive somatic cells of mammals. The other is reactivated during the early pre-implantation stages of embryogeneis and subsequently becomes heterochromatic.

Detailed cytological examination of the X chromosome cycle in human and mouse embryos has been carried out in our laboratory (Semenova-Tian-Shanskaya 1977, 1978). Both X chromosomes were found to be isochromatic (euchromatic), i.e. genetically active, in the gonocytes on their way to the germinal ridges. Repopulation of the latter coincides with heterochromatization of one X chromosome, which remains dormant in the oogonia. During the pre-leptotene stage the dormant X chromosome is re-activated and both X chromosomes are isochromatic and transcriptionally active from that stage onwards. These cytological findings are in line with numerous biochemical studies, unambiguously indicating genetic activity of both X chromosomes from the onset of meiosis and throughout oogenesis up to germinal vesicle breakdown shortly before ovulation. XX mice possess twice as many X chromosome gene products in their oocytes as do XO females (Burgoyne and Biggers 1976). Not surprisingly, the eggs of XO females are characterized by a reduced capability for normal fertilization and subsequent development both *in vivo* and *in vitro*. The absence of one X chromosome in male embryos (YO karyotype, i.e. nullisomy for X chromosome genes) is usually manifested as a very early embryolethal (see Chapter 8). However, while YO embryos from XO females are usually eliminated at the six- to eight-cell stage (Burgoyne and Biggers 1976), similar embryos from female mice with an inverted X chromosome (both X chromosomes were active in oogenesis but one was lost because of the high non-disjunction rate) may proceed a little further (Evans and Burgoyne 1983; Burgoyne, personal communication).

Androgenetic haploids of YO karyotype have especially poor viability and do not survive beyond a few cleavages (Tarkowski 1977). All these observations suggest that the viability of YO embryos in mice to some extent correlates with the volume of X chromosome gene products stored during oogenesis.

The same relationship seems to be true both for the gonosomes and for the autosomes. After the maternal gene products have been exhausted, subsequent stages of development are controlled by genetic information from the embryonic genome itself. For example, not a single mutation known so far in mice (either genetic or chromosomal) can induce developmental arrest before the two-cell stage but any later stage may be affected. Mouse embryos of $c^{25H} c^{25H}$ genotype stop cleaving at the three to six blastomeres stage, and though they survive for the subsequent 24–28 h, they do not resume cleavage and reveal pathological changes of their nuclei (see Chapter 7 and Nadijcka *et al*. 1979). What happens to the DNA synthesis of these specimens remains unclear. T-locus deletions (T^{Hp}, T^{Or}, T^{Orl}) affect development of homozygotes at a somewhat later stage (see Chapter 12). Cleavage in T^{Hp}/T^{Hp} specimens is blocked at the morula stage and in T^{Orl}/T^{Orl} at the blastocyst stage.

Whether *c*-locus gene products are more crucial for early embryogenesis, or whether their store is more easily exhausted because of reduced level or lower stability is completely obscure as yet.

Cleavage arrest is often encountered in autosomal monosomies. The absence of most autosomes is not detrimental for cleaving mouse eggs up to the blastocyst stage, but some specific types of monosomy cannot support cleavage beyond four cleavage divisions (see Chapter 7). Early death of Ms17 embryos is potentiated by chromosome 17 loss in oogenesis (a clear-cut maternal effect), as well as by particular combinations of Rb translocations (Baranov *et al.* 1980). The maternal effect of chromosome 17 loss is already evident at the third or fourth cleavage division, i.e. much earlier than the maternal effect of the T^{Hp} deletion in chromosome 17 (Johnson 1975; see also Chapter 12). We shall come back to the developmental manifestations of Ms17 and T^{Hp} later in this chapter. It should be stressed now however that gene products of this particular chromosome and especially of some of its loci are indispensable for early cleavage. Thus the number of cytoplasmic factors stored in the ooplasm of the egg may determine the time that the corresponding genes switch on. Early cleavage arrest typical of some other monosomies could also be attributed to the storage of specific gene products in mature eggs. The insufficiency of gene dosage compensation mechanisms in monosomic embryos is quite consistent with this suggestion, which is also supported by the developmental analysis of some particular lethal mutations. For instance, the *Os* mutation in mice (see Section 12.6.3) is always lethal in homozygotes at the late blastocyst stage because of a complete mitotic block (Magnuson and Epstein 1984). The death of *Os/Os* specimens at the late blastocyst stage shows that up to the fifth cleavage division at least, all functions of the mitotic spindle are controlled by stored maternal products, while the corresponding embryonic genes start to control these processes only at the late blastocyst stage. Similar considerations seem to be valid for other early lethals in mice. Expression of these mutations somewhat resembles functional nullisomy for particular genes. One dose of the normal allele is sufficient to prevent the lethal effect of *Os*, A^y, t^{12}, T^{hp}, or T^{Or} mutations, while the lethal effect of a double dose of the t^{12} allele can be cured by the presence of a single normal gene dose in triploid embryos, $t^{12}t^{12} +$ (McGrath and Hillman 1982). Both c-albino deletions of chromosome 7 and the cluster of *T*-deletions (T^{hp}, T^{Or}, T^{Or}) of chromosome 17 behave as early embryonic lethals only in homozygotes, which correspond to the nullisomic condition of these segments. Relatively late expression of dominant lethals induced by X-irradiation or by radiomimetic agents could be attributed to the presence of at least one normal copy of the particular gene or autosomal segment, which provides sufficient genetic information to support normal cleavage and blastulation, but which is not enough to control the more complex post-implantation events.

Thus cleavage divisions of mammalian embryos are regulated both by cytoplasmic ovum factors and by information generated by activated embryonic genes. Early cleavages are principally governed by maternal gene products stored during oogenesis, and these are subsequently complemented by the products of so-called 'early genes' or 'cell-cycle genes' of the embryo itself (Dyban 1984). The clusters of cell cycle genes are not randomly dispersed throughout the genome but are confined to specific chromosomes. The X chromosome, autosomes 7, 17, 2, and 5 are the most plausible homes for these early genes (see Chapters 11 and 12). These early genes are not specific regulators of mitotic phases but their products may be indispensable for the mitotic spindle and for cytotomy. They may also meet some crucial metabolic requirements of embryonic cells.

13.1.5 Compaction and morula formation

Morphogenetic processes concerned with morula formation, compaction, and subsequent blastulation are probably initiated from the second or third cleavage divisions onwards. Two- to four-cell stage blastomeres are still totipotent, and if isolated may still give rise to a whole embryo (Tarkowski and Wroblewska 1967; Tarkowski 1970; Willadsen 1979, 1980; Willadsen, Lehn-Jensen, Fehilly, and Newcomb 1981). Single or paired blastomeres of the 8–16 cell stage, if aggregated or injected to produce chimaeras, can participate in the formation of trophoblast as well as all inner cell mass (ICM) derivatives (Kelly 1975; Gardner 1975; Johnson, Handyside, and Braude 1977).

The subsequent behaviour and hence the fate of blastomeres is chiefly determined by their position in the morula, as well as by their interactions during the preceding interphase. The microvilli of the cell membrane play a crucial role in these interactions. Their location and relative density on the outer surface varies from the four-cell stage onwards, in relation to cell cycle phase. The descendants of the blastomere that was the first to cleave at the two-cell stage are more likely to form intercellular contacts and hence to be enrolled into the ICM (Graham and Deussen 1978; Graham and Lehtonen 1979; Reeve and Ziomek 1981). Significant variations in the cleavage rate of daughter blastomeres isolated at the two-cell stage were found in *in vitro* studies carried out in our laboratory (Dyban and Sekirina 1981). Thus, in spite of their identical karyotype, the nuclei of daughter blastomeres trapped in different cytoplasmic environments display different cleavage abilities, which are important for their subsequent fate.

The formation of the morula corresponds to the eight-cell stage, during which compaction and polarization of the blastomeres take place. By the end of the eight-cell stage (60 h after fertilization) the blastomeres change their shape, the cells flatten against each other to minimize intercellular space and obscure cell boundaries, and gap and tight junctions appear (Ducibella 1976; Johnson 1981). Stage-specific antigens are thought to appear on the outer surface of the blastomeres. They are believed to play a major part in compaction and subsequent blastomere behaviour (Ducibella 1976). Whether their production is regulated at the transcriptional, translational, or post-translational level remains unknown. The participation of stage-specific antigens of exclusively maternal origin was suggested to be sufficient for normal compaction (Ducibella 1976), but this remains to be proved.

Compaction can be prevented with antibodies against these stage-specific antigens, though cleavage itself is not affected (Johnson *et al.* 1977a). Compaction processes depend on Ca^{2+} concentration. A special membrane protein, uvomorulin, probably binds Ca^{2+} ions and thus plays an important role in compaction (Hyafil, Morello, Babinet, and Jacobs 1980). If cultured in Ca^{2+}-free medium, the compacted morula undergoes decompaction. The blastomeres revert to a round shape and continue to divide mitotically. The addition of Ca^{2+} ions induces normal compaction and blastocyst formation.

Compaction of haploid embryos occurs one cleavage later than in diploids (at about the 16-cell stage). Decompaction and recompaction of haploid parthenogenones was observed (Kaufman 1983a; see also Chapter 1).

Polarization of the blastomeres is another morphogenetic phenomenon known to occur during the formation of the morula. It involves nuclear movements, redistribution of cell organelles, and other morphogenetic changes important for subsequent development (Johnson 1981; Johnson, Pratt, and Handyside 1981; Johnson *et al.* 1984). The typical radial symmetry of the early eight-cell embryo is more or less lost during compaction, and is replaced by an epithelial type of organization in the compacted morula, which soon differentiates into a polar outer layer of trophectodermal cells (TE) and the non-polar cells of the inner cell mass (ICM) (Johnson and Ziomek 1981; Reeve and Ziomek 1981; Ziomek *et al.* 1982). α-amanitin-induced inhibition of transcriptional activity at the two-cell stage does not seem to affect compaction and blastomere polarization up to the end of the eight cell stage (Johnson and Pratt 1983; Johnson *et al.* 1984). These observations support the notion that cell interactions and positional information acquired in the course of cleavage are controlled by proteins regulated at the post-translational level.

All informational requirements of mouse embryos during the third and fourth cell cycles appear to be satisfied primarily by stored maternal information, and also partly by very transient gene activity at the start of the second cycle (Johnson and Pratt 1983; Johnson *et al.* 1984).

This circumstance may explain the absence of chromosomal aberrations exclusively affecting compaction and blastomere polarization. Embryos that survive to the eight-cell stage usually undergo compaction. Both polarization and compaction could be delayed or even completely prevented by genome aberrations or gene mutations. Affected blastomeres sometimes do not take part in compaction but are extruded. Probably not only maternal information stored in the egg, but also activity of specific embryonic genes are necessary for polarization and compaction.

13.1.6 Cavitation and blastocyst formation

Transformation of compacted morulae into blastocysts occurs chiefly by fluid accumulation in the interblastomeric spaces. Because of tight junctions the secreted fluid cannot leak out, and it gradually transforms the intercellular space into a blastocoele cavity. Formation of the blastocyst is accomplished by differentiation of initially totipotent cells into TE and ICM. The two primary cell populations differ from one other in their protein synthesis as well as in their membrane properties. Transformation of morulae into blastocysts requires specific mRNA synthesis during the fifth cleavage division (transition from 16 to 32 blastomeres) (Braude 1979; Johnson 1981). Both ICM and TE precursors are characterized by specific protein synthesis before blastulation.

Fluid accumulation in the blastocoele is an energy-dependent process involving special mechanisms of ion transportation. The role of the embryonic genome in this process is still rather obscure. It is known that cavitation does not depend on the number of cleavages but is determined by the biological age of the embryo. Embryos with artificially postponed cytotomy but a normal rate of DNA synthesis start fluid accumulation at the proper time. As few as two blastomeres of the appropriate age in contact with one another may start fluid accumulation and form pseudoblastocysts or trophoblastic vesicles (Tarkowski 1970). Thus cavitation does not depend on the number of cell divisions but correlates more closely with the number of rounds of DNA replication and with the nucleo -cytoplasmic ratio (Smith and McLaren 1977).

Triploid embryos start cavitation earlier than normal diploids, in spite of their reduced blastomere number. This may be attributed to the corresponding shift of nucleo-cytoplasmic ratio (McGrath and Hillman 1982). On the other hand, compaction and cavitation in experimental haploids usually occur at a more advanced embryonic age and at a much higher mean blastomere count (see Chapter 1; also Dyban and Noniashvili 1985a,b).

The activity of a whole embryonic genome and metabolic cooperation between blastomeres may be an indispensable prerequisites for both compaction and cavitation (Dyban 1984). A threshold level of certain substances may have to be reached to switch on these morphogenetic processes. A decreased DNA content will require more time to generate the appropriate quantity of the trigger substances, and conversely a higher level of DNA will support their synthesis more effectively and thus may induce compaction to start earlier, even at a lower cell number.

On the other hand both compaction and cavitation correlate with cell size. Haploid embryos with delayed cavitation possess rather small cells compared to triploid or tetraploid embryos.

A good many chromosomal aberrations induce embryonic death before or soon after blastulation, but none prevent blastulation without blocking cleavage or compaction. It looks likely therefore that all chromosomal aberrations with early lethal manifestations primarily affect cell cycles and thus induce arrest of cleavage. None of these affected embryos would reach the biological age appropriate to cavitation and formation of the blastocyst.

13.1.7 Commitment of cells to trophectoderm and inner cell mass (foundation of two tissue lineages)

This process begins at the sixth cell cycle and probably terminates by the eighth cell cycle. Embryos at this stage consist of about 200 cells and are rather difficult to flush from the uterine crypts. They are therefore poorly studied cytogenetically. A good many chromosomal aberrations, primarily autosomal monosomies, induce embryonic death at this stage, i.e. during the transition from blastocyst to egg cylinder embryo. Not only monosomics but also haploid and diploid androgenones, haploid gynogenones, and haploid parthenogenones (see Chapter 1), as well as some gene mutations (see Chapter 12; also Magnuson and Epstein 1981; Epstein and Magnuson 1982; Magnuson 1983)), behave as embryonic lethals at this particular stage.

Thus numerous abnormalities of the chromosome set, as well as abnormal nuclear activity in parthenogenones, androgenones, and gynogenones, result in embryonic death around the time of implantation. Strictly coordinated and integrated functions of numerous embryonic genes on most if not all of the autosomes is likely to be required in the peri-implantation period. It cannot however be excluded that embryonic death at this stage may also be due to reduced cleavage rate, as well as to blastocyst abnormalities initiated somewhat earlier in development.

It has already been pointed out that normal blastocyst formation depends on the number of blastomeres at the start of cavitation (Tarkowski and Wroblewska 1967). Only embryos composed of 22–25 blastomeres at the start of cavitation were able to generate an ICM and thus transform into normal blastocysts. The embryos with a smaller number of blastomeres gave rise to trophoblastic vesicles composed only of trophectodermal cells (Snow 1973). These are able to induce a decidual reaction but die during implantation. The blastocyst must contain some critical number of ICM cells in order to induce proper interactions between ICM and TE cells, indispensable for TE cell proliferation and hence generation of secondary giant cells and ectoplacental cone growth. Without these events, implantation fails (Ansell and Snow 1975; Snow, Aitken, and Ansell 1976).

The death of gynogenetic, androgenetic, and parthenogenetic embryos may be primarily due to defects in these TE and ICM interactions, with resulting underdevelopment of TE derivatives and concomitant implantation failure. (See Chapter 1.)

13.2 Chromosomal control of early embryonic development in mammals

Much if not all maternal mRNA ceases to be functional after the mid two-cell stage, i.e. 27–30 h after fertilization (Johnson 1981). Unlike mouse embryos, the embryos of other animals (for example amphibians) have by this time (27–30 h) completed gastrulation. Maternal translational products are detectable in the ooplasm of mammalian eggs up to the blastocyst stage and even a little later, and so can participate in the morphogenetic processes of blastulation.

According to the hypothesis of Dyban (1974, 1979, 1984), early development in mammals is regulated by so-called 'early' genes, which are not randomly scattered throughout the genome but on the contrary are clustered in a few autosomes. This does not however mean that genes of other autosomes are completely dormant at these early stages. Genes on most autosomes are transcribed at this time but only 'early' genes control the synthesis of proteins indispensable for the cell cycle machinery and for embryonic survival.

Genes coding for actin and tubulin, and for other parts of the mitotic apparatus as well as for cell membrane proteins, are likely candidates for these hypothetical 'early' genes. The 'early' genes are embryonic counterparts of the genes active during oogenesis that are responsible for the accumulation of similar genetic products in the maturing egg. Thus the primary function of early 'genes' is to increase the reliability of the developmental machinery responsible for normal cell cycles.

The hypothesis postulates that the activity of 'early' genes in some way compensates for the progressive loss of maternal gene products (egg cytoplasmic factors). Transcriptional–translational products of 'early' genes become indispensable for normal cleavage and embryonic survival, since the equivalent gene products of maternal (egg) genome origin are already exhausted.

Thus both cytoplasmic factors (products of the maternal genome stored in pre-ovulatory eggs) and embryonic genome products are involved in the genetic control of pre-implantation development in mammals. Hence the control mechanisms of early embryogenesis in mammals are based on very complex and precisely balanced cooperation between maternal genetic information transcribed during oogenesis, and genetic information from 'early' genes of the embryo itself, as well as complex epigenetic factors (positional information, cytoplasmic properties, blastomere number, etc.)

The blastocyst stage is indispensable for the proper implantation of embryos, and thus might be considered as a highly specific adaptive stage in mammalian development. The blastocyst stage cannot however be accepted as the morphological equivalent of the gastrula stage in lower vertebrates; the latter has more in common with the egg cylinder stage in mammals, when all three primary germ layers are already formed (McLaren 1979, 1982, 1983). These considerations are in good agreement with the cytogenetic data that demonstrate exceptionally high vulnerability of egg cylinder stage embryos to most chromosomal and genome imbalances. The egg cylinder stage in mammals, just like the gastrula stage in other animals, is probably fully under the genetic control of the embryonic genome (Dyban 1984). An alternative view is that the embryonic genome exerts complete genetic control from the two-cell stage onwards (Epstein 1981c,d, 1983b; Magnuson and Epstein 1981a, 1984). Many more experimental studies are needed to clarify which of these two alternatives is correct.

13.2.1 Mechanisms of aneuploidy expression in early mammalian embryos

The mechanisms responsible for the phenotypic expression of aneuploidy are very complex. No one has yet analysed the causal chain linking the primary lesion in an individual gene to its phenotypic expression in the whole organism. Any type of aneuploidy (trisomy and monosomy included) involves genetic imbalance of many thousands of genes, both structural and regulatory. The actual mechanisms of aneuploidy expression must therefore be very complicated, and are mostly obscure in spite of many thorough studies (Ts21 in humans, for instance). Several attempts to clarify the developmental puzzle of trisomy have recently been undertaken. One suggestion is that the phenotypic effects of trisomy should be subdivided into primary and secondary. Primary effects would be limited to the genes of the extra chromosome (gene dosage effects). Secondary effects would be subdivided into two further groups, (a) biochemical alterations caused by gene dosage effects, and (b) indirect effects due to defective regulation of genes in other chromosomes (Epstein 1981, 1983b). This view is very useful for analysing the post-natal effects of aneuploidy, but does not help in elucidating the effects of aneuploidy during embryogenesis.

Routine cytological studies of aneuploid cells are not very informative from this point of view but more detailed studies of aneuploid cells cultured *in vitro* have succeeded in revealing a specific complex of minor cytological faults and biochemical alterations that can be assembled into the so-called 'cell syndrome' (Kuliev 1972, 1976; Grinberg 1982). The significance of this for understanding the pathogenesis of abnormalities and embryonic death remains rather obscure, while data on the proliferative capacities and disturbances of cell cycles in aneuploid cells are still rather controversial. Impairment of growth kinetics of cultured aneuploid cells from human abortuses have repeatedly been reported by some investigators (Schneider and Epstein 1972) but not by others (Gropp *et al.* 1983). No major differences in the cell cycle or proliferative kinetics between normal cells and those trisomic for chromosomes 1, 12, or 19 have been reported from *in vitro* studies (Gropp *et al*. 1983), though a reduction in the proliferative capacity of thymus cells in Ts19 embryos has been noted *in vivo* (Bersu *et al*. 1982).

Two more recent experimental approaches seem to be more rewarding in understanding the developmental lesions caused by chromosomal aberrations. Both involve the artificial production of primary or secondary chimaeras (McLaren 1976c), and both clearly demonstrate that many though not all trisomic and monosomic cells can be rescued for extended periods of time if they are combined with normal diploid cells. This has been clearly shown for aggregation chimaeras $2n \leftrightarrow$ Ts17; $2n \leftrightarrow$ Ts15; $2n \leftrightarrow$ Ts16. These were found to be viable post-natally, with a substantial proportion of trisomic cells in all tissues and organs of adult animals (see Chapter 6; also Epstein 1984 for review). The secondary chimaeras were produced by the injection of Ts blood stem cells into lethally irradiated adult mice. Cells with Ts12,14, and 19 were capable of completely restoring haemopoiesis with functionally normal red and white blood cells, while Ts 13 and Ts 16 cells failed to support efficient haemopoiesis in irradiated mice (Gropp *et al*. 1983; see also Chapter 6).

It seems probable therefore that at the cellular level the defects caused by aneuploidy are to a large extent and at least temporarily alleviated by complementation with normal diploid cells.

What is actually 'cured' by complementation of diploid and aneuploid cells remains in the field of pure speculation. It should be recalled, however, that the most reliable and constant changes in genetically imbalanced cells occur at the biochemical level and concern gene dosage effects evident in a number of aneuploid cells in mouse and in humans.

It has been repeatedly postulated, but to our knowledge never experimentally proved, that imbalance of so-called 'regulatory' genes is responsible for the expression of aneuploidy (Vogel 1973; Korochkin 1981; Krone and Wolf 1977). The triplication of the structural SOD-1 gene in 21q22 region of human chromosome 21 is consistent with dosage effects but with an otherwise quite normal phenotype, while the triplication of an adjacent rather small chromosomal segment causes typical Down's syndrome (Epstein *et al*. 1982). The existence of these 'regulatory' genes has already been proven for *E. coli* and *Drosophila*. They have also been seriously suspected in mice and humans.

Much more work is needed to prove unequivocally the existence of regulatory genes in mammals, to localize them in chromosomes, and to understand their action in normal and abnormal embryogenesis.

The role of topographic interactions among chromosomes in interphase nuclei as a possible chief regulator of gene orchestration should also be thoroughly studied (see Section 9.2 of this chapter; also Markert 1982). The hypothesis of Baranov (1983a,b,c) seeks to explain the early lethal effects of some types of aneuploidy (primarily nullisomy and monosomy) by the failure of proper interchromosomal interactions in the early stages of embryogenesis.

13.2.2 The hypothesis of interchromosomal relationships and homologous chromosome activation in the early stages of mammalian embryogenesis

The hypothesis is based on two independent lines of evidence:

1. Structural and functional relationships of chromosomes in early embryogenesis of mammals.

2. Developmental characteristics of embryos with monosomy and nullisomy [see Chapters 7 and 13 (Section 13.1.1)]

Basic information on the karyology of early mammalian embryos (chiefly the laboratory mouse) as well as on the establishment of a definite topographic (stereological) structure of the nuclei during the first few cleavages is given in Section 13.1.1.1. In spite of the obvious paucity of this information, and its contradictory nature, it appears that specific topographic location of homologous chromosomes in interphase nuclei has already been established at the earliest developmental stages and these interactions seem to be of great functional significance (Ashley and Pocock 1981).

The mechanisms underlying the specificity of gene activation as well as the nature of the activating substances are still completely unknown. Some of these activators are synthesized and stored in the oocyte (see Section 13.1.1.1; also Longo and Kunkle 1978; Gurdon 1981). Whether these substances recognize homologous genes at random or whether there operates some mechanism of preferential recognition of paternal and maternal genes is still unclear (Chapman, West, and Adler 1977). Some preliminary studies and indirect genetic experiments indicate the existence of multiple mechanisms responsible for gene activation.

The existence of a special activator mechanism differentiating between paternal and maternal chromosomes is suggested by studies on the 'imprinting phenomenon' originally postulated for the mammalian X chromosome (Cooper 1971) and extended later to the autosomes (Johnson 1975). The T^{hp} mutation—a deletion of chromosome 17 (see previous chapter)— behaves as a pre-natal lethal in heterozygous ($T^{hp}/+$) embryos if it is inherited through the oocyte genome, but it is fully viable post-natally if it is transmitted through the sperm. This remarkable phenomenon, suggesting unequal participation of maternal and paternal chromosomes in the genetic control of early development, has been repeatedly shown in mice heterozygous for reciprocal or Robertsonian translocations (Lyon and Glenister 1977; Searle and Beechey 1978). In reciprocal crosses of heterozygotes for reciprocal translocations, a proportion of zygotes result from the fusion of complementary unbalanced gametes, with disomy and nullisomy of some translocated segments or the whole chromosome (Searle *et al.* 1971b). The resulting embryo will be genetically balanced, though both homologous chromosomes (or segments) are inherited from the same parent and none from the other. One may expect therefore to obtain viable progeny if both homologous chromosomes are properly activated, and conversely to get developmental failure if the activation mechanism is somehow different for male and female chromosomes. The latter situation seems to hold for most chromosomes and their segments studied so far (Lyon and Glenister 1977; Searle and Beechey 1978). The progeny survived if the double chromosome dose was received from the ovum, but died if it was transmitted through the sperm. Unexpectedly, however, defective complementation has been detected in chromosomes 2, 7, and 8, all of which are known to carry 'early' genes (see Section 13.1.1 and Chapter 12). These chromosomes could not support normal development if both homologues were received through the female gamete (Searle and Beechey 1978).

Maternal chromosomes seem to participate somewhat more actively in genetic control of early development, but not all paternal chromosomes remain equally 'imprinted' after fertilization, since some paternal genes are already activated during early cleavage (see Chapter 12). Much more biochemical and embryological study is needed to elucidate the exact mechanism underlying this remarkable phenomenon. Nonetheless it seems quite probable that the products of maternal genes are in some way involved in the proper activation of paternal genes, and at least some early activated paternal genes may stimulate the activity of maternal chromosomes. Examples of these complex interactions of homologous and non-homologous chromosomes are often encountered in more advanced stages of development, such as the phenomenon of allelic exclusion during immunoglobulin synthesis (McLean 1975; Wabl and Steinberg 1982), transection phenomena in *Drosophila*, gene regulation through physical interaction of homologous chromosomes (Ashburner 1981), and reactivation of genes in somatic heterokaryons after the loss of some chromosomes (Ruddle 1981). These and many other facts scattered through the literature (Lewin 1981) favour the assumption of chromosome 'cross-talk' in the functionally active nucleus. The language of chromosomes and the actual meaning of their talk are quite poorly understood at present.

Monosomy for at least some autosomes (namely 2, 5, and 17) is already evident during cleavage and may block development by the morula stage. The embryolethal stages of these monosomies are very close to or even coincide with these for the lethal alleles c^{25H}, T^{hp}, and t^{12} (see Chapter 12). Two of these mutations (c^{25H} and T^{hp}) are cytologically visible deletions of chromosomes 7 and 17, respectively. Death after a small number of cleavage divisions has been shown for YO embryos (nullisomy of X chromosome) and for haploid embryos with segmental deficiency of chromosome 15 (see Chapter 1). In all these cases, early embryonic death resulted from the complete absence (nullisomy) of embryonic gene products essential for normal cleavage. Embryos with monosomy 2, 5, or 17 possess at least one copy of all genes carried in the hemizygous chromosome, yet their development too does not proceed beyond the first few cleavage divisions. Thus, phenotypically these types of monosomy could mimic at least some nullisomics in mice.

The data on chromosomal interactions during early development, and the early death of monosomic embryos as well as of embryos deficient for different parts of chromosomes 7 or 17 have stimulated one of the authors (Baranov 1983a,b,c) to suggest the existence of a hypothetical mechanism of mutual activation of homologous chromosomes during the early stages of embryogenesis.

According to this hypothesis the main developmental consequences of autosomal monosomy are due to complete or partial inactivation of genetic loci in the remaining non-paired chromosome or to significant delay in their activation. For autosomes, the loss of the maternal partner seems to be more detrimental to future development than the loss of the paternal chromosome, but for some chromosomes (2, 7, and 8) the opposite seems to hold.

Preliminary data on autosomal monosomies seem consistent with this interpretation (Baranov 1983a). The viability of Ms17 embryos generated through chromosome loss in oogenesis seems to be affected much more severely and at earlier stages of development than is the viability of the same monosomics generated through chromosome 17 loss in male meiosis (see Chapter 7). Advanced development of some haploid embryos (see Chapter 1) is not in significant conflict with the above hypothesis if one assumes that any stimuli that can induce parthenogenic development in ovulated eggs can at least partly substitute for the activation effects of paternal chromosomes.

The hypothesis of mutual activation of homologous chromosomes is also supported by developmental analysis of the segmental deficiencies of chromosome 17. Absence of the rather small CD-region of this chromosome has been found to be expressed at the same developmental stage as monosomy for the whole chromosome 17 (Baranov 1983b). Keeping in mind the activation hypothesis this might be interpreted as an argument for a special activation (or inactivation) centre in this particular region of chromosome 17.

It was in the same segment of chromosome 17, proximal to the T138 breakpoint (see Fig. 36, Chapter 9), that Johnson postulated the existence of a special activation centre required for the activation of chromosome 17 during embryogenesis (Johnson 1975). It is also this region of chromosome 17 that appears to be involved in the cleavage and blastulation of mouse eggs (Babiarz 1983). Thus one may suspect that chromosome 17, like the X chromosome, possesses its own centre of activation (or inactivation), which is primarily involved in switching on (or off) a developmental programme encoded by genes of the same chromosome or elsewhere in the genome.

This hypothesis could be tested either by direct measurement of the gene products of the chromosomes involved in the monosomy by means of the twin technique (Epstein 1981b), or by comparing the effects of monosomy and nullisomy for the same chromosome.

One further suggestion on the developmental characteristics of aneuploidy in mammals has recently been put forward by Markert (1982), according to which the programme of gene regulation during early development is based mainly on the topographic interactions of interphase chromosomes, the non-random arrangements of which are controlled by the activity of special genes or by repeated-sequence regulator DNA. This highly sensitive mechanism of chromosome interactions might be easily upset by various chromosomal aberrations, all of which, but especially aneuploidy, could severely interfere with the normal timing and level of gene transcription.

This interesting hypothesis cannot explain why aneuploid cells can give rise to teratomas with derivatives of all embryonic germ layers and advanced differentiation. It is hard to see how aneuploid cells could easily change their developmental programme in teratomas or in chimaeric embryos, nor does it seem likely that chromosome position in the embryonic interphase nuclei of aneuploid cells differs from that in the same cells transplanted ectopically or included in experimental chimaeras.

Markert's suggestion also fails to explain the early death of monosomic compared to trisomic embryos. On the other hand, his views on possible topographic interactions of interphase chromosomes during early stages of embryogenesis are somewhat in line with Baranov's hypothesis (see above). Application of the molecular DNA probe technique to cytogenetic and cytological studies will help to elucidate the position of homologous chromosomes in interphase nuclei.

A hypothesis postulating imprinting of parental chromosomes and the necessity of interactions between chromosomes of the maternal and paternal pronuclei in one-cell mouse embryos for the successful accomplishment of embryogenesis has also been put forward by other groups of authors (Surani and Barton 1983; McGrath and Solter 1984; see also Chapter 1). The method of pronuclear transplantation used in these studies could be useful as an experimental test of the hypothesis of homologous chromosome interactions and activation during early embryogenesis in mice as well as in other mammals.

13.2.3 Chromosomal control at post-implantation stages

Expression of chromosomal aberrations and especially autosomal trisomies during post-implantation embryogenesis has been already dealt with in Chapters 7 and 9. A few further comments regarding the developmental effects of chromosomal aberrations and their implications for the understanding of chromosome functions during post-implantation development will be made here.

First, one should compare developmental profiles of mouse embryos with trisomy and monosomy for each chromosome. It is evident (Baranov 1983d) that no autosomal trisomy is lethal before implantation and that the lethal effects of trisomy do not depend on the size of the extra chromosome. There is however unambiguous correlation of survival profiles of mouse embryos with trisomy and monosomy for each particular chromosome (Baranov 1983a). For example, monosomies 2, 5, and 17 are expressed at early developmental stages, causing a significant lag in growth, and death by the morula stage. Embryos with corresponding trisomies have especially severe retardation soon after implantation and die significantly earlier than other aneuploids. In contrast, monosomies 1, 12, 14, and 16 do not affect pre-implantation development; trisomies for the same chromosomes do not interfere with major organogenesis and are compatible with survival up to the late fetal period and sometimes even to term (see Chapter 6). This direct relationship is probably not obligatory for all autosomes, though it seems to hold for most. A similar correlation between trisomy and monosomy is known for aneuploids in humans and *Drosophila* (Vogel 1973). This so-called 'phenotypic parallelism' may be considered as an additional argument in favour of a recent suggestion (see Section 13.2.2) that genes responsible for early development are not uniformly dispersed among all autosomes but are organized in a few functional clusters (Dyban 1974; see Section 13.2.1). The interaction of these clusters in time and space provides coordination and integration of the numerous morphogenetic processes taking place in each embryonic anlage and in the embryo as a whole. The greater the number of these 'early' gene blocks carried by the chromosome, the earlier the stage of embryogenesis that should be affected by its imbalance. This general principle turns out to be valid both before and after implantation, which suggests that some common developmental mechanisms exist, responsible for normal development as well as for the abnormal expression of trisomy and monosomy. Mutual activation of homologous chromosomes could be considered as one such mechanism (see Section 13.2.2). If so (Baranov 1983a,b,c), the extra chromosome should remain inactive (switched off) during early development of trisomies. This is partly supported by the absence of harmful effects of trisomy before implantation, while at more advanced stages the genes of the extra chromosome become activated and the gene dosage effect appears. The chromosomes carrying 'early' genes remain completely inactive in the monosomic condition and so are lethal by the morula stage, while in excess they are switched on and impair development soon after implantation. Thus the hypothesis of mutual activation might explain both the early lethality of monosomics and the direct relationship between the survival profiles of monosomies and the corresponding trisomies (Baranov 1983a).

Another problem of interest relevant to cytogenetic studies of post-implantation embryos concerns the teratological implications. Numerous data of clinical and experimental cytogenetics show that chromosomal aberrations may be correlated with quite specific abnormalities of different organs. This suggests that each chromosome or even each segment may participate in the developmental control of different morphogenetic processes during organ formation. Although at first sight plausible, this view is hard to substantiate. We still know too little of the mechanisms either of teratogenesis or of gene action to make this approach helpful. Since aneuploidy involves imbalance of many thousands of genes, the possibility of finding out the primary effects of chromosomal imbalance seems remote. Embryonic death at any particular stage, as well as any particular malformations, could be provoked by a range of different developmental mechanisms operating at different stages of embryogenesis (Dyban 1974). The death of aneuploid embryos at implantation does not imply that the particular chromosome is in any way involved in implantation. Similarly, death during organogenesis does not imply the involvement of the corresponding autosome in placentation or organogenesis.

Mammalian development at the gastrula stage, which corresponds to the egg cylinder stage in the mouse, is controlled by a complex interaction of genes and epigenetic factors. These processes involve many thousands of genes on all chromosomes. Some of these genes may play a major role in crucial steps of morphogenesis, the involvement of other genes in developmental control may be rather modest, but the transcriptional activity of every chromosome is indispensable at this stage (see Korochkin 1981).

Only with this reservation in mind should one consider suggestions for critical phases of lethal and teratogenic effects of chromosomal abnormalities in mouse development (Gropp 1978, 1981a,b; Gropp et al. 1983). The overall data (see Chapters 1–3, 6,7) establish the existence of at least three main critical stages in the development of mouse aneuploids (Gropp 1981; Gropp et al. 1983). The earliest coincides with implantation and is critical for monosomic embryos, most of which are eliminated just before or during this stage. The second stage roughly corresponds to major organogenesis. Failure of yolk sac vascularization and abnormalities of allantoic growth are two major causes of the massive death of chromosomally abnormal embryos from the tenth to twelfth days of gestation. The origin of gross malformations (neural tube defects, cranio-facial anomalies, cardiovascular disorders) fall into the second critical stage. The majority of mouse trisomies (2–11, 15, 17) as well as some structural aberrations and triploidy comprise this highly polymorphic group.

Late fetal damage associated with general hypoplasia, oedema (mostly due to the insufficiency of the definitive placenta), and numerous malformations are responsible for the third, long-lasting critical stage, with most mouse embryos trisomic for chromosomes 12, 13, 14, 16, 18, and 19 being eliminated a few days before birth or soon thereafter. Severe abnormalities of the cardiovascular system and failure of the chorio-allantoic placenta are considered as major causes of fetal death, while major hypoplasia and inability to breathe are responsible for the elimination of chromosomally abnormal newborns.

Thus, implantation, major organogenesis, and birth itself should be taken as the most crucial developmental checkpoints, providing very efficient elimination of chromosomally abnormal specimens in mammals. This does not however mean that the developing embryo is really more sensitive to chromosomal imbalance at these particular stages.

Many chromosomal trisomies in mice are distinguished by specific or semi-specific malformations and so provide opportunity for the detailed step-by-step analysis of abnormal developmental mechanisms (Gropp 1982). Of special interest are comparative studies of phenotypically rather similar anomalies caused by exogenous teratogens and by specific chromosomal disorders. Exencephaly induced by an excess of vitamin A and by Ts12 or Ts14 may be taken as an example (see Chapter 6; also Putz and Morris-Kay 1981).

It should not be forgotten however that exencephaly may result from a primary defect of the neural plate, from excessive growth of the margins of the neural plate, from impairment of the chordo-mesodermal anlage, from abnormal growth of the bones at the base of the skull, and so on. Over five different developmental mechanisms are implicated in the origin of cleft palate (Baranov and Dyban 1977).

Many more studies are needed, with the assistance of various methods of experimental embryology as well as other experimental approaches, to identify developmental failures induced by specific, semi-specific, and non-specific manifestations of chromosomal imbalance.

It should also be pointed out that detailed analysis of developmental effects caused by chromosomal aberrations is much more difficult at post-implantation than at pre-implantation stages. Modern cytogenetics and developmental biology in mammals is not yet in a position to penetrate fully the intrinsic mechanisms underlying the chromosomal control of mammalian development.

Concluding remarks

The ultimate goal of our monograph is to review the basic data dealing with developmental effects of different chromosomal aberrations in mammalian embryogenesis. It seems justifiable on the basis of this evidence to conclude that the developmental cytogenetics of mammals still lags behind the leading experimental approaches such as molecular biology, for instance, in elucidating the intrinsic genetic mechanisms governing ontogenesis. On the other hand the last 10–15 years have witnessed striking advances in cytogenetic studies. These achievements have mainly been due to the invention of appropriate methods of chromosome analysis for early stages of ontogenesis, including gametogenesis, fertilization, and pre- and post-natal development. Combined with modern high resolution banding techniques, these methods provide reliable identification of homologous chromosomes as well as structural rearrangements throughout the life cycle.

Experimental embryology now has techniques for the deliberate production of whole genome imbalance, as well as convenient biological tools for the study of the developmental effects of trisomy, monosomy, or nullisomy for each individual chromosome in the laboratory mouse.

Hence experimental cytogenetics of mammals is now free of any substantial barriers to the comprehensive study of developmental manifestations of any genome or chromosomal abnormality. We have all the prerequisites needed to overcome the problems hindering our understanding of chromosomal activity during early embryogenesis in mammals.

An urgent task of modern developmental cytogenetics is the detailed analysis of biological effects due to autosomal monosomy and nullisomy. Experimental data on the developmental effects of autosomal monosomies are still rather fragmentary and autosomal nullisomies still remain practically 'terra incognita'. However, there are several feasible routes to induce both monosomy and nullisomy experimentally and convenient biological tools (mice with appropriate chromosomal translocations) at the disposal of many cytogenetic laboratories, with which to launch efficient production and detailed analysis of these chromosomal abnormalities. The same is true for the developmental effects of partial deletions both in autosomes and in the X chromosome.

Methods of experimental embryology combined with other experimental approaches should be widely applied to the study of developmental mechanisms interrupted by aneuploidy. Two recently invented methods for non-invasive identification of early mouse embryos of known karyotype are of especial value for such studies. The split-embryo technique permits chromosomal analysis of the twin embryos at the blastocyst stage (Epstein *et al.* 1978), while an alternative technique used in the Department of Embryology allows efficient selection at the four-cell stage and onwards (Severova and Dyban 1984). Both methods could help substantially in clarifying embryo-lethal and phenocritical stages of monosomies and other aberrations, and thus help to distinguish between specific and non-specific reactions induced by aneuploidy.

Pre-implantation stages offer more opportunities for current developmental genetics than post-implantation stages. Pre-implantation stages are easily accessible for various *in vivo* and *in vitro* methods of experimental embryology. Morphogenetic processes in pre-implantation embryos are simple compared to the very complex multicellular interactions after implantation. Moreover, these early stages are much more thoroughly studied from genetic, biochemical, and experimental embryological points of view. Hence combined cytogenetic, experimental embryological, cytological, and molecular biological approaches might be very productive in understanding chromosomal functions during early development.

Experimental chimaeras offer a further promising approach to developmental cytogenetics (Epstein 1984). Aneuploid–diploid chimaeras made during cleavage are very convenient biological tools for studying metabolic cooperation between cells of different karyotype, and hence provide opportunities for extensive studies of the expression of chromosomal anomalies both at the cellular and organismal levels. Abnormalities of metabolic intercellular cooperation seem to be a plausible mechanism for developmental failure in parthenogenetic, gynogenetic, and androgenetic embryos. Studies of the expression of chromosomal anomalies at the molecular, subcellular, cellular, and organismal levels are of high priority.

Differential gene activity during the early development of mammals as well as in other animals seems firmly established. There are many different experimental methods to investigate this crucial problem of genetics and developmental biology. The cytogenetic approach on its own does not seem to be very productive in this field in comparison with, for instance, methods of molecular genetics. However, the combination of the two techniques might be extremely efficient for the study of gene activity in differentiation. Recombinant DNA techniques widely used at present for gene mapping studies might also be effectively applied for the chromosomal assignment of 'early' genes transcribed during early stages of embryogenesis, with the subsequent production of gene libraries and analysis of their functions.

Hence the marriage of cytogenetics and molecular genetics might be a very productive strategy for modern developmental genetics.

Each phenotypic character depends on the action of many genes, and the effect of each gene is dispersed among many phenotypic characters. The same is true for gene expression in early mammalian embryos. The genes, however, are not completely isolated from each other either structurally or functionally, and each chromosome should be considered as an integrated unit of many thousands of individual genes. Absence or excess of the whole chromosome would mean the imbalance of many structural genes. Thus we can hardly distinguish the developmental effects of any particular gene from the thousands of neighbouring genes carried on the same chromosome. Cytogenetic tools however afford a unique opportunity to elucidate the developmental effects of the whole gene complex of some particular chromosome, and thus to study chromosome interactions in embryonic development.

It has been suggested (Dyban 1979) that genes responsible for morphogenetic processes related in time and coordinated in space are inherited as a single gene block, regulated as one integral system mainly by means of heterochromatization.

Exploration of the developmental effects of monosomy and nullisomy, as well as of deletions and inversion, may provide experimental verification of this hypothesis, and may yield information on the biological significance of gene linkages and the function of gene blocks carried in a definite order along a chromosome.

One other urgent question of current developmental cytogenetics concerns heterochromatization of the whole chromosome or its segments. Present techniques allow X chromosome heterochromatization to be studied in early mammalian embryos, in normal embryos and in combination with autosomal imbalance. The precise mechanism of X chromosome inactivation still remains unknown and its correlation with trisomy or monosomy of particular autosomes has not been studied.

Another promising branch of developmental cytogenetics deals with allelic interactions in early embryos with trisomy, tetrasomy, triploidy and tetraploidy. Of special interest in this respect are experiments on the characteristics of mutant genes with very early expression. Some indirect evidence suggests that the expression of chromosomal aberrations depends on genetic background, i.e. on the presence of modifier genes. This problem deserves special analysis in experiments in which both genetic and non-genetic (external) factors are precisely specified, and combined with various known chromosomal anomalies.

In view of recent findings favouring complementary interactions between maternal and paternal pronuclei as an indispensable prerequisite of normal embryogenesis (Surani *et al.* 1984; McGrath and Solter 1984), it seems urgent to study gene reprogramming from early embryogenesis onwards. This task is closely related to the problem of chromosome interactions in early development. The necessity of interchromosomal interactions for adequate homologous chromosome activation both in time and in space during early mammalian embryogenesis has been postulated (Baranov 1983a,b,c,d), and all methodological prerequisites now exist for testing this hypothesis.

Developmental cytogenetics of mammals has not only theoretical but also practical applications both in medicine and in veterinary practice.

There are still a good many scattered facts and islands of fragmentary knowledge in the developmental cytogenetics of mammalian embryogenesis. This substantially restrains or makes premature any fundamental generalizations or theoretical considerations. Much more experimental effort is needed to clarify the developmental effects of monosomy, nullisomy, and deletions on mammalian embryogenesis.

Successful advance of any science depends on its methodological outfit and on the precision of its tasks, as well as on the enthusiasm of its practitioners. The presence of new reliable methods for the chromosomal analysis of mammalian embryos, the existence of convenient biological tools and experimental embryological approaches for the deliberate production of mammalian embryos with any desired type of chromosomal imbalance, as well as the ample opportunities existing for their detailed analysis both *in vivo* and *in vitro* by means of modern methods of developmental biology and molecular biology — all these factors should guarantee a rapid advance of our knowledge in the field of developmental cytogenetics of mammals.

References

Abramczuk, J. and Sawicki, W. (1975a). Pronuclear synthesis of DNA in fertilized and parthenogenetically activated mouse eggs. *Exp. Cell Res.* **92**, 361–72. *Aging gametes. Their biology and pathology* (ed. R.J. Blandau). Karger, Basel.

Adamson, E.D. and Gardner R.L. (1979). Control of early development. *Br. Med. Bull.* **35**, 113–119.

Adolph, S. and Klein, J. (1981). Robertsonian variations in *Mus musculus* from Central Europe, Spain and Scotland. *J. Hered.* **72**, 219–221.

Alfert, M. (1950). A cytochemical study of oogenesis and cleavage in the mouse. *J. Cell. comp. Physiol.* **36**, 381–409.

Amoroso, E.C. and Parkes, A.S. (1947). Effect on embryonic development of X-irradiation of rabbit spermatozoa *in vitro*. *Proc. R. Soc. Lond.* **184**, 57.

Anderson, E., Hoppe, P.C., and Lee, G.S. (1984). The karyotype and ultrastructural characteristics of spontaneous pre-implantation mouse parthenotes. *Gamete Res.* **9**, 451–67.

Andrews, T., Dunlop, W., and Roberts, D.E. (1984). Cytogenetic studies in spontaneous abortuses. *Hum. Genet.* **66** (1), 77–84.

Angell, R.R., Aitken, R.J., Van Look, P.F.A., Lumdsden, M.A., and Templeton, A.A. (1983). Chromosomal abnormalities in human embryos after *in vitro* fertilization. *Nature* **303**, 336–8.

Ansell, J.D. and Snow, M.H.L. (1975). The development of trophoblast *in vitro* from blastocysts containing varying amounts of inner cell mass. *J. Embryol. exp. Morphol.* **33**, 177–85.

Arkhangelskaya, I.B. (1985). Peculiarities of *in vitro* development of cell fragments isolated microsurgically from zygotes and early blastomeres of mice. In *Mechanisms of normal and abnormal embryogenesis in mammals*, pp. 47–50. Nauka, Leningrad (in Russian).

Artzt, K., Shin, H.-S., and Bennett, L. (1982). Gene mapping within the T/t complex of the mouse. II. Anomalous position of the H-2 complex in t-haplotypes. *Cell* **28**, 471–6.

—— and Cheng, C. (1982). Biochemical studies of T/t antigens. *Oncodev. Biol. Med.* **4**, 53–8.

Ashburner, M. (1981). Chromatin and gene expression. *Hoppe-Seylers Z. physiol. Chem.* **362**, 379–87.

Ashley, T. (1983). Sex vesicle loss: a possible explanation of the excess of XO over XXY conceptuses in mice and men. *Hum. Genet.* **65**, 209–10.

—— and Pocock, N.A. (1981). A proposed model of chromosomal organization in nuclei at fertilization. *Genetica* **55**, 161–9.

Astaurov, B.L. (1948). Experimental studies of merogony and androgenesis and their significance for development theory and heredity. *Usp. sovrem. Biol.* **25**, (1), 49–88 (in Russian with English summary).

—— (1960). Differential X-ray sensitivity of nucleus and cytoplasm as a consequence of their functional specialization. *Byull. Mosk. Obsch. Ispyt. Prir. Otd. Biol.* (Moscow Soc. Nat. Sci.) **63**, 35–40.

—— (1968). Problems of individual development. (Results and perspective). *Zh. obsch. Biol.* **29**, 139–52 (in Russian with English summary).

—— (1969). Experimental polyploidy in animals. *Ann. Rev. Genet.* **3**, 99–126.

—— (1972). Genetics and the problems of ontogenesis. *Ontogenez* **6**, 547–65 (in Russian with English summary).

—— (1974). *Heredity and development*. Nauka, Moscow.

Austin, C.R. (1961). *The mammalian egg*. Blackwell, Oxford.

—— (1969). Anomalies of fertilization leading to triploidy. *J. Cell comp. Physiol.* **56**, (Suppl. 1), 1–15.

—— (1960). Variations and anomalies in fertilization. *Fertilization* **2**, 437–66.

—— and Braden, A.W.H. (1953). An investigation of polyspermy in the rat and rabbit. *Aust. J. biol. Sci.* **6**, 674.

——, —— (1954a). Anomalies in rat, mouse and rabbit egg. *Aust. J. biol. Sci.* **7**, 537–42.

——, —— (1954b). Induction and inhibition of the second polar division in the rat egg and subsequent fertilization. *Aust. J. biol. Sci.* **7**, 195–210.

Ayme, S., Mattei, J.F., Mattei, M.G., and Giraud, F. (1980). Anomalies chromosomiques, facteurs de risque actuellement connus. *J. Genet. hum.* **28**, 155–78.

Babiarz, B.S. (1983) Deletion mapping of the T/t complex: evidence for a second region of critical embryonic genes. *Devl Biol.* **95**, 342–51.

Bachvarova, R. (1974). Incorporation of tritiated adenosine into mouse ovum RNA. *Devl Biol.* **40**, 50–8.

—— (1981). Synthesis, turnover and stability of heterogeneous RNA in growing mouse oocytes. *Devl Biol.* **86**, 384–92.

—— and De Leon, V. (1980). Polyadenylated RNA of mouse ova and loss of maternal RNA in early development. *Devl Biol.* **74**, 1–8.

—— Burns, J., Spiegelman, I., Choy, J., and Chaganti, R.S.K. (1982). Morphology and transcriptional activity of mouse oocyte chromosomes. *Chromosoma (Berl.)* **86**, 181–96.

Baghdassarian, A., Boyard, F., Borgaonkar, D.S., Arnold, E.A., Solez, K., and Migeon, C.J. (1975). Testicular function in XYY men. *Johns Hopkins med. J.* **136**, 15–24.

Baker, B.S. (1976). The genetic control of meiosis. *Ann. Rev. Genet.* **10**, 53–134.

Baker, T.G. (1972). Oogenesis and ovulation. In *Germ cells and fertilization* (ed. C.R. Austin and R.V. Short), pp. 14–45. Cambridge University Press, Cambridge.

Balakier, H. and Tarkowski, A. (1976). Diploid parthenogenetic mouse embryos produced by heat-shock and cytochalasin B. *J. Embryol. exp. Morphol.* **35**, 25–39.

——, —— (1980). The role of germinal vesicle karyoplasm in the development of male pronucleus in the mouse. *Exp. Cell Res.* **128**, 79–85.

Baranov, V.S. (1974). Trisomy of autosomes in newborns. Experiments in mice heterozygous for Rb6Bnr translocation. *Bull. exp. Biol. Med. USSR* **127** (5), 99–103 (in Russian with English summary).

—— (1976a). Morphogenetic and cytogenetic analysis of chromosomal embryopathies in mice. Doctoral Thesis, Leningrad (in Russian).

—— (1976b). Spontaneous polyploidy in laboratory mice. An embryological and cytological analysis. *Ontogenez* 7, 229–38 (in Russian with English summary).

—— (1979). A new marker reciprocal translocations T(7.14)2 Iem and T(7.14.15)3 Iem in laboratory mice. *Genetika* 15, 1651–60 (in Russian with English summary).

—— (1980). Mice with Robertsonian translocations in experimental biology and medicine, *Genetica* (*The Netherlands*) 52/53, 23–32.

—— (1981). Rb(2.6)4 Iem—a new marker Robertsonian translocation in laboratory mice. *Tsitologiiya* 13, 820–9 (in Russian with English summary).

—— (1982a). Influence of structural aberrations of the chromosomes 7 and 14 on embryogenesis of laboratory mice. *Ontogenez* 13 (4), 360–7 (in Russian with English summary).

—— (1982b). Genetic and cytogenetic marking for neurological Quaking (qk) mutation in laboratory mouse *Mus musculus. Genetika* 18 (2), 235–40 (in Russian with English summary).

—— (1983a). Chromosomal control of early embryonic development in mice. I. Experiments on embryos with autosomal monosomy. *Genetica* (*The Netherlands*) 61, 165–77.

—— (1983b). Chromosomal control of early embryonic development in mice. II. Experiments on embryos with structural aberrations of autosomes 7,9,14 and 17. *Genet. Res.* 41, 227–39.

—— (1983c). Chromosomal control of early embryonic development in mammals. *Ontogenez* 14 (6), 573–88. (in Russian with English summary).

—— (1983d). The effect of monosomy of some autosomes on pre-implantation development in laboratory mice. *Ontogenez* 14 (1), 73–8l (in Russian with English summary).

—— (1985). Peculiarities of t^{12} mutation expression in mouse embryos with structural aberrations of chromosome 17. *Genetica* 21 (10), 1685–92 (in Russian with English summary).

—— and Chebotar, N.A. (1980). Mutagenic effect of pyrimethamine injection before ovulation on pre-implantational rat embryos. *Cytol. genet.* 6, 20–4 (in Russian with English summary).

—— and Dyban A.P. (1968). Analysis of spermatogenesis and embryogenic damage in mice heterozygous for T6-translocation. *Genetica* 4, (12), 70–83 (in Russian with English summary).

——, —— (1970). Correlation of embryonic damages with types of chromosomal aberrations, arising in meiosis of mice with T-6 translocation. *Ontogenez* 1, (3), 248–261 (in Russian with English summary).

——, —— (1971a). New marker Robertsonian translocation (centric fusion of autosomes) in laboratory mice. *Tsitologiya* 13 (7), 820–9 (in Russian with English summary).

——, —— (1971b). Embryogenesis and peculiarities of karyotype in mouse embryos with centric fusion of chromosomes (Robertsonian translocation)T1Iem. *Ontogenez* 2 (2), 164–176 (in Russian with English summary).

——, —— (1972a). Frequency of spontaneous chromosomal aberrations in early post-implantation development of mice. *Bull. exp. Biol. Med. USSR* 74 (4), 107–10 (in Russian with English summary).

——, —— (1972b). Embryonic damage caused by autosomal trisomy arising in the progeny of mice with Robertsonian translocation (centric fusion of autosomes) T1

Iem. *Arkh. Anat. Histol. Embryol.* **63** (8), 67–77 (in Russian with English summary).

——, —— (1972c). The effect of maternal genotype on the frequency of trisomy of autosomes in embryogenesis. *Bull. exp. Biol. Med. USSR* **74** (12), 92–6 (in Russian with English summary).

——, —— (1973). Cytogenetic analysis of early embryogenesis in mice, heterozygous for Robertsonian translocation T1 Ald. *Ontogenez* **4** (6), 587–94 (in Russian with English summary).

——, —— (1976). The prezygotic selection of male gametes in laboratory mice. *Tsitologiya* **18** (5), 556–9 (in Russian with English summary).

——, —— (1977). Developmental analysis of some embryonic anlagen by means of specific teratogens. In *External factors in individual development,* pp. 325–35. Nauka, Moscow (in Russian).

—— and Udalova, L.D. (1974). Identification of chromosomes involved in Rb1 Iem translocation in mice. *Tsitologiya* **16** (9), 1169–71 (in Russian with English summary).

——, —— (1975). Autosomal trisomy in mice, heterozygous for Robertsonian translocations. *Arkh. Anat. Histol. Embryol.* **69** (11), 63–74.

——, Dyban, A.P., and Chebotar, N.A. (1980). Peculiarities of pre-implantation mouse development in case of monosomy by autosome 17. *Ontogenez* **11** (2), 148–59 (in Russian with English summary).

——, Gregorova, S., and Forejt, J. (1981). Fertility and cytogenetic analysis of early embryogenesis in mice with chromosomal translocation T(16.17)43 H. *Genetika* **17**, 1454–9 (in Russian with English summary).

——, Waisman, B.L., and Udalova L.D. (1982). Some morphological peculiarities and state of rRNA synthesis in the mouse embryos with trisomy for chromosome 19. *Ontogenez* **13** (1), 46–54. (in Russian with English summary).

Barlow, P., Owen, D.A.J., and Graham, Ch. (1972). DNA synthesis in the pre-implantation mouse embryo. *J. Embryol. exp. Morphol.* **27**, 431–445.

—— and Sullivan, F.M. (1975). Behavioural teratology. In *Teratology, trends and applications* (ed. C.L. Berry and D.E. Poswillo), pp. 103–20. Springer, Berlin.

Basler, A. (1978). Timing of meiotic stages in oocytes of the Syrian hamster (*Mesocricetus auratus*) and analysis of induced chromosome aberrations. *Hum. Genet.* **42**, 67–77.

Beatty, R.A. (1957). *Parthenogenesis and polyploidy in mammalian development.* Cambridge University Press, Cambridge.

—— (1970). The genetics of mammalian gametes. *Biol. Rev.* **45**, 73–120.

—— (1974). Genetic aspects of spermatozoa. In *Physiology and genetics of reproduction* (ed. E.M. Coutinho and F. Fuchs). Plenum Press, New York.

—— (1978). The origin of human triploidy: an integration of quantitative and qualitative evidence. *Ann. Hum. Genet.* **42**, 299–311.

—— and Coulter, V.J. (1978). Digynic triploidy in rabbit blastocysts after delayed insemination. *Genet. Res.* **32**, 9–18.

—— and Fechheimer, N.L. (1972). Diploid spermatozoa in rabbit males and their experimental separation from haploid spermatozoa. *Biol. Reprod.* **7**, 267–77.

—— and Fischberg, M. (1949). Spontaneous and induced triploidy in pre-implantation mouse eggs. *Nature* **163**, 807–8.

——, —— (1951a). Heteroploidy in mammals. I. Spontaneous heteroploidy in preimplantation mouse eggs. *J. Genet.* **50**, 345–59.

——, —— (1951b). Cell number in haploid, diploid and polyploid mouse embryos. *J. exp. Biol.* **28**, 541–52.

——, —— (1952). Heteroploidy in mammals. III. Induction of tetraploidy in preimplantation mouse eggs. *J. Genet.* **50**, 471–5.

——, Lim, M.C., and Coulter, V.J. (1975). A quantitative study of the second meiotic metaphase in male mice *Mus musculus*. *Cytogenet. Cell Genet.* **15**, 256–75.

Beechey, C.V. and Searle, A.G. (1979). *Mouse News Letter* **60**, 44–5.

——, Kirk, M., and Searle, A.G. (1980). A reciprocal translocation induced in the oocyte and affecting fertility in male mice. *Cytogenet. Cell Genet.* **27**, 129–46.

Benirschke, K. (1974). Chromosomal errors and reproductive failure. In *Physiology and genetics of reproduction* Part A. (ed. E.M. Coutinho and F. Fuchs) pp. 73–91. Plenum, New York.

——, Bogart, M.H., McClure, H.M., and Nelson-Rees, W.A. (1974). Fluorescence of the trisomic chimpanzee chromosomes. *J. med. Primatol.* **3**, 311–14.

Bennett, D. (1975). The T-locus in the mouse: a review. *Cell* **6**, 441–54.

—— (1981). T/t locus, its role in embryogenesis and its relation to classical histocompatibility system. *Prog. Allergy* **29**, 35–53.

——, Goldberg, E., Dunn, L., and Boyse, E. (1972). Serological detection of a cell-surface antigen specified by the T/Brachyury/mutant gene in the house mouse. *Proc. nat. Acad. Sci. U.S.A.* **69**, 2076–80.

Bensaude, O., Babinet, C., Morange, M., and Jacob, F. (1983). Heat-shock proteins first major products of zygotic gene activity in mouse embryos. *Nature* **305**, 331–2.

Berger, R. 1972 Monosomies G. *Nouv. Presse. med.* **1**, 1033–34.

Bergstrom, S. (1978). Experimentally delayed implantation. In *Methods in mammalian reproduction* (ed. J.C. Daniel) pp. 419–35. Academic Press, London.

Bernstein, R.M. and Mukherjee, B.B. (1972). Control of nuclear RNA synthesis in 2-cell and 4-cell mouse embryos. *Nature* **238**, 457–9.

——, —— (1973). Cytoplasmic control of nuclear activity in preimplantation mouse embryos. *Devl Biol.* **34**, 47–66.

Berry, R.J., Beechey, C.V., and Searle, A.G. (1973). Cytogenetic radiosensitivity and chiasma frequency in wild and laboratory mouse. *Mutat. Res.* **19**, 127–31.

Bersu, E.T. (1984). Morphologic development of the fetal trisomy 19 mouse. *Teratology* **29**, 117–29.

——, Crandall, C., and White B. (1982). Growth characteristics of the murine trisomy 19 thymus. *Teratology* **26**, 85–94.

——, ——, —— (1983). Erythropoiesis in the fetal trisomy 19 mouse. I. Characterization of erythrocyte populations in peripheral blood. *Teratology* **27**, 271–6.

Biggers, J.D. and Borland, R.M. (1976). Physiological aspects of growth and development of preimplantation mammalian embryos. *A. Rev. Physiol.* **38**, 95–119.

—— and Stern, S. (1973). Metabolism of the preimplantation mammalian embryo. *Adv. reprod. physiol. Sci. Lond.* **6**, 1–59.

——, Whitten W.K., and Whittingham, D.G. (1971). The culture of mouse embryos *in vitro*. In *Methods of mammalian embryology* (ed. J.C. Daniel) pp. 86–116. W.H. Freeman, San Francisco.

Blandau, R.J. (1961). Biology of eggs and implantation. In *Sex and internal secretions* Vol. 2, pp. 797–882. Williams and Wilkins, Baltimore.

—— (1971). Observing ovulation and egg transport. In *Methods in mammalian embryology* (ed. J.C. Daniel), pp. 1–14. W.H. Freeman, San Francisco.

304   References

Bochkov, N.P. (1971). The incidence of spontaneous chromosomal diseases in man. In *Human chromosomes and irradiation*, pp. 126–41. Atomizdat, Moscow.

——, Zakharov, A.F., and Ivanov, V.I. (1984). *Medical genetics (physicians manual)*. Meditsina, Moscow (in Russian).

Boer, P. de (1973). Fertile tertiary trisomy in the mouse. *Cytogenet. Cell Genet.* **12**, 435–42.

—— (1974). A meiotic study of two translocations and a tertiary trisomy in the mouse (*Mus musculus*). Ph.D. thesis, Wageningen.

—— and van Beek, M.E.A.B. (1982). Meiosis of T70H translocation trisomic male mice. Meiotic configuration and segregation. *Chromosoma (Berl.)* **87**, 303–13.

—— and Branje, H.E.B. (1979). Association of the extra chromosome of tertiary trisomic male mice with the sex chromosomes during the first meiotic prophase and its significance for impairment of spermatogenesis. *Chromosoma (Berl.)* **73**, 369–79.

—— and Gissen, M. (1974). The location of the positions of the breakpoints involved in the T26H and T70H mouse translocations with the aid of Giemsa-banding. *Can. J. Genet. Cytol.* **16**, 783–8.

—— and Groen, A. (1974). Fertility and meiotic behaviour of male T70H tertiary trisomics of the mouse. A case of preferential telomeric meiotic pairing in a mammal. *Cytogenet. Cell Genet.* **13**, 485–510.

—— and de Maar, P.H.M. (1976). A histological study of embryonic death caused by heterozygosity for the T26H reciprocal mouse translocation. *J. Embryol. exp. Morphol.* **35** (3), 595–606.

—— and Hoeven, F.A. (1980). The use of translocation-derived marker-bivalent for studying the origin of meiotic instability in female mice. *Cytogenet. Cell Genet.* **26**, 49–58.

——, ——, and van der Chardon, J.A.P. (1976). The production, morphology, karyotypes and transport of spermatozoa from tertiary trisomic male mice and the consequences for egg fertilization. *J. Reprod. Fert.* **60**, 257–65.

—— and Searle, A.G. (1980). Workshop on chromosomal aspects of male sterility in mammals. Summary and synthesis. *J. Reprod. Fert.* **60**, 257–65.

—— and Speed, R.M. (1982). Meiosis of T70H translocation trisomic male mice. II. Meiotic rate, spermatocyte interactions and fertility. *Chromosoma (Berl.)* **87**, 315–25.

Bolton, V.N., Oades, P.J., and Johnson, M.H. (1984). The relationship between cleavage, DNA replication, and gene expression in the mouse 2-cell embryo. *J. Embryol. exp. Morphol.* **79**, 139–63.

Bomsel-Helmreich, O. (1965). Heteroploidy and embryonic death. *CIBA Foundation Symposium: preimplantation stages of pregnancy* pp. 246–67. Churchill, London.

—— (1967). Triploidie expérimentale chez quelques mammifères. Thèse doctorale des Sciences naturelles. Paris, Institute Nationale des Recherches Agronomiques.

—— (1976). Experimental heteroploidy in mammals. In *Current topics in pathology* (ed. A. Gropp and K. Benirschke) pp. 155–73. Springer, Berlin.

Bond, L.J. and Chandley, A.C. (1983). *Aneuploidy*. Oxford University Press, Oxford.

Bonnevie, J. (1934). Embryological analysis of gene manifestation in Little and Bagg's abnormal mouse tribes. *J. exp. Zool.* **67**, 443–520

—— (1940). Tatsachen der genetischen Entwicklungsphysiologie. *Handbuch der Erbbiologie des Menschen* Bd. 3, pp. 154–172.

Borsuk, E. (1982). Preimplantation development of gynogenetic diploid mouse embryos. *J. Embryol. exp. Morphol.* **69**, 215–222.

Boué, J.G. and Boué, A. (1970). Les aberrations chromosomiques dans les avortements spontanés humans. *Presse méd.* **78**, 635–41.

——, —— (1971). Avortements et chromosomes. *Sci. Prog.* **3**, 22–31.

——, —— (1974). Chromosome abnormalities and abortion. In *Physiology and genetics of reproduction. Part. B* (ed. E.M. Coutinho and F. Fuchs) pp. 317–39. Plenum, New York.

——, —— (1976). Chromosomal anomalies in early spontaneous abortion. In *Current topics in pathology* (ed. A. Gropp and K. Benirschke) pp. 193–208. Springer, Heidelberg.

——, —— (1978). Prenatal diagnosis in 100 structural rearrangements of the chromosomes. *Cytogenet. Cell Genet.* **20**, 214–25.

——, —— and Lasar, P. (1975). The epidemiology of human spontaneous abortions with chromosomal anomalies. In *Aging gametes*, pp. 330–48. Karger, Basel.

——, ——, Philippe, L., and Gueguen, S. (1972). Sur les durées de developpement et de retention de 716 zygotes, produits d'avortements spontanés précocés. *C. r. hebd. Séanc. Acad. Sci. Paris* **D272**, 2992–6.

——, Philippe, E., Giroud, A., and Boué, A. (1976). Phenotypic expression of lethal chromosomal anomalies in human abortuses. *Teratology* **14**, 3–28.

Bowman, P., and McLaren, A. (1970a). Cell number in early embryos from strains of mice selected for large and small body size. *Genet. Res.* **15**, 261–3.

——, —— (1970). Cleavage rate of mouse embryos *in vivo* and *in vitro*. *J. Embryol. Exp. Morphol.* **24**, 203–7.

Braden, A.W.H. (1957). Variation between strains in the incidence of various abnormalities of egg maturation and fertilization in the mouse. *J. Genet.* **55**, 476–86.

—— (1958). Variation between strains of mice in phenomena associated with sperm penetration and fertilization. *J. Genet.* **56**, 1–11.

—— (1972). T-locus in mice; segregation – distortion and sterility in the male. In *The genetics of the spermatozoon* (ed. R.A. Beatty and S. Gluecksohn-Waelsch) pp. 289–305. Bogtrykkeict Forum, Copenhagen.

—— and Austin, C.R. (1954). Fertilization of the mouse egg and the effect of delayed coitus and of hot-shock-treatment. *Aust. J. biol. Sci.* **7**, 552–65.

—— and Gluecksohn-Waelsch, S. (1958). Further studies of the effect of the t-locus in the house mouse on male fertility. *J. exp. Zool.* **138**, 341–52.

Braude, P.R. (1979a). Control of protein synthesis during blastocyst formation in the mouse. *Devl Biol.* **68**, 440–52.

—— (1979b). Time-dependent effects of amanitin on blastocyst formation in the mouse. *J. Embryol. exp. Morphol.* **52**, 193–202.

——, Pelham, H., Flach, G., and Lobatto, R. (1979). Post-transcriptional control in the early mouse embryo. *Nature* **282**, 102–5.

Breckon, G. and Savage, J.R.K. (1982). Homozygous deficiency: syrian hamsters with only 42 chromosomes. *Cytogenet. Cell Genet.* **33**, 285–94.

Brenneke, H. (1937). Strahlenschädigung an Mäuse and Rattensperma, beobachtet an der Frühentwicklung der Eier. *Strahlen therapie* **60** (S), 214–38.

Brinster, R.L. (1967). Protein content of the mouse embryo during the first five days of development *J. Reprod. Fert.* **13**, 413–20.

—— (1971a). Uptake and incorporation of amino acids by the preimplantation mouse embryo *J. Reprod. Fert.* **27**, 329–38.

—— (1971b). Protein synthesis and enzyme constitution of the preimplantation mammalian embryo. In *Regulation of mammalian reproduction* (ed. S. Segal, N. Clozier, and P. Corfman) pp. 250–80.

—— (1973). Parental glucose phosphate isomerase activity in three-day mouse embryos. *Biochem. Genet.* **9**, 187–91.

——, Wiebold, J.L., and Brunner, S. (1976). Protein metabolism in preimplanted mouse ova. *Devl Biol.* **51**, 215–24.

Brodsky, V.Y. and Urivaeva, I. (1981). *Cell polyploidization, proliferation and differentiation.* Nauka, Moscow.

Brook, J.D., Gosden, R.G., and Chandley, A.S. (1984). Maternal aging and aneuploid embryos. Evidence from the mouse that biological but not chronological age is the important influence. *Hum. Genet.* **66**, 41–5.

Brooks, R.F. (1977). Continuous protein synthesis is required to maintain the probability of entry into S phase. *Cell* **12**, 311–17.

Bruere, A.M. (1975). Further evidence of normal fertility and the formation of balanced gametes in sheep with one or more different Robertsonian translocations. *J. Reprod. Fert.* **45**, 323–31.

——, Scott, I.S., and Henderson, L.M. (1981). Aneuploid spermatocyte frequency in domestic sheep heterozygous for three Robertsonian translocations. *J. Reprod. Fert.* **63**, 61–6.

Burgoyne, P.S. (1978). The role of sex chromosomes in mammalian germ cell differentiation. *Ann. Biol. Anim. Biochem. Biophys.* **18** (2B), 317–25.

Burgoyne, P.S. and Baker, T.G. (1981). Oocyte depletion in XO mice and their XX sibs from 12 to 200 days *post partum. J. Reprod. Fert.* **61**, 207–12.

—— and Biggers, J.D. (1976). The consequences of X-dosage deficiency in the germ line: impaired development *in vitro* of preimplantation embryos from XO mice. *Devl Biol.* **51**, 109–17.

—— and Diddle, F.G. (1980). Spermatocyte loss in XYY mice. *Cytogenet. Cell Genet.* **28** (1–2), 143–4.

——, Evans, E.P., and Holland, K. (1983a). XO monosomy is associated with reduced birthweight and lowered weight gain in the mouse. *J. Reprod. Fert.* **68** (2), 381–5.

——, Tam, P.P.L., and Evans, E.P. (1983b). Retarded development of XO conceptuses during early pregnancy in the mouse. *J. Reprod. Fert.* **68** (2), 387–93.

Butcher, R.L. and Fugo, N.W. (1967). Overripeness and the mammalian ova. II. Delayed ovulation and chromosome anomalies. *Fert. Steril.* **18**, 297–302.

Cacheiro, N.L.A., Russell, L.B., and Swartout, M.S. (1974). Translocations, the predominant cause of total sterility in sons of mice treated with mutagens. *Genetics* **76**, 73–91.

Calarco, P.G. and Brown, E.H. (1968). Cytological and ultrastructural comparisons of t^{12}/t^{12} and normal mouse morulae. *J. exp. Zool.* **168**, 169–86.

—— and Pedersen, R.A. (1976). Ultrastructural observations of lethal yellow (A^y/A^y) mouse embryos. *J. Embryol. exp. Morphol.* **35**, 73–80.

Capanna, E., Gropp, A., Winking, H., Noack, G., and Civitelli, M.V. (1976). Robertsonian metacentrics in the mouse. *Chromosoma* **58**, 341–53.

Carothers, A.D. and Beatty, R.A. (1975). The recognition and incidence of haploid and polyploid spermatozoa in man, rabbit and mouse. *J. Reprod. Fert.* **44**, 487–500.

Carr, D.H. (1965). Chromosome studies in spontaneous abortions. *Obstet. Gynecol.* **26**, 308–26.

—— (1967a). Chromosomal anomalies as a cause of spontaneous abortions. *Am. J. Obstet. Gynecol.* **97**, 283–93.

—— (1967b). Cytogenetics of abortion. In *Comparative aspects of reproductive failure* (ed. K. Benirschke) pp. 96–117. Springer, Berlin.

—— (1969). Lethal chromosome errors. In *Comparative mammalian cytogenetics* (ed. K. Benirschke) pp. 68–90. Springer, Berlin.

—— (1971a). Genetic basis of abortion. *A. Rev. Genet.* **5**, 65–80.

—— (1971b). Chromosome studies in selected spontaneous abortions. Polyploidy in man. *J. med. Genet.* **8**, 164–74.

—— (1972). Chromosomal anomalies in human foetuses. *Res. Reprod.* **4**, 3–4.

—— (1983). Cytogenetics of human reproductive wastage. In *Issues and reviews in teratology* (ed. H. Kalter) Vol.1, pp. 33–72. Plenum Press, New York.

——, Haggar, R.A., and Hart, A.G. (1968). Germ cells in the ovaries of XO female infants. *Am. J. clin. Pathol.* **49**, 521–6.

Carter, T.C., Lyon, M.F., and Phillips, R.J.S. (1955). Gene tagged chromosome translocations in eleven stocks of mice. *J. Genet.* **53**, 154–66.

Cascio, S.M. and Wasserman, P.M. (1982). Programme of early development in the mammals: post-transcriptional control of a class of proteins synthesized by mouse oocytes and early embryos. *Devl Biol.* **89**, 397–408.

Cassidy, S.B., Whitworth, T., Sanders, D., Lorber, C.A., and Engel, E. (1977). Five month extrauterine survival in a female triploid (69 XXX) child. *Ann. Genet.* **20**, 277–9.

Cattanach, B.M. (1962). XO-mice. *Genet. Res.* **3**, 487–90.

—— (1964). Autosomal trisomy in the mouse. *Cytogenetics* **3**, 159–66.

—— (1967). A test of distributive pairing between two specific nonhomologous chromosomes in the mouse. *Cytogenetics* **6**, 67–77.

—— (1975). Control of chromosome inactivation. *A. Rev. Genet.* **9**, 1–18.

—— and Moseley, H, (1973). Nondisjunction and reduced fertility caused by the tobacco mouse metacentric chromosomes. *Cytogenet. Cell Genet.* **12**, 267–87.

—— and Pollard, C.E. (1969). An XYY sex-chromosome constitution in the mouse. *Cytogenetics* **8**, 80–6.

——, Williams, C.E., and Batley H. (1972). Identification of the linkage groups carried by the metacentric chromosomes of the tobacco mouse (*Mus poschiavinus).* *Cytogenetics* **11**, 412–23.

—— and Savage, G.R.K. (1976). A new Robertsonian translocation. *Mouse News Letter* **54**, 38.

Centerwall, W.R. and Benirschke, K. (1973). Male tortoise-shell and calico (T–C) cats. Animal models of sex-chromosome mosaics, aneuploids, polyploids and chimaerics. *J. Hered.* **64**, 272–8.

——, —— (1975). An animal model for the XXY Kleinfelter's syndrome in man: tortoise-shell and calico male cats. *Am. J. vet. Res.* **36**, 1275–80.

Chan, F.P.H., Ball, J.K., and Sergovich, F.R. (1979). Trisomy 15 in murine thymomas induced by chemical carcinogens, X-irradiation and an endogenous murine leukemia virus. *J. nat. Cancer Inst.* **62** (3), 605–610.

Chandley, A.C. (1979). The chromosomal basis of human infertility. *Br. med. Bull.* **35**, 181–6.

—— (1980). Genetic causes of abnormal human gamete formation. In *Research on fertility and sterility* (ed. J. Cortes-Prierto) pp. 381–92. MJR Press, New York.

—— (1981). Male infertility and meiosis in man. In *Oligozoospermia: recent progress in andrology* (ed. G. Frajese *et al.*) pp. 247–65. Raven Press, New York.

—— (1982). The origin of aneuploidy. In *Human genetics, Part B: medical aspects* pp. 337–47. Alan Liss, New York.

—— (1984). Infertility and chromosome abnormality. *Oxford Rev. reprod. Biol.* **6**, 1–46.

——, Christie, S., Fletcher, I., Frackiewics, A., and Jacobs, P.A. (1972). Translocation heterozygosity and associated subfertility in man. *Cytogenetics* **11**, 516–33.

——, Edmond, P., Christie, S., Gowans, L., Fletcher, J., Frackiewics, A., and Newton, M. (1973). Cytogenetics and infertility in man. I. Karyotype and seminal analysis. *Ann. Hum. Genet.* **39**, 231–54.

Chang, M.C. (1954). Development of parthenogenetic rabbit blastocysts induced by low temperature storage of unfertilized ova. *J. exp. Zool.* **125**, 127–50.

—— (1957). Natural occurrence and artificial induction of parthenogenetic cleavage of ferret ova. *Anat. Rec.* **126**, 187–90.

—— and Fernandez-Cano F.C. (1958). Effects of delayed fertilization on the development of pronucleus. *Anat. Rec.* **132**, 307–19.

Chapman, V.M. and Wudl, L. (1975). The expression of β-glucuronidase during mouse embryogenesis. In *Isozymes* (ed. C. Market) pp. 57–67. Academic Press, New York.

——, West, J.D., and Adler, D.A. (1977). Genetics of early mammalian embryogenesis. In *Concepts in mammalian embryogenesis* pp. 95–135. MIT Press, Cambridge, Mass.

——, Whitten, W., and Ruddle, F. (1971). Expression of paternal glucose phosphate Isomerase (GPI – 1) in preimplantation stages of mouse embryos. *Devl Biol.* **26**, 153–8.

——, Adler, D., Labarva, C., and Wudl, L. (1976). Genetic variation of β-glucuronidase expression during early embryogenesis in mammals. *CIBA Foundation Symposium: embryogenesis in mammals,* Vol. 40, pp. 115–24. North Holland, Amsterdam.

Chebotar, N.A. (1976a). Frequency of spontaneous non-disjunction in metaphase-II oocytes of CBA mice. *Tsitologiya* **18** (7), 897–9 (in Russian with English summary).

—— (1976b). Influence of rat age on the frequency of spontaneous chromosome aberrations. *Genetika* **12** (6), 158–61 (in Russian with English summary).

—— (1978). Cytogenetical analysis of the 1st cleavage division in rat embryos. *Tsitologiya* **20** (1), 102–4 (in Russian with English summary).

—— (1980). Cytogenetic analysis of the preovulatory stages of oogenesis and the first cleavage division in the CBA strain mice. *Tsitologiya* **22** (6), 710–14 (in Russian with English summary).

Chen, H.Y., Brinster, R.L., and Merz, E.A. (1980). Changes in protein synthesis following fertilization of the mouse egg ovum. *J. exp. Zool.* **212**, 355–60.

Church, R.B. and Schultz, G.A. (1964). Differential gene activity in the pre- and postimplantation mammalian embryo. *Curr. Top. dev. Biol.* **8**, 179–202.

——, Shea B., and Tervit R. (1974). Chromosomal analysis of bovine follicular oocytes after incubation *in vitro*. *J. Anim. Sci.* **89**, 142–55.

Chyi-Chyang, L., Tsushida, W.C., and Morris S.A. (1971). Spontaneous meiotic chromosome abnormalities in male mice *Mus musculus*. *Can. J. Genet. Cytol.* **13**, 95–100.

Clegg, K.B. and Piko, L. (1977). Size and specific activity of the UTP pool and overall rates of RNA synthesis in early mouse embryos. *Devl Biol.* **58**. 76–95.

——, —— (1983). Quantitative aspects of RNA synthesis and polyadenylation in 1-cell and 2-cell mouse embryos. *J. Embryol. exp. Morphol.* **74**, 169–172.

Clough, E., Ryle, W.C.D., Hare, D.T., Helly R.L., and Patterson D.F. (1970). An XXY sex chromosome constitution in a dog with testicular trypoplasis and congenital heart disease. *Cytogenetics* **9**, 71–7.

Committee on Standardized Genetic Nomenclature for Mice (1972). Standard karyotype of the mouse *Mus musculus. J. Hered.* **63**, 69.

Conen, P.E. and Glass, J.H. (1963). 45/XO Turner's syndrome in the newborn: report of two cases. *J. clin. Endocrinol.* **23**, 1–10.

Cooper, D.W. (1971). Directed genetic change model for X-chromosome inactivation in eutherian mammals. *Nature* **230**, 292–4.

Copeland, N.G., Jenkins, N.A., and Lee, B.K. (1983). Association of the Lethal Yellow (A^y) coat color mutation with an ectopic MuLV provirus. *Hereditas* **98**, A155.

Court-Brown, W.M. (1969). Chromosome aneuploidy in man and its frequency with special reference to mental subnormality and criminal behaviour. In *International review of experimental pathology* Vol. 7, pp. 180–254. Academic Press, New York.

Cox, D.R., Epstein, L.B., and Epstein, C.J. (1980). Genes coding for sensitivity to interferon (IfRec) and soluble superoxide dismutase (SOD-1) are linked in mouse and man and map to mouse chromosome 16. *Proc. nat. Acad. Sci. USA* **77** (4), 2168–72.

——, Smith, S.A., Epstein, L.B., and Epstein, C.J. (1984). Mouse trisomy 16 as an animal model of human trisomy 21 (Down's Syndrome): formation of viable trisomy 16- diploid mouse chimaeras. *Devl Biol.* **101**, 416–24.

Cozad, K.M. and Warner, C.M. (1982). Detection of H-2 antigens on 8-cell mouse embryos. *J. exp. Zool.* **221**, 213–17.

Cure, S., Boué, A., and Boué, J. (1973). Consequence of chromosomal anomalies on cell multiplication In *Les accidents chromosomiques de la reproduction* (ed. A. Boué and C. Thibault) pp. 95–109. Coll. Inst. Santé Rech. Méd., Paris.

Cuthbertson, K.S.R. (1982). Parthenogenetic activation of mouse oocytes *in vitro* with ethanol and benzyl alcohol. *J. Exp. Zool.* **226**, 311–14.

Czolowska, R., Modlinski, J., and Tarkowski, A.K. (1984). Behaviour of thymocyte nuclei in nonactivated and activated mouse oocytes. *J. Cell Sci.* **69**, 19–34.

Daentl, D.L., and Epstein, C.J. (1971). Developmental interrelationships of uridine uptake, nucleotide formation and incorporation into RNA by early mammalian embryos. *Devl Biol.* **24**, 428–42.

Dagg, C.P. (1960). Sensitive stages for the production of developmental abnormalities in mice with 5-fluorouracil. *Am. J. Anat.* **106**, 89–96.

Dalcq, A. and Pasteels, J. (1955). Détermination photometrique de la teneur relative en DNA des noyaux dans les oeufs en segmentation du rat et de la souris. *Exp. Cell Res.* Suppl. 3, 72–97.

Das, R.K. and Karr, R.N. (1981). A 41 XYY mouse. *Experientia* **37** (8), 821–2.

Davidenkova, E.F., Verlinskaya, D.K., and Tysiachnuk, S.F. (1973). *Clinical syndromes in sex-chromosome aberrations*. Meditsina, Leningrad (in Russian).

——, ——, —— (1974). Chromosomal diseases in clinic and the function of heterochromatin in the extra chromosomes. *Genetica* **10** (1), 152–62 (in Russian with English summary).

Davidson, E.H. (1976). *Gene activity in early development*. Academic Press, New York.

Deckers, J.F.M. and van der Kroon, P.H.W. (1981). Some characteristics of the XO mouse (*Mus musculus* L.). I Vitality: growth and metabolism. *Genetica* **55**, 179–85.

——, ——, and Douglas, L. Th. (1981). Some characteristics of the XO mouse (*Mus musculus* L.) II. Reproduction: fertility and gametic segregation. *Genetica* **57**, 3–11.

DeFries, J.C. (1967). Quantitative genetics and behavior. Overview and perspective. In *Behavior—genetic analysis* (ed. J. Hirsch), pp. 322–39. McGraw-Hill, New York.

Detlaff, T. A. (1964). Cell divisions, duration of interkinetic stages and differentiation in early stages of embryonic development. *Adv. Morphogen.* **3**, 323–62.

—— (1977). Some temporal and temperature correlations of embryonic development in poikilotermal animals. In *Problems in experimental biology* pp. 269–87. Nauka, Moscow.

—— and Detlaff, M.A. (1967). On relative dimensionless characteristics of the development duration in embryogenesis. *Arch. Biol.* **72**, 1–16.

Dev, V.G., Miller, D.A., and Miller, O.J. (1973). Chromosome markers in *Mus musculus*: strain difference in C-banding. *Genetics* **75**, 663–70.

——, Tantravahi, R., Miller, D.A., and Miller, O.J. (1977). Nucleolus organizers in *Mus musculus* subspecies and in the rat–mouse cell line. *Genetics* **86**, 389–98.

Dickson, A. (1967). Variations in development of mouse blastocysts. *J. Anat.* **101**, 263–7.

Dolch, K.M. and Chrisman, C.L. (1981). Cytogenetic analysis of preimplantation blastocysts from prepubertal gilts treated with gonadotropins. *Am. J. vet. Res.* **42**, 344–6.

Domarcus, H.V. (1983). Hypotelorism in trisomy 1 producing mice. *Anat. Rec.* **206**, 307–312.

Domon, M. (1983). A changing pattern of the cell cycle during the first two cleavage divisions of the mouse. *Devl Growth Differ.* **25** (6), 537–45.

Donahue, R.P. (1972). Cytogenetic analysis of the first cleavage division in mouse embryos. *Proc. nat. Acad. Sci. USA* **69**, 74–7.

—— and Karp, L.E. (1973). Chromosomal anomalies after fertilization of aged, postovulatory mouse oocytes. *Am. J. Hum. Genet.* **25**, 24.

Donkelaar, P.J., Geyshbevts, L.G.M., and Dederen, P.J.W. (1979). Stages in the prenatal development of the Chinese Hamster *Cricetus griseus*. *Anat. Embryol.* **156**, 1–28.

Ducibella, T. (1976). Surface changes of the developing trophoblast cell. In *Development in mammals* (ed. M.N. Johnson) Vol. 1, pp. 5–30. North Holland, Amsterdam.

—— and Anderson, E. (1975). Cell shape and membrane changes in the eight cell mouse embryo: prerequisites for morphogenesis of the blastocyst. *Devl Biol.* **47**, 45–58.

Dyban, A.P. (1959). *Pathological embryology of man*. Medgiz, Leningrad (in Russian).

—— (1967). Some current problems of experimental teratology. *Vestn. Acad. Med. nauk USSR* **1**, 18–30 (in Russian).

—— (1970a). Studies of early embryogenesis in mammals with induced chromosomal aberrations. *Arkh. Anat. Histol. Embriol.* **58** (1), 7–18 (in Russian with English summary).

—— (1970b). Embryonic differentiation and cell interactions in induced chromosomal aberrations in mammals. *Cell interactions in differentiation and growth* pp. 126–39. Nauka, Moscow (in Russian).

—— (1970c). Method for meiotic and mitotic chromosome preparations from mammalian testes. *Tsitologiya* **12** (5), 687–8 (in Russian with English summary).

—— (1972). Cytogenetical aspects of normal and abnormal embryogenesis in mammals. In *Problems in developmental genetics* pp. 62–85. Nauka, Moscow (in Russian).

—— (1973a). Cytogenetics of early embryonic development in mammals. *Vestn. Acad. Med. nauk USSR* **1**, 18–29 (in Russian with English summary).

—— (1973b). Medical aspects of experimental embryology. *Vestn. Acad. Med. nauk SSSR* **6**, 38–47 (in Russian with English summary).

—— (1974). New methodological approaches and perspectives of experimental cytogenetics of mammalian embryogenesis. *Ontogenez* **5** (6), 568–81 (in Russian with English summary).

—— (1976a). Experimental embryological and cytogenetical approaches to some problems of behavioral genetics in mammals. In *Memory in the mechanisms of normal and abnormal reactions*, pp. 220–39. Meditsina, Leningrad (in Russian).

—— (1976b). Experimental meditsina, embryological and cytogenetical aspects in memory investigation. In *Mechanisms of memory modulation*, pp. 183–92. Nauka, Leningrad (in Russian).

—— (1977). Theoretical and practical applications of experimental teratological studies. *Ontogenez* **8** (6), 582–98 (in Russian with English summary).

—— (1979). Differential activity of chromosomes in early embryogenesis of mammals. In *Itogi nauki i tekhniki, obshchaya genetika* Vol. 6, pp. 41–76. VINIITI, Moscow.

—— (1984). Studies of cellular, chromosomal and molecular mechanisms in very early mammalian embryogenesis. *Soviet science research Part B. Developmental biology*. Gordon & Breach, New York. (In press.)

—— and Baranov, V.S. (1977). Oogenesis in mammals. In *Modern problems of oogenesis* pp. 200–33. Nauka, Moscow (in Russian).

——, —— (1978). *Cytogenetics of mammalian development.* Nauka, Moscow (in Russian).

—— and Chebotar, N.A. (1975). Spontaneous chromosomal aberrations in oogenesis of laboratory rats. *Bull. exp. Biol. Med. USSR* **8**, 103–6 (in Russian with English summary).

—— and Khozhai, L.I. (1980). Parthenogenetic development of ovulated mouse ova induced by ethyl alcohol. *Byull. eksp. Biol. Med.* **89**, 487–9 (in Russian with English summary).

—— and Noniashvili, E.M. (1985a). The factors determining the incidence and the rate of parthenogenetic development in mouse eggs activated by ethyl alcohol treatment. *Ontogenez* **6**, 582–97 (in Russian with English summary).

——, —— (1986). Induced parthenogenesis in mammals. Cytogenetical aspects. In *Developmental biology and the control of animal heredity* (ed. T. Tuzpaev and V. Strunnikov) pp. 105–22. Nauka, Moscow (in Russian).

—— and Sekirina, G.G. (1981). Preimplantation development in monozygotic twins. Experiments on mouse embryos. *Ontogenez* **12** (2), 130–9 (in Russian with English summary).

312   References

—— and Sorokin, A.V. (1983). A comparison of the size of paternal and maternal homologous chromosomes during the first two cleavage divisions in mice. *Ontogenez* **14** (3), 238–46 (in Russian with English summary).

—— and Udalova, L.D. (1967). Studies of chromosome aberrations at early stages of embryogenesis in mammals I. Experiments with 2, 4-diamino-5, 7-chlorphenyl-6-ethylpyrimidine in rats. *Genetica* **3** (4), 52–65 (in Russian with English summary).

——, —— (1974). On polymorphism of C-heterochromatin in mice of different strains. *Bull. exp. Biol. Med. USSR* **127** (8), 106–8 (in Russian with English summary).

——, Puchkov, V.F., Baranov, V.S., Samoshkina, N.A., and Chebotar, N.A. (1975). Laboratory mammals: mouse *Mus musculus*, rat *Rattus norvegicus*, rabbit *Oryctolagus cuniculus*, hamster *Cricetus auratus*. In *Animal models in developmental biology*, pp. 505–66. Nauka, Moscow (in Russian).

——, Samoshkina, N.A., Patkin, E.L., and Chebotar, N.A. (1976). Radioautographical and cytophotometrical analysis of DNA synthesis in pre-implantation embryos of mice and rats. *Ontogenez* **7** (5), 450–9 (in Russian with English summary).

——, Udalova, L.D., Akimova, I.M., and Chebotar, N.A. (1971). Spontaneous heteroploidy at different stages of embryogenesis in rats. *Arkh. Anat. Histol. Embriol.* **60** (2), 22–38 (in Russian with English summary).

——, Waisman, B.L., and Golinsky, G.F. (1983). Production of viable karyo- and cytoplasts from fertilized mouse eggs by centrifugation in ficoll gradient. *Ontogenez* **4**, 420–6 (in Russian).

Dyban, P.A. (1981). Morphological characteristics of spontaneous ovarian teratomas of LT/Sv and LT/Sv × BJ mice, *Exp. Oncol.* **3**, 44–50 (in Russian with English summary).

—— (1982). Abnormal interactions between oocyte growth and proliferative activity of follicular epithelium in the ovaries of LT/Sv mutant mice, *Ontogenez* **13** (6), 650–5 (in Russian with English summary).

Eaton, G.J. and Green, M.M. (1963). Giant cell differentiation and lethality of homozygous yellow mouse embryos. *Genetica* **34**, 156–61.

Edwards, J.H., Yuncken, C., Rushton, D.L., Richards, S., and Mittwoch, U. (1967). Three cases of triploidy in man. *Cytogenetics* **6**, 81–104.

Edwards, R.G. (1954a). Colchicine-induced heteroploidy in early mouse embryos. *Nature* **174**, 276–7.

—— (1954b). The experimental induction of pseudogamy in early mouse embryos. *Experientia* **10**, 499–500.

—— (1956). The experimental induction of gynogenesis in the mouse III. Treatment of spermatozoa with trypaflavine, toluidine blue or nitrogen mustard. *Proc. R. Soc. Lond.* **B149**, 117–25.

—— (1957a). The experimental induction of gynogenesis in the mouse I. Irradiation of the sperm by X-rays. *Proc. R. Soc. Lond.* **B146**, 469–87.

—— (1957b). The experimental induction of gynogenesis in the mouse, II. Ultra-violet irradiation of the sperm. *Proc. R. Soc. Lond.* **B146**, 488–504.

—— (1958a). Colchicine-induced heteroploidy in the mouse, I. The induction of triploidy by treatment of the gametes. *J. exp. Zool.* **137**, 317–48.

—— (1958b). The number of cells and cleavages in haploid, polyploid and other heteroploid mouse embryos at 3.5 days gestation. *J. exp. Zool.* **138**, 189–207.

—— (1961). Induced heteroploidy in mice: effects of diacetylmethylcolchicine on eggs at fertilization. *Exp. Cell Res.* **24**, 615–17.

—— and Fowler, R.E. (1970). The genetics of human preimplantation development. In *Modern trends in human genetics* (ed. H. Emery) Vol.1, pp. 181–215. Appleton-Century-Crofts, New York.

—— and Gates, A.H. (1969). Timing of the stages of the maturation divisions, ovulation, fertilization, and the first cleavage of eggs of adult mice treated with gonadotrophins. *J. Endocrinol.* **18**, 292–304.

Eglitis, M.A. (1980). Formation of tetraploid mouse blastocysts after blastomere fusion with polyethylene glycol. *J. exp. Zool.* **213**, 309–13.

—— and Wiley, L.M (1981). Tetraploidy and early development: effects on developmental timing and embryonic metabolism. *J. Embryol. exp. Morphol.* **66**, 91–108.

Ehrman, L. and Parson, P. (1981). *Behaviour, genetics and evolution* (2nd edn). McGraw-Hill, New York.

Eicher, E.M. (1973). Translocation trisomic mice, production by female but not male translocation carriers. *Science* **180**, 181.

—— (1979). Murine ovarian teratomas and parthenotes as cytogenetic tools. *Cytogenet. Cell Genet.* **20**, 232–9.

—— and Washburn, L.L. (1978). Assignment of genes to regions in mouse chromosomes. *Proc. nat. Acad. Sci. USA* **75**, 946–50.

Ellem, K.A. and Gwatkin, R.B. (1968). Patterns of nucleic acid synthesis in the early mouse embryo. *Devl Biol.* **18**, 311–30.

Endo, S., Takagi, N., and Sasaki, M. (1982). The late replicating X-chromosome in digynous mouse triploid embryos. *Devl Genet.* **3**, 165–76.

Engel, W. and Franke, W. (1976). Maternal storage in mammalian oocyte. In *Current topics in pathology* (ed. A. Gropp and K. Benirschke) Vol. 62, pp. 29–52. Springer, Berlin.

——, Zenzes M.T., and Schmid, M. (1977). Activation of mouse ribosomal RNA genes at the 2-cell stage. *Hum. Genet.* **38**, 57–63.

Epiphanova, O.I. (1973). Regulatory mechanisms of cell cycles studies by means of transcriptional and translational inhibitors. In *Cell cycle* pp. 72–103. Nauka, Moscow (in Russian).

Eppig, J.J. (1978). Granulosa cell deficient follicles. Occurence, structure and relationship to ovarian teratocarcinogenesis in strain LT/SV mice. *Differentiation* **12**, 11–120.

—— (1981). Preimplantation embryonic development of spontaneous mouse parthenotes after oocyte meiotic maturation *in vitro*. *Gamete Res.* **4**, 3–13.

—— (1982). The relationship between parthenogenetic embryonic development and cumulus cell–oocyte intercellular coupling during oocyte meiotic maturation. *Gamete Res.* **5**, 229–37.

Epstein, C.J. (1975). Gene expression and macromolecular synthesis during preimplantation embryonic development. *Biol. Reprod.* **12**, 82–105.

—— (1975). The genetic activity of early mammalian embryos. In *Current topics in pathology* (ed. A. Gropp and K. Benirschke) Vol.62, pp. 53–8. Springer, Berlin.

—— (1981a). Animal models for autosomal trisomy. In *Trisomy 21* (Down's syndrome): *research perspectives* (ed. F. de la Cruz and P.S. Gerald). Plenum Press, New York and University Park Press, Baltimore.

—— (1981b). Inactivation of the X chromosome. In *The biology of normal human growth* (ed. M. Ritzen) p. 79–90. Raven Press, New York.

—— (1981c). The effects of chromosomal aneuploidy on early development: experimental approaches. In *Fertilization and embryonic development* in vitro (ed. L. Mastroianni, J.D. Biggers, and A.W. Sadler) pp. 273–81. Plenum Press, New York.

—— (1983a). The X-chromosome in development. In *Cytogenetics of the mammalian X chromosome. Part A. basic mechanisms of X chromosome behavior*, pp. 51–65. A.R. Liss, New York.

—— (1983b). Sex chromosome expression in embryonic development. *Differentiation* **23** (Suppl.), 31–4.

—— (1984). The mouse trisomies: experimental systems for the study of aneuploidy. In *Issues and reviews in teratology* (ed.H.Kalter) Vol. 3, pp. 171–217. Plenum Press, New York.

—— and Magnuson, T. (1982). Genetic control of early mammalian development. In *Human genetics, Part A. The unfolding genome* pp. 327–38. A.R. Liss, New York.

—— and Smith, *S.A.* (1973). Amino acid uptake and protein synthesis in preimplantation mouse embryos. *Devl Biol*. **33**, 171–84.

——, —— (1974). Electrophoretic analysis of protein synthesis in preimplantation mouse embryos. *Devl Biol*. **33**, 171–185.

——, —— (1978). Handling of actinomycin D by preimplantation mouse embryos. *Exp. Cell Res*. **111**, 117–26.

—— and Travis, B. (1979). Preimplantation lethality of monosomy for mouse chromosome 19. *Nature* **280**, 144–5.

——, Smith, S., and Travis, B. (1980). Expression of H-Y antigen on preimplantation mouse embryos. *Tissue Antigens* **15**, 63–7.

——, Epstein, L.B., Cox, D., and Weil, J. (1981). Functional implication of gene dosage effects in trisomy 21. In *Trisomy 21 human genetics* (ed. C.R. Burgio, M. Fraccaro, L. Tiepolo, and U. Wolf) Suppl.2, pp. 155–71. Springer, Berlin.

——, ——, Weil, J., and Cox, D. (1982b). Trisomy 21: mechanisms and models. *Ann. N.Y. Acad. Sci*. 110–18.

——, Smith, S., Travis, B., and Tucker, G. (1978). Both X-chromosomes function before visible X-chromosome inactivation in female mouse embryos. *Nature* **274**, 500–3.

——, Tucker, J., Travis, B., and Gropp, A. (1977). Gene dosage for isocitrate dehydrogenase in mouse embryos trisomic for chromosome 1. *Nature* **267**, 615–16.

——, Smith, S.A., Zamora, T., Sawicki, J.A., Magnuson, T.R., and Cox D. (1982a). Production of viable adult trisomy 17 ↔ diploid mouse chimeras. *Proc. nat. Acad. Sci. USA* **79**, 4376–80.

Erickson, R.P. (1978). t-alleles and the possibility of post-meiotic gene expression during mammalian spermatogenesis. *Fed. Proc*. **37**, 2517–21.

——, Betlach, C.I., and Epstein, C.J. (1974a). Ribonucleic acid and protein metabolism of t^{12}/t^{12} embryos and T/t^{12} spermatozoa. *Differentiation* **4**, 281–90.

——, Eicher, E.M., and Glucksohn-Waelsch, S. (1974b). Demonstration in the mouse of X-ray induced deletions for a known enzyme structural locus. *Nature* **248**, 416–18.

——, Lewis, S. and Slusser, K. (1983). Deletion mapping of the t-complex of chromosome 17 of the mouse. *Nature* **274**, 163–4.

——, Siekeritz, P., Jacobs, K., and Gluecksohn-Waelsch, S. (1974c). Chemical and immunological studies of liver microsomes from mouse mutants with ultrastructurally abnormal hepatic endoplasmic reticulum. *Biochem. Genet.* **12**, 81–95.

Evans, E.P. (1976). Male sterility and double heterozygosity for Robertsonian translocations in mouse. *Chromosomes Today* **5**, 75–81.

—— Beechey, C.V., and Burtenshaw, M.D. (1978). Meiosis and fertility in XYY mice. *Cytogenet. Cell Genet.* **20**, 249–63.

—— and Burgoyne P. (1983). The chromosome unbalance observed in early preimplantation embryos from mice heterozygous for paracentric inversion of the X-chromosome, In Eighth International Chromosome Conference, Lübeck, Abstracts 2–19.

——, Breckon, C., and Ford, C.E. (1964). An air-drying method for meiotic preparation from mammalian testes. *Cytogenetics* **3**, 285–94.

—— Brown, B.B., and Burtenshaw, M.D. (1980). Research note. *Mouse News Letter* **63**, 30.

——, Burtenshaw, M.D., and Ford, C.E. (1972). Chromosomes of mouse embryos and newborn young: preparation from membranes and tail tips. *Stain Technol.* **47**, 229–34.

——, Lyon, M.F., and Daglish, M. (1967). A mouse translocation giving a metacentric marker chromosome. *Cytogenetics* **8**, 105–19.

——, Lyon, M.F., Ford, C.E., and Searle, A.G. (1969). A 39, X/41 XYY mosaic mouse. *Cytogenetics* **8**, 87–98.

Evans, H.J. (1967). The nucleolus, virus infection and trisomy in man. *Nature* **214**, 361–2.

Evans, M.J. and Kaufman, M.N. (1983). Pluripotential cell growth directly from normal mouse embryos. *Cancer Surv.* **2** (1), 185–207.

Fabricant, J.D. and Schneider, E.L. (1978). Studies on the genetic and immunologic components of the maternal age effect. *Devl Biol.* **66**, 337–43.

Fedorov, V.K., Dyban, A.P., Baranov, V.S., Dmitriev, J.S., and Udalova L.D. (1973). Study of reproductive function and higher nervous activity in mice with T6 autosome trisomy. *Sechenov physiol. J. USSR*, **53** (5), 704–12.

Fischberg, M. and Beatty, R.A. (1951). Spontaneous heteroploidy in mouse embryos up to midterm. *J. exp. Zool.* **118**, 321–26.

——, —— (1952). Heteroploidy in mammals. II. Induction of triploidy in preimplantation mouse eggs. *J. Genet.* **50**, 455–70.

Flach, G., Johnson, M.H., Braude, P.R., Taylor, R.A., and Bolton, V.N. (1982). The transition from maternal to embryonic control in the 2-cell mouse embryo. *EMBO J.* **1**, 681–6.

Ford, C.E. (1969). Meiosis in mammals. In *Comparative mammalian cytogenetics* (ed. K. Benirschke) pp. 91–106. Springer, Berlin.

—— (1970). The population cytogenetics of other mammalian species. In *Human population cytogenetics.* Pfizer Medical Monographs Vol. 8, pp. 222–30. Edinburgh University Press, Edinburgh.

—— (1971). Genome unbalance and reproductive wastage in man and mammals. *Nord Med.* **86**, 1545–8.

—— (1972). Gross genome unbalance in mouse spermatozoa: does it influence the capacity to fertilize? In *The genetics of the spermatozoon*, pp. 359–69. Proceedings International Symposium, Edinburgh.

—— (1975). The time in development at which gross genome unbalance is expressed. In *The early development of mammals. Second Symposium of the British Society of Developmental Biology*, pp. 285–304. Cambridge University Press, London.

—— and Clegg, H.M. (1969). Reciprocal translocations. *Br. med. Bull.* **25**, 110–114.

—— and Evans, E.P. (1973a). Robertsonian translocations in mice: segregational irregularities in male heterozygotes and zygotic unbalance. *Chromosomes Today* **4**, 387–97.

—— and Evans, E.P. (1973b). Non-expression of genome unbalance in haplophase and early diplophase of the mouse. In *Chromosomal errors in relation to reproductive failure* (ed. A. Boué and C. Thibault) pp. 271–85. Coll. Inst. Santé Rech. Méd, Paris.

Forejt, J. (1972). Giemsa specific centromeric heterochromatin in three inbred mouse strains. *Folia biol.* **18**, 213–15.

—— (1973). Centromeric heterochromatin polymorphism in the house mouse. Evidence from inbred strains and natural populations. *Chromosoma* **43**, 187–201.

—— (1974). Nonrandom association between a specific autosome and the X-chromosome in meiosis of the male mouse: possible consequence of the homologous centromeres separation. *Cytogenet. Cell Genet.* **13**, 369–83.

—— (1979). Meiotic studies of translocations causing male sterility in the mouse II. Double heterozygotes for Robertsonian translocations. *Cytogenet. Cell Genet.* **23**, 163–70.

—— (1981). Hybrid sterility gene located in the T/t–H-2 supergene on chromosome 17. In *Current trends in histocompatibility* (ed. R.A. Reisfeld and S. Ferrone) pp. 103–31. Plenum Press, New York.

—— and Gregorova, S. (1977). Meiotic studies of translocations causing male sterility in the mouse I. Autosomal reciprocal translocations *Cytogenet. Cell Genet.* **19**, 159–79.

——, Capkova, J., and Gregorova, S. (1980). T(16.17)43 H translocation as a tool in analysis of the proximal part of chromosome 17 (including T-t gene complex) of the mouse. *Genet. Res.* **35**, 165–77.

——, Gregorova, S., and Goetz, P.(1982). XY pair associates with the synaptonemal complex of autosomal male-sterile translocations in pachytene spermatocytes of the mouse (*Mus musculus*). *Chromosoma* **82**, 41–53.

Fourcroy J.L. (1982). RNA-synthesis in immature mouse oocyte development. *J. exp. Zool.* **219**, 257–66.

Franklin, M. and Schmickel, R.D. (1971). Group G-deletion syndrome. *J. med. Genet.* **8**, 341–5.

Fraser, G.R. (1963). Parental origin of the sex chromosomes in the XO and XXY in man. *Ann. hum. Genet.* **26**, 297–320.

Fraser, L.R. (1970). Differing requirements for capacitation *in vitro* of mouse spermatozoa from two strains. *J. Reprod. Fert.* **49**, 83–7.

—— and Drury, L.M. (1976). Mouse sperm genotype and the rate of egg penetration *in vitro. J. exp. Zool.* **197**, 13–20.

—— and Maudlin, I. (1979). The analysis of aneuploidy in first-cleavage mouse embryos fertilized *in vitro* and *in vivo. Envir. Health Perspect.* **31**, 141–9.

Fredga, K. (1968). The idiogram of water vole, a favourable animal for cytogenetic research. *Chromosoma* **25**, 75–89.

Fryns, J.P., van de Kerckhove, A., Godderis, P., and van den Berghe, H. (1978).

Maternal origin and unusually long survival in a case of full triploidy. *Clin. Genet.* **13**, 115.

Fujimoto, S., Phalavan, N., and Dukelow, M.R. (1974). Chromosome abnormalities in rabbit preimplantation blastocyst induced by superovulation. *J. Reprod. Fert.* **40**, 177–81.

——, Passantino, T.J., Koenczoel, L., and Segal, S.J. (1975). A simplified method for chromosome preparations of the rat preimplantation embryos. *Tsitologiya* **40**, 469–75.

Fuller, J.T. and Thompson, W.R. (1960). *Behaviour genetics.* John Wiley, New York and London.

Funaki, K. (1981). Active development in preimplantation stages of giant digynic triploids in the Chinese hamster. *Proc. Jap. Acad.* **57B**, 18–22.

—— and Mikamo, K. (1980). Giant diploid oocytes as a cause of digynic triploidy in mammals. *Cytogenet. Cell Genet.* **28**, 158–68.

Fundele, R., Bucher, T., Gropp, A., and Winking, H. (1981). Enzyme pattern in trisomy 19 of the mouse. *Devl Genet.* **2**, 291–303.

Gaillard, J.A. (1974). Differentiation and organization in teratomas. In *Neoplasia and cell differentiation*, pp. 319–49. Karger, Basel.

Gamow, E. and Prescott, D. (1970). The cell cycle during early embryogenesis of the mouse. *Exp. Cell Res.* **59**, 117–23.

Gardner, R.L. (1972). An investigation of inner cell mass and trophoblast tissues following their isolation from the mouse blastocyst. *J. Embryol. exp. Morphol.* **28**, 279–312.

—— (1974). Origin and properties of trophoblast. In *Immunology of trophoblast* (R.G. Edwards, C.W.S. Home, and M.H. Johnson) pp. 48–65. Cambridge University Press Cambridge.

—— (1975). Analysis of determination and differentiation in the early mammalian embryo using intra- and interspecific chimeras. In *The developmental biology of reproduction* (ed. C.L. Markert) pp. 207–238. Academic Press, New York.

—— (1976). Gene A^y/A^y in early mammalian development (General discussion). In *Embryogenesis in mammals* (ed. A. McLaren). *CIBA Foundation Symposium*, Vol. 40, pp. 232–3. North Holland Press, London.

—— and Lyon, M.F. (1971). X-chromosome inactivation studied by injection of a single cell into the mouse blastocyst. *Nature* **231**, 385–6.

—— and Papaioannou, V.E. (1975). Differentiation in the trophectoderm and inner cell mass. In *Early development of mammals. Second Symposium of the British Society of Developmental Biology* pp. 107–32. Cambridge University Press, London.

Gartler, S.M. and Andine, R.J. (1976). Mammalian X-chromosome inactivation. In *Advances in human genetics.* Vol. 7, pp. 99–140. Academic Press, New York.

——, Andina, R., and Ceki, N. (1975). Ontogony of X-chromosome inactivation in the female germ line. *Exp. Cell Res.* **91**, 454–60.

Gates, H., Doyle, L., and Noyes, R. (1961). A physiological basis of heterosis in hybrid mouse fetuses. *Am. Zool.* **1**, 449–50.

Gehring, W.J. (1976). Development genetics of *Drosophila*. *A. Rev. Genet.* **10**, 209–52.

Georgiev, G.P. (1971). The structure and functional activity of chromatin. In *Advances in modern genetics* pp. 28–46. Nauka, Moscow.

Gill, T.J. III Siew, S., and Kunz, H.W. (1983). Major histocompatibility complex (MHC) linked genes affecting development. *J. exp. Zool.* **228**, 325–45.

Ginsberg, L. and Hillman, N. (1973). ATP metabolism in t^n/t^n mouse embryos. *J. Embryol. exp. Morphol.* **33**, 715–23.

Gluecksohn-Waelsch, S. (1953). Lethal factors in development. *Q. Rev. Biol.* **28**, 115–35.

—— (1964). Genetic control of mammalian differentiation. In *Genetics today*, pp. 211–19. Pergamon Press, New York.

—— (1979). Genetic control of morphogenetic and biochemical differentiation: lethal albino deletions in the mouse. *Cell* **16**, 225–37.

—— and Erickson, R. (1970). The T-locus of the mouse: implications for mechanisms of development. *Curr. Top. dev. Biol.* **5**, 281–316.

Goldbard, S.B., Verbanac, K.M., and Warner, C.M. (1982a). Role of the H-2 complex in preimplantation mouse embryo development. *Biol. Reprod.* **26**, 591–6.

——, ——, —— (1982b). Genetic analysis of H-2 linked gene(s) affecting early mouse embryo development. *J. Immunogen.* **9**, 77–82.

Golbus, M.S., Calarco, P.G., and Epstein C.J. (1973). The effects of inhibitors of RNA synthesis (α-amanitin and actinomycin D) on preimplantation mouse embryogenesis. *J. exp. Zool.* **186**, 207–16.

Goldfeder, A. (1965). Biological properties and radiosensitivity of tumours. Determination of the cell-cycle and time of synthesis of deoxyribonucleic acid using tritiated thymidine and autoradiography. *Nature* **207**, 612–14.

Golubovskaya, I.N. (1979). Genetic control of meiosis. *Int. Rev. Cytol.* **58**, 247–90.

Gosden, R.G. (1973). Chromosomal anomalies of preimplantation mouse embryos in relation to maternal age. *J. Reprod. Fert.* **35**, 351–4.

Graham, C.F. (1970). Parthenogenetic mouse blastocysts. *Nature* **226**, 165–7.

—— (1971). Virus assisted fusion of embryonic cells. Karolinska Symposium on Research Methods in Reproductive Physiology, Stockholm, pp. 154–87.

—— Cell cycle during mammalian development. In *The cell cycle in development and differentiation* (ed. M. Balls and F. Billett) pp. 293–310. Cambridge University Press, Cambridge.

—— (1973b). Nucleic acid metabolism during early mammalian development. In *The regulation of mammalian reproduction* (ed. S.I. Segal. R. Crozier, P. Corfman, and F. Condliffe) pp. 286–98. Charles Thomas, Springfield.

—— (1974a). The production of parthenogenetic mammalian embryos and their use in biological research. *Biol. Rev.* **49**, 399–422.

—— (1974b). The necessary conditions for gene expression during early mammalian development. 31st Symposium, Society of Developmental Biology pp. 202–17.

—— (1977). Teratocarcinom cells and normal mouse embryogenesis. In *Concepts in mammalian embryogenesis* (ed. M.J. Sherman) pp. 315–94. MIT Press, Cambridge, Massachusetts.

—— and Deussen, Z.A. (1974). *In vitro* activation of mouse eggs. *J. Embryol. exp. Morphol.* **31**, 497–512.

——, —— (1978). Features of cell lineage in preimplantation mouse development. *J. Embryol. exp. Morphol.* **48**, 53–72.

—— and Lehtonen, F. (1979). Formation and consequences of cell patterns in preimplantation mouse development. *J. Embryol. exp. Morphol.* **49**, 277–94.

——, McBurney, M.W., and Iles, S.A. (1975). Teratomas from haploid and diploid parthenogenetic mouse embryos. In *Teratomas and differentiation* (ed. M.J. Sherman and D. Solter) pp. 33–49. Academic Press, New York.

Granholm, N.H. and Johnson, P.M. (1978). Enhanced identification of lethal yellow (A^y/A^y) mouse embryos by means of delayed development of four-cell stages. *J. exp. Zool.* **205**, 327–9.

Green, M.C. (1966). Mutant genes and linkages. In *Biology of the laboratory mouse* pp. 87–150. McGraw-Hill, Toronto.

—— (1973). The laboratory mouse, *Mus musculus*. In *Handbook of genetics* (ed. R.C. King) Vol. 4, pp. 203–41. Plenum Press, New York.

Greenwood, R.D. and Sommer, A.M. (1971). Monosomy G: a case report and review of the literature. *J. med. Genet.* **8**, 496–500.

Gregorova, S., Baranov, V.S., and Forejt, J. (1981). Partial trisomy (including T-t gene complex) of the chromosome 17 of the mouse. The effect on male fertility and the transmission to progeny. *Fol. biol.* **27**, 171–7.

Griffen, A.B. (1967). A case of tertiary trisomy in the mouse and its implications for the cytological classification of trisomies in other mammals. *Can J. Genet. Cytol.* **9**, 503–10.

—— and Bunker, M.C. (1964). Three cases of trisomy in the mouse. *Proc. nat. Acad. Sci. USA* **52**, 1194–8.

Grinberg, K.N. (1969). Chromosomal diseases. Autosomal aberrations. In *Basic studies in human cytogenetics* pp. 310–410. Meditsina, Moscow (in Russian).

—— (1982). Cellular genetics in the study of hereditary diseases. *Vestn. Acad. Med. nauk. USSR* **6**, 76–81 (in Russian with English summary).

Gropp, A. (1973a). Fetal mortality due to aneuploidy and irregular meiotic segregation in the mouse. In *Les accidents chromosomiques de la reproduction* (ed. A Boué and C. Thibault) pp. 255–69. Coll. Inst. Santé Rech. Méd., Paris.

—— (1973b). Reproductive failure due to fetal aneuploidy in mice. Proceedings of the Seventh International Congress on Fertility and Sterility, Tokyo–Kyoto, pp. 326–30.

—— (1975). Chromosomal animal model of human disease. Fetal trisomy and developmental failure. In *Teratology* (ed. C. Berry and D.E. Poswillo) pp. 17–31. Springer, Berlin.

—— (1978). Relevance of phases of development for expression of abnormality. Perspectives drawn from experimentally induced chromosome aberrations. In *Abnormal fetal growth: biological bases and consequences* (ed. F. Naftolin) pp. 85–100. Dahlem Konferenzen, Berlin.

—— (1981a). Chromosomaberrationen, Geschwulste und Enwicklungsstorungen. *Klin. Wochenschr.* **59**, 965–75.

—— (1981b). Clinical and experimental pathology of fetal wastage. In *Human reproduction, Proceedings of third world congress, Berlin* (ed. K. Semm and L. Mettler), pp. 208–16. Excerpta Medica, Amsterdam.

—— (1982). Value of animal model for trisomy. *Virchow Arch. Pathol. Anat. Physiol.* **395**, 117–31.

—— and Grohe, G. (1981). Strain background dependence of expression of chromosome triplication in the mouse embryo. *Hereditas*, **94**, 7–8 (Abstract).

—— and Kolbus, U. (1974). Exercephaly in the syndrome of trisomy 12 of the foetal mouse. *Nature* **249**, 145–7.

—— and Winking, H. (1981). Robertsonian translocations: cytology, meiosis, segregation patterns and biological consequences of heterozygosity. *Symp. Zool. Soc. Lond.* **47**, 141–81.

—— and Zech, L. (1973). Identification of metacentric marker chromosomes in the mouse by use of banding techniques. In *Nobel symposia chromosome identification* (ed. T. Caspersson and L. Zech) pp. 118–23. Academic Press, New York.

——, Giers D., and Kolbus, U. (1974) Trisomy in fetal backcross progeny of male and female metacentric heterozygotes of the mouse. *Cytogenet. Cell Genet.*, **13**, 511–35.

—— Kolbus, U., and Giers, D. (1975). Systematic approach to the study of trisomy in the mouse. *Cytogenet. Cell Genet.* **14**, 42–62.

——, Putz, B., and Zimmermann, U. (1976). Autosomal monosomy and trisomy causing developmental failure. In *Developmental biology and pathology* (ed. A. Gropp and K. Benirschke) pp. 177–92. Springer, Berlin.

——, Tettenborn, V., and Leonard, A. (1970a). Identification of acrocentric chromosomes involved in the formation of 'fusion' metacentrics in mice. Proposal for nomenclature of *Mus poschiavinus* metacentrics. *Experientia* **26**, 1018–9.

——, ——, and Lehmann, E. von. (1970b). Chromosomenvariation vom Robertson'schen Typus bei der Tabakmaus, M. Poschiavinus und ihren Hybriden mit der Laboratoriumsmaus. *Cytogenetics* **9**, 9–23.

——, Winking, H., Herbst, E.W., and Claussen, C.P. (1983). Murine trisomy: developmental profiles of the embryo, and isolation of trisomic cellular systems. *J. exp. Zool.* **228** (2), 253–70.

——, Winking, H., Redi, C., Capanna, E., Britton-Davidian, J., and Noack G. (1982). Robertsonian karyotype variation in wild house mice from Rhaeto-Lombardia. *Cytogenet. Cell Genet.* **34**, 67–77.

Gropp, D., Gropp, A., and Winking, H. (1981). Research note. *Mouse News Letter* **64**, 70.

Gross, P.R. (1967a). The control of protein synthesis in embryonic development and differentiation. In *Current topics in developmental biology* Vol. 2, pp. 1–46. Academic Press, New York.

—— (1967b). RNA metabolism in embryonic development and differentiation. I. Fertilization. *New Engl. J. Med.* **22**, 1239–47.

—— (1967c). RNA metabolism in embryonic development and differentiation. II. Biosynthetic patterns and their regulation. *New Engl. J. Med.* **23**, 1297–305.

Grüneberg, H. (1966). The case for somatic crossing over in the mouse. *Genet Res.* **7**, 57–75.

Gurdon, J.B. (1974). *The control of gene expression in animal development*. Oxford and Harvard University Presses, Oxford and Cambridge, Mass.

—— (1977). Egg cytoplasm and gene control in development. *Proc. R. Soc. Lond.* **B198**, 211–47.

—— (1981). Concepts of gene control in development. In *Developmental biology using purified genes* pp. 1–10. Academic Press, New York.

Guttman, R., Naftali, G., and Nevo, E. (1975). Aggression patterns in three chromosomal forms of the mole rat *Spalax ehrenbergi*. *Anim. Behav.* **23** (3), 485–93.

Hadorn, D. (1961). *Developmental genetics and lethal factors*. Methuen, London.

Hagemeier, A., Smit, E.M.E., Govers, F., and Both, N.J. (1982). Trisomy 15 and other non-random chromosome changes in Rauscher murine leukemia virus-induced leukemia cell lines. *J. nat. Cancer Inst.* **69** (4), 945–52.

Hamerton, J.L. (1968). Robertsonian translocations in man: evidence for prezygotic selection. *Cytogenetics* **7**, 260–76.

—— (1970). Robertsonian translocation. Evidence on segregation from family studies. *Pfizer Medical Monographs*, Bd 5, pp. 63–80. Edinburgh University Press, Edinburgh.

—— (1971). *Human cytogenetics*. Academic Press, New York.

——, Canning, N., Ray, M., and Smith, S. (1975). A cytogenetic survey of 14060 newborn infants. I. Incidence of chromosome abnormalities. *Clin. Genet.* **8**, 223–43.

Handyside, A.H. (1980). Distribution of antibody and lectin-binding sites on dissociated blastomeres from mouse morulae: evidence for polarization at compaction. *J. Embryol. exp. Morphol.* **60**, 99–115.

Hansmann, I. (1974). Chromosome aberrations in metaphase-II oocytes: stage sensitivity in mouse oogenesis to amethopterin and cyclophosphamide. *Mutat. Res.* **22**, 175–91.

—— (1983). Factors and mechanisms involved in nondisjunction and X-chromosome loss. In *Cytogenetics of the mammalian X chromosome. Part A. Basic mechanisms of X chromosome behavior* pp. 131–70. A.R. Liss, New York.

—— and Röhrborn, G. (1973). Chromosome aberrations in preimplantation stages of mice after treatment with triazoquinone. *Humangenetik* **18**, 101–9.

——, Gebauer, J., Bihl, L., and Grimm, T. (1978). Onset of nucleolus organizer activity in early mouse embryogenesis and evidence for its regulation. *Exp. Cell. Res.* **114**, 263–8.

—— and El-Nahass, E. (1979). Incidence of nondisjunction in mouse oocytes. *Cytogenet. Cell Genet.* **24**, 115–21.

—— and Probeck, H.D. Chromosomal imbalance in ovulated oocytes from Syrian hamsters (*Mesocricetus auratus*) and Chinese hamsters (*Cricetus griseus*). *Cytogenet. Cell. Genet.* **23**, 70–6.

—— and Jederny, J. (1983). The genetic basis of non-disjunction. Increased incidence of hyperploidy in oocytes from F_1 hybrid mice. *Hum. Genet.* **65** (1), 56–60.

Harman, M.T. and Kirgris, H.D. (1938). The development and atresia of the Graafian follicle and the division of intra-ovarian ova in the guinea pig. *Am. J. Anat.* **63**, 79–99.

Harper, M.I., Fosten, M., and Monk M. (1982). Preferential paternal X inactivation in extra-embryonic tissues of early mouse embryos. *J. Embryol. exp. Morphol.* **67**, 127–35.

Hassold, T.J., Matsuyama, A., Newlands, I.M., Matsuura, J.S., Jacobs, P.A., Manuel, B., and Tsuei, J.A. (1978). Cytogenetic studies of spontaneous abortions in Hawaii. *Ann. hum. Genet.* **41**, 443–54.

Henderson, S.A. and Edwards, R. (1968). Chiasma frequency and maternal age in mammals. *Nature* **218**, 22–8.

——, Eicher, E.M., Yu, M.T., and Atwood, K.C. (1976). The chromosomal location. Variation in ribosomal RNA gene number in mouse chromosomes. *Cytogenet. Cell Genet.* **15**, 307–16.

Herbert, M.C. and Graham, C.F. (1974). Cell determination and biochemical differentiation of the early mammalian embryo. *Current topics in developmental biology* Vol.8, pp. 151–78. Academic Press, London.

Herbst, E.W., Pluznik, D.H., Gropp, A., and Uthgenannt, H. (1981). Trisomic hemopoietic stem cells of fetal origin restore hemopoiesis in lethally irradiated mice *Science* **211**, 1175–7.

Herrup, K. (1983). Role of staggerer gene in determining cell number in cerebellar cortex. *Dev. Brain Res.* **11**, 267–79.

—— and Mullen, R.J. (1979). Staggerer chimeras: intrinsic nature of Purkinje cells defects and implications for normal cerebellar development. *Brain Res.* **178** 443–57.

Hertig, A.T. (1967). The overall problem in man. In *Comparative aspects of reproductive failure* (ed. K. Benirschke) pp. 11–40. Springer, Berlin.

Hertwig, P. (1938). Unterschiede in der Entwicklungsfähigkeit von F_1-Mäusen nach Röntgenbestrahlung von Spermatogonien fertigen und unfertigen Spermatozoen. *Biol. Zentrallol.* **59**, 273–301.

Herzog, A. and Höhn, H. (1971). Autosomale Trisomie bei der brachygnathie des Rindes. *Cytogenetics* **10**, 347–55.

Heyner, S. and Hunziker, R. (1979). Differential expression of alloantigens of the major histocompatibility complex on unfertilized and fertilized mouse eggs. *Devl Genet.* **1**, 69–76.

Hillman, N. (1972). Autoradiographic studies of t^{12}/t^{12} mouse embryos. *Am. J. Anat.* **134**, 411–24.

—— (1975). Studies of the T-locus. In *The early development of mammals* (ed. M. Balls and A.E. Wild) pp. 189–206. Cambridge University Press, Cambridge.

—— and Tasca, R.J. (1969). Ultrastructural and autoradiographic studies of mouse cleavage stages. *Am. J. Anat.* **126**, 151–74.

——, —— (1973). Synthesis of RNA in t^{12}/t^{12} mouse embryos. *J. Reprod. Fert.* **33**, 501–6.

——, Hillman, R., and Wileman, G. (1970). Ultrastructural studies of cleavage stage t^{12}/t^{12} mouse embryos. *Am. J. Anat.* **128**, 311–40.

Hodel, C. and Egl F. (1965) Ullrich-Turner-Syndrom bei Neugehorenen mit Aortenisthmus-stenose und Vena cava superior sinistra. *Ann. Paediat.* **204**, 387–96.

Hoffman, M.A., Dmitriev, N.J., and Lopatina, N.G. (1980). Neuron density in different brain regions with trisomy and translocation of T6 autosome. *Sechenov Physiol. J. USSR* **66**, 594–6 (in Russian).

——, Eremeiev, N.S., Glushchenko, T.G., and Dmitrieva, N.J. (1979). RNA content in neural and glial cells in brains of mice with trisomy and translocation of T6 autosome. *Sechenov Physiol. J. USSR* **65**, 1271–5 (in Russian).

Hofsaess, F.K. and Meachem, T.N. (1971). Chromosome abnormalities in early rabbit embryos. *J. exp. Zool.* **177**, 9–11.

Hollander, W.F. and Waggie, K.S. (1977). Gnome and other effects of a small translocation in the mouse. *J. Heredity* **68**, 41–7.

Hongell, A. and Gropp, A. (1982). Trisomy 13 in the mouse. *Teratology* **26**, 95–104.

Hoppe, P.C. and Illmensee, K. (1977). Microsurgically produced homozygous-diploid uniparental mice. *Proc. nat. Acad. Sci. USA* **74**, 5657–61.

——, —— (1982). Full term development after transplantation of parthenogenetic embryonic nuclei into fertilized mouse eggs. *Proc. nat. Acad. Sci. USA* **79**, 1912–16.

Howe, C.C. and Solter, D. (1979). Cytoplasmic and nuclear protein synthesis in preimplantation mouse embryos. *J. Embryol. exp. Morphol.* **52**, 209–25.

Hsu, T.C. (1979). *Human and mammalian cytogenetics. An historical perspective.* Springer, Berlin.

—— and Mead, R.A. (1969). Mechanisms of chromosomal changes in mammalian speciation. In *Comparative mammalian cytogenetics* (ed. K. Benirschke) pp. 8–17. Springer, Berlin.

Hulten, M. and Pearson, P.L. (1971). Fluorescent evidence for spermatocytes with two chromosomes in an XYY male. *Ann. hum. Genet.* **34**, 273–6.

Hyafil, F., Morello, D., Babinet, C., and Jacobs, F. (1980). A cell surface glycoprotein involved in the compaction of embryonal carcinoma cells and cleavage stage embryos. *Cell* **31**, 927–34.

Ignatieva, T.M. (1980). *Early development in fishes and amphibians.* Nauka, Moscow (in Russian).

Iles, S.A. and Evans, E.P. (1977). Karyotype analysis of teratocarcinomas and embryoid bodies of C3H mice. *J. Embryol. exp. Morphol.* **38**, 77–92.

Issa, M., Blank, C.E., and Atherton, G.W. (1969). The temporal appearance of sex chromatin and of late-replicating X-chromosomes in blastocysts of domestic rabbit. *Cytogenetics* **8**, 219–37.

Ivett, J.L., Tice, R.R., and Bender, M, A. (1978). Y—two X's? An XXY genotype in Chinese hamster, *C.griseus. J. Hered.* **69** (2), 128–9.

Iwakawa, Y., Nozaki, M., and Matsushiro, A. (1982). Analysis of molecular mechanisms of preimplantation development of mouse embryos; effects of t^{12} mutation and tunicamysin. *A. Rep. Inst. Virus Res. Kyoto Univ.* **25**, 60–1.

Izquierdo, L. and Roblero, L. (1965). The incorporation of labelled nucleosides by mouse morulae. *Experientia* **21**, 531–3.

Jacobs, P.A. (1982). Pregnancy losses and birth defects. In *Reproduction in mammals* Vol.2. *Embryonic and fetal development* (ed. C.R. Austin and R.V. Short) pp. 142–58. Cambridge University Press, Cambridge.

——, Wilson, C.M., Sprenkle, J.A., Rosenshern, N.B., and Migeon, B.N. (1980). Mechanisms of origin of complete hydatiform moles. *Nature* **286**, 714–16.

——, Angell, R.R., Buchman, I.M., Hassold, T.J. Matsuyama, A.M., and Manuel B. (1978). The origin of human triploids. *Ann. hum. Genet.* **42**, 49–57.

Jagiello, G. (1981). Reproduction in Down's syndrome. In *Trisomy 21 (Down's syndrome)—research perspectives* (ed. F.F. De la Cruz and P.S. Gerald) pp. 151–62. University Park Press, Baltimore.

—— and Ducayen M. (1973). Meiosis of ova from polyovular (C58/j) and polycystic (C57/j) strains of mice. *Fertil. Steril.* **24**, 10–23.

—— and Fang, J.S. (1979). Analysis of diplotene chiasma frequencies in mouse oocytes and spermatocytes in relation to aging and sexual dimorphism. *Cytogenet. Cell Genet.* **23**, 53–60.

—— and Lin, J.S. (1973). An assessment of the effects of mercury on the meiosis of mouse ova. *Mutat. Res.* **17**, 93–101.

—— Ducayen, M., and Lin, J.S. (1972). Meiosis suppression by caffeine in female mice. *Mol. gen. Genet.* **118**, 209.

——, Karnecki, J., and Ryon, R.J. (1968). Superovulation with pituitary gonadotrophin method for obtaining meiotic metaphase figures in human ova. *Lancet* **81**, 178–81.

——, Miller, W.A., Ducayen, M.B., and Lin, J.S. (1974). Chiasma frequency and disjunctional behaviour of ewe and cow oocytes matured *in vitro. Biol. Reprod.* **10**, 354–63.

——, Ducayen, M.B., Miller, W.A., Lin, J.S., and Fang, S.A. (1973). A cytogenetic analysis of oocytes from *Macaca mulatta* and *Nemestrina* matured *in vitro. Humangenetik* **18**, 117–22.

Jameela, S.S. and Murthy, D.K. (1979). Spontaneous occurrence of Robertsonian centric fusion in mouse. *Curr. Sci. (Ind).* **48**, 199–200.

Johnson, D.K. (1974). Hairpin-tail: a case of post-reductional gene action in the mouse eggs? *Genetics* **76**, 795–805.

—— (1975). Further observations on hairpin (T^{Hp}) mutation in the mouse. *Genet. Res.* **24**, 207–13.

Johnson, L.V. and Calarco, P.G. (1980). Electrophoretic analysis of cell surface proteins of preimplantation mouse embryos. *Devl Biol.* **77**, 224–7.

—— and Granholm, N.H. (1978). *In vitro* analysis of pre- and early postimplantation development of lethal yellow (A^Y/A^Y) mouse embryos. *J. exp. Zool.* **204**, 381–90.

Johnson, M.H. (1981). The molecular and cellular basis of preimplantation mouse development. *Biol. Rev.* **56**, 463–98.

—— and Pratt, H.P.M. (1983). Cytoplasmic localization and cell interactions in the formation of the mouse blastocyst. In *Time, space and pattern in embryonic development* (ed. N. Jeffrey and N. Raff) pp. 287–312. Alan Liss, New York.

—— and Ziomek, C.A. (1981). The foundation of two distinct cell lineages within the mouse morula. *Cell* **24**, 71–80.

——, Handyside, A.N., and Braude, P.R. (1977). Control mechanisms in early mammalian development. In *Development in mammals* (ed. M. Johnson) Vol.2, pp. 67–99. North Holland, Amsterdam.

——, McConnell, D., and Van Blerkom, G. (1984). Programmed development in the mouse embryo. *J. Embryol. exp. Morphol.* **83** (Suppl.), 197–231.

——, Pratt, N.P.M., and Handyside, A.N. (1981). The generation and recognition of positional in the preimplantation mouse embryo. In *Cellular and molecular aspects of implantation* (ed. S. Glasser and D. Bullock) pp. 55–74. Plenum Press, London.

Jonasson J., Therkelsen A.J., Lauritsen J.G., and Lindsten J. (1972). Origin of triploidy in human abortuses. *Hereditas* **71**, 168–71.

Joseph, A., and Thomas, M. (1982). Cytogenetic investigations in 150 cases with complaints of sterility or primary amenorrhea. *Hum. Genet.* **61**, 105–9.

Juberg, R.C., and Davis, L.M. (1970). Etiology of nondisjunction: lack of evidence for genetic control. *Cytogenetics* **9**, 284–94.

Jude, U.H., Winking, H., and Gropp, A. (1980). Research note. *Mouse News Letter* **63**, 23.

Kafiani, K.A. and Kostomarova, A.A. (1978). *Informational macromolecules in early development of animals*. Nauka, Moscow (in Russian).

Kajii, T. and Niikawa, N. (1977). Origin of triploidy and tetraploidy in man: two cases with chromosome markers. *Cytogenet. Cell Genet.* **18**, 109–25.

—— and Ohama, K. (1977). Androgenetic origin of hydatiform mole. *Nature* **268**, 633–4.

——, Ferrier, A., Niikawa, N., Takahara, H., Ohama, K., and Avirachan, S. (1980). Anatomic and chromosomal anomalies in 639 spontaneous abortuses. *Hum. Genet.* **55**, 87–98.

Kaleta, E. (1977). Influence of genetic factors on the fertilization of mouse ova *in vitro*. *J. Reprod. Fert.* **51**, 357–81.

Kalter, H. (1968). *Teratology of central nervous system*. University of Chicago Press, Chicago.

Kanda, N. (1973). A new differential technique for staining the heteropycnotic X-chromosome in female mice. *Exp. Cell Res.* **80**, 463–7.

Karp, G.C., Manes, C., and Huahn, W.E. (1974). Ribosome production and protein synthesis in the preimplantation rabbit embryo. *Differentiation* **2**, 65–75.

Kato, Y., Yamazaki, K., Kimura, S., Hayasaka, M., and Kondo S. (1982). Patterns of cell proliferation in the preimplantation mammalian embryo. In *Genetic approaches in developmental neurobiology* pp. 3–19. University of Tokyo Press, Tokyo and Springer Verlag, Heidelberg.

Kaufman, M.H. (1972). Non-random segregation during mammalian oogenesis. *Nature* **238**, 465–6.

—— (1973a). Parthenogenesis in the mouse. *Nature* **242**, 475–6.

—— (1973b). Analysis of the first cleavage division to determine the sex ratio and incidence of chromosome anomalies at conception in the mouse. *J. Reprod. Fert.* **35**, 67–72.

—— (1975). The experimental induction of parthenogenesis in the mouse. In *The early development of mammals* (ed. M. Balls and A.E. Wild) Vol. 2, pp. 25–44. Cambridge British Society Developmental Biology Symposium.

—— (1978a). The experimental production of mammalian parthenogenetic embryos. In *Methods in mammalian reproduction* (ed. J.C. Daniel Jr.) pp. 21–49. Academic Press, New York.

—— (1978b). Chromosome analysis of early postimplantation presumptive haploid parthenogenetic mouse embryos. *J. Embryol. exp. Morphol.* **45**, 85–91.

—— (1981). Parthenogenesis: a system facilitating understanding of factors that influence early mammalian development. In *Progress in anatomy* (ed. R.J. Harrison and R.L. Holmes) pp. 1–34. Cambridge University Press, London.

—— (1982a). The chromosome complement of single-pronuclear haploid mouse embryos following activation by ethanol treatment. *J. Embryol. exp. Morphol.* **71**, 139–54.

—— (1982b). Two examples of monoamniotic monozygotic twinning in diploid parthenogenetic mouse embryos. *J. exp. Zool.* **224**, 277–82.

—— (1983a). *Early mammalian development: parthenogenetic studies.* Cambridge University Press, Cambridge.

—— (1983b). Ethanol-induced chromosomal anomalies at conception in mice. *Nature* **304**, 258–261.

—— and Sachs, L. (1975). The early development of haploid and aneuploid parthenogenetic embryos. *J. Embryol exp. Morphol.* **34**, 643–55.

——, Barton, S.C., and Surani, M.A.H. (1977a). Normal postimplantation development of mouse parthenogenetic embryos to the forelimb bud stage. *Nature* **265**, 53–5.

——, Guc-Cubrilo, M., and Lyon, M.F. (1978). X chromosome inactivation in diploid parthenogenetic mouse embryos. *Nature* **271** (5645), 547–9.

——, Evans, M.J., Robertson E.J., and Bradley, A. (1984). Influence of injected pluripotential (EK) cells on haploid and diploid parthenogenetic development. *J. Embryol. exp. Morphol.* **80**, 75–86.

——, Robertson, E.J., Handyside, A.H., and Evans, M.J. (1983). Establishment of pluripotential cell lines from haploid mouse embryos. *J. Embryol. exp. Morphol.* **73**, 249–61.

Kelly, S.J. (1975). Potency of early cleavage blastomeres of the mouse. In *The early development of mammals* (ed. M. Balls and A.E. Wild) pp. 97–105. Cambridge University Press, London.

Kemler, R., Babinet, C., Condamine, H.M, Gachelin, G., Guenet, J.L, and Jacob, F. (1976). Embryonal carcinoma antigen and the T/t locus in the mouse. *Proc. nat. Acad. Sci. USA* **73**, 4080–4.

Kemphues, K.Y., Raff, E.C., Raff, R., and Kaufman, T.C. (1980). Mutation in a testis-specific B-tubulin in *Drosophila*: analysis of its effects on meiosis and location of the gene. *Cell* **21**, 445–51.

Keneklis, T.P. and Odartchenko, N. (1974). Autoradiographic visualisation of paternal chromosomes in mouse eggs. *Nature* **247**, 215–16.

Kessler S. and Moos, R.H. (1973). Behavioural aspects of chromosomal disorders. *A. Rev. Med.* **24**, 89–103.

Khlebodarova, R.M., Serov, O.L., and Korochkin, L.I. (1976). Expression of alleles at the Pgd locus in early rat development. *Dokl. Acad. nauk USSR Ser. Biol.* **224**, 428–9 (in Russian).

Khozhai, L.I. (1981). Karyology of initial embryogenesis and parthenogenetic development in mice. Ph.D. thesis, Leningrad (in Russian).

Kidder, G.M. and Pedersen, R.A. (1982). Turnover of embryonic messenger RNA in preimplantation mouse embryos. *J. Embryol. exp. Morphol.* **67**, 37–49.

Kierszenbaum, A.L. and Tres, L.L. (1975). Structural and transcriptional features of the mouse spermatid genome. *J. Cell Biol.* **65**, 258–63.

King, W.A., Linares, T., and Gustavsson, I. (1981). Cytogenetics of preimplantation embryos sired by bull heterozygous for the 1/29 translocation. *Hereditas* **94**, 219–24.

Klein, I. (1971). Cytological identification of chromosomes, carrying on the IXth linkage group (including H-2) in the mouse. *Proc. nat. Acad. Sci. USA* **681**, 1594–7.

—— and Rasha, K. (1968). Deficiency of 'ribosomal' DNA in t^{12} mutant of the mouse. Proceedings of the XII International Congress on Genetics, pp. 149–50.

Knowland, J. and Graham, C. (1972). RNA synthesis at the two-cell stage of mouse development. *J. Embryol. exp. Morphol.* **27**, 167–76.

Köller, P.C. (1944). Segmental interchange in mice. *Genetics* **29**, 247–63.

Kolombet, L.V. (1977). The sequence of LDH-loci activation in early development of mouse embryos. *Ontogenez* **8** (3), 269–74 (in Russian with English summary).

Komar, A. (1973). Parthenogenetic development of mouse eggs activated by heat-shock *J. Reprod. Fert.* **35**, 433–43.

Konyukhov, B.V. (1969). *Biological models of inherited diseases of man*. Meditsina, Moscow.

—— (1975). Genetical control of morphogenesis. *Uspehy sovr. Biol.* (Advances in modern biology) **80**, 185–203 (in Russian with English summary).

—— (1980). *Developmental genetics of vertebrates*. Nauka, Moscow (in Russian).

Korochkin, L.I. (1981). *Gene interactions in development*. Springer, Berlin.

Koski, M.A., Dixon, L.K., and Fahrion, N. (1977). Olfactory mediated choice behaviour in mice: development and genetic analysis. *Behav. Biol.* **18**, 324–32.

Krafka, I. (1939). Parthenogenetic cleavage in the human ovary. *Anat. Rec.* **75**, 19–21.

Kratzer, P.G. (1983). Expression of maternally and embryologically derived hypoxanthine phosphoribosyl transferase (HPRT) activity in mouse eggs and early embryos. *Genetics* **104**, 685–9.

—— and Gartler, S.M. (1978). HPRT activity changes in preimplantation mouse embryos. *Nature* **274**, 503–4.

Krco, C.J. and Goldberg, E.H. (1976). H-Y (male) antigen: detection on 8-cell embryos. *Science* **193**, 1134–5.

——, —— (1977). Major histocompatibility antigens on preimplantation mouse embryos. *Transpl. Proc.* **9**, 1367–70.

Krishna, M. and Generoso, W.M. (1977). Timing of sperm penetration, pronuclear formation, pronuclear DNA synthesis and first cleavage in naturally ovulated mouse eggs. *J. exp. Zool.* **202**, 245–5.

Krone, W. and Wolf, U. (1977). Chromosome variation and gene action. *Hereditas* **86**, 31–6.

Krushinsky, L.V. (1977). *Biological aspects of reasoning ability*. Moscow University Publishing House, Moscow (in Russian).

——, Dyban, A.P., Baranov, V.S., Poletaeva, I.I., and Romanova, L.G. (1976). Capacity for extrapolation in laboratory mice with Robertsonian translocations of chromosomes. *Dokl. Akad. nauk. USSR, Ser. Biol.* **231**(3), 759–61 (in Russian).

——, ——, ——, ——, —— (1982). Characteristics of higher nervous activity in mice with Robertsonian translocations of chromosomes. *Zh. vyss. Nervn. Deyat. I.P. Pavlova* **32** (3), 446–53 (in Russian).

——, ——, ——, ——, ——, and Popova, N.V. (1978). Paths of physiological and genetical investigations of extrapolatory ability in mice. *Zh. vyss. Nervn. Deyat. I.P. Pavlova* (Journal of Higher Nervous Activity) **28** (5), 903–12 (in Russian).

——, Astaurova, N.B., Kuznetzova, L.M., Ochinskaya, E.I., Poletaeva, I.I., Romanova, L.G., and Sotskaya M.N. (1975). The role of genetical factors in determining the extrapolatory ability in animals. In *Actual problems in behaviour genetics* (ed. V.K. Fedmov and V.V. Ponomarenko). Nauka, Leningrad (in Russian).

——, Dyban, A.P., Baranov, V.S., Poletaeva I.I., Romanova, L.G., Forejt, J., and Gregorova, S. (1981). Investigation of extrapolatory ability in laboratory mice with partial autosomal trisomy. *Dokl. Akad. nauk. USSR Ser. Biol.* **260**, 1497–9 (in Russian).

Krzanowska, H. (1970). Relation between fertilization rate and penetration of eggs by supplementary spermatozoa in different mouse strains and crosses. *J. Reprod. Fert.* **22**, 199–204.

—— (1974). The passage of abnormal spermatozoa through the uterotubal junction of the mouse. *J. Reprod. Fert.* **38**, 81–90.

Kucharenko, V.I., Kuliev, A.M., Grinberg, K.N., and Teskich, V.V. (1975). Complex studies of human cell lines with abnormal karyotype II. Parameters of mitotic cycle. *Genetica* **11** (8), 139–46.

Kulazenko, V.P. (1976). Morphogenetic and cytogenetic damages in spontaneous abortuses of man. Doctoral thesis, Moscow (in Russian).

——, Kuliev, A.M., and Rozovsky, I.S. (1975). Morphogenetic changes in triploid embryos of man. *Arkh. Anat. Histol. Embryol.* **71**, 5–16 (in Russian with English summary).

Kuliev, A.M. (1972). Phenotype–karyotype interrelationship in embryogenesis of man. *Genetica* **8**, 140–53 (in Russian with English summary).

—— (1974). Lethal factors in human embryogenesis. In *Lectures in medical genetics* pp. 160–75. Meditsina, Moscow (in Russian).

—— (1976). Phenotypical peculiarities of chromosomal embryolethals in man. Doctoral thesis, Moscow (in Russian).

——, Grinberg, K.N., Vasileiski, S.S., Stepanova, L.G., and Urovskaya, L.G. (1972). Cell lines with abnormal karyotypes. *Genetika* **8** (8), 146–55 (in Russian with English summary).

Lauritsen, J.G. (1982). The cytogenetics of spontaneous abortion. *Res. Reprod.* **14** (3), 3–4.

Lazyuk, G.I. and Lurie, I.V. (1979). Multiple inborn malformations. In *Teratology of man* pp. 262–336. Medizina, Moscow (in Russian).

——, Lurie, I.V., and Cherstvoi, E.D. (1983). *Inherited syndromes of multiple inborn malformations.* Meditsina, Moscow (in Russian).

Ledbetter, D., Riccardi, V.M., and Airhart, S.D. (1981). Deletion of chromosome 15 as a cause of the Prader–Willi syndrome. *New Engl. J. Med.* **304**, 325–8.

Leipoldt, M. and Engel, W. (1983). Hidden breaks in ribosomal RNA of phylogenetically tetraploid fish and their possible role in the diploidization process. *Biochem. Genet.* **21**, 819–41.

Lennartz, R.J., Schunilfeder, N., and Maurer, W. (1966). Dauer der DNS-Synthesephase bei Ascitestumoren der Maus unterschiedlicher Ploidie. *Naturwissenschaften* **53**, 21–2.

Leonard, A. and Deknudt, G. (1967a). Chromosome rearrangements induced in the mouse by embryonic X-irradiation. I. Pronuclear stage. *Mutat. Res.* **4**, 689–97.

——, —— (1967b). A new marker for chromosome studies in mouse. *Nature* **214**, 504–5.

—— and Leonard, E.D. (1975). Aging and chromosome aberrations in male mammalian germ cells. *Exp. Gerontol.* **10**, 309–11.

——, Deknudt, G., and Linden, G. (1971). Ovulation and prenatal losses in different strains of mice. *Exp. Anim.* **6**, 1–6.

Levey, I.L., Stull, G.B., and Brinster, R.L. (1978). Poly(A) and synthesis of polyadenylated RNA in the preimplantation mouse embryo. *Devl Biol.* **64**, 140–8.

Lewandowski, R.C. Jnr. and Yuntz, J.J. (1975). New chromosomal syndromes. *Am. J. Dis. Child.* **129**, 515–29.

Lewin, R. (1981). Do chromosomes cross-talk? *Science* **214**, 1334–5.

Lewis, S.E. (1978). Developmental analysis of lethal effects of homozygosity for the c^{25H} deletion in the mouse. *Devl Biol.* **65**, 553–7.

Lifschytz, E. and Lindsley, D.L. (1972). The role of X-chromosome inactivation in spermatogenesis. *Proc. nat. Acad. Sci. USA* **69**, 182–6.

—— and Hareven, D. (1982). Heterochromatin markers: arrangements of obligatory heterochromatin histone genes and multisite gene families in the interphase nucleus of *Drosophila melanogaster. Chromosoma (Berlin)* **86**, 443–55.

Lindzey, G. and Thiessen, D. (1970). *Contribution to behavioural–genetic analysis. The mouse as a prototype.* Academic Press, New York.

Long, S.E. (1977). Cytogenetic examination of preimplantation blastocysts of ewes mated to rams heterozygous for the Massey I(t$_j$) translocation. *Cytogenet. Cell Genet.* **18**, 82–9.

——, Berepubo, N. and Amaitar, A. (1980). 37 XO chromosome complement in a kitten. *J. Small Anim. Pract.* **21** (11), 627–31.

—— and Williams, C.V. (1980). Frequency of chromosomal abnormalities in early embryos of the domestic sheep (*Ovis aries*). *J. Reprod. Fert.* **58**, 197–201.

Longo, F.J. and Kunkle, M. (1978). Transformation of sperm nuclei upon insemination. *Curr. Top. devl. Biol.* **12**, 149–84.

Lopatina, N.G., Dmitriev, J.S., Baranov, V.S., Udalova, L.D., and Pigarevski, P.V. (1976). Investigation of some higher nervous activity peculiarities in mice with chromosomal translocation T6 involving 14 and 15 pairs of autosomes. In *Mechanism of memory modulation.* (ed. N.P. Bekhtereva) pp. 199–202. Nauka, Leningrad.

Lu, Te-yu and Markert, C.L. (1980). Manufacture of diploid/tetraploid chimeric mice. *Proc. nat. Acad. Sci. USA* **776**, 6012–17.

Lubimova-Kerkis, T.I., Kuzin, B.A., and Korochkin, L.I. (1981). LDH activities in the eggs and early embryos of BALB/c, C57BL/6J mice and their reciprocal hybrids. *Genetika* **17** (1), 115–27 (in Russian with English summary).

Luciani, J.M., Guichaoua, M.R., Mattei, A., and Morazzani, M.R. (1984). Pachytene analysis of a man with a 13q, 14q, translocation and infertility. Behaviour of the trivalent and non-random association with sex-vesicle. *Cytogenet. Cell Genet.* **38** (1), 14–22.

Lurie, I.V. and Savenko, L.A. (1978). Genetical analysis in families with reciprocal translocations. *Genetika* **9** (11), 159–164 (in Russian with English summary).

Luthardt, F.W. (1976). Cytogenetic analysis of oocytes and early preimplantation embryos from XO mice. *Devl Biol.* **54**, 73–81.

—— and Donahue, R.P. (1973). Pronuclear DNA synthesis in mouse eggs. *Exp. Cell Res.* **82**, 143–51.

——, —— (1975). DNA synthesis in developing two-cell mouse embryos. *Devl Biol.* **44**, 210–20.

——, Palmer, C.G., and Yu, P.L. (1973). Chiasma and univalent frequencies in aging female mice. *Cytogenet. Cell Genet.* **12**, 68–79.

Lyapunova, E.A., Vorontzov, N.N., Korobitsyna, E.Y., Ivanitskaya, Y.M., Borisov, Y.M., Yakimenko, L.V., and Dovgal, V.Y. (1980). A Robertsonian translocation in *Ellobius talpinus*. *Genetica* **52/53**, 239–48 (in Russian).

Lyon, M.F. (1966). X-chromosome inactivation in mammals. In *Advances in teratology* pp. 25–54. Logos Press, London.

—— (1969). A true hermaphrodite mouse presumed to be XO/XY mosaic. *Cytogenetics* **8**, 326–31.

—— (1972). X-chromosome inactivation and developmental patterns in mammals *Biol. Rev.*, **47** 1–35.

—— (1974). Mechanisms and evolutionary origins of variable X-chromosome activity in mammals. *Proc. R. Soc., Lond.* **B187**, 243–260.

—— and Glenister, P.H. (1977). Factors affecting the observed number of young resulting from adjacent-2 disjunction in mice carrying a translocation. *Genet. Res.* **29**, 83–93.

—— and Hawker, S.G. (1973). Reproductive lifespan in irradiated and unirradiated chromosomally XO mice. *Genet. Res.* **21**, 185–94.

—— and Meredith, R. (1966). Autosomal translocations, causing male sterility and viable aneuploidy in the mouse. *Cytogenetics* **5**, 335–54.

——, Sayers, I.M., and Evans, E.P. (1978). *Mouse News Letter* **58**, 44.

——, Evans, E.P., Jarvis, S.E., and Sayers, I. (1979). t-haplotypes of the mouse may involve a change in intercalary DNA. *Nature* **279** (5708), 38–42.

Maclean N., Harnden, D.G., Brown W.M., Bond, I., and Mantle, D.J. (1964). Sex chromosome abnormalities in newborn babies. *Lancet* **i**, 286–8.

McBurney, M.W., Bramwell, S.R., Deussen, Z.A., and Graham, C.F. (1975). Development of parthenogenetic and fertilized mouse embryos in the uterus and in extrauterine sites. *J. Embryol. exp. Morphol.* **34**, 387–406.

McClure, H.M., Belden, K.H., Pieper, W.A., and Jacobson C.B. (1960). Autosomal trisomy in a chimpanzee: resemblance to Down's syndrome. *Science* **165**, 1010–12.

McFeely, R.A. (1967). Chromosome abnormalities in early embryos of the pig. *J. Reprod. Fert.* **13**, 578–81.

—— (1969). Aneuploidy, polyploidy and structural rearrangement of chromosomes in mammals other than man. In *Comparative mammalian cytogenetics* pp. 434–44. Springer, Berlin.

McGaughey, R.W. and Chang, M.C. (1969). Inhibition of fertilization and production of heteroploidy in eggs of mice treated with colchicine. *J. exp. Zool.* **171**, 465–79.

McGrath, J. and Hillman, N. (1982). The phenotypic expression of $t^6/t^6/t^6$ genotype. *J. Embryol. exp. Morphol.* **69**, 107–13.

——, ——, (1982). The effect of experimentally induced triploidy on the lethal expression of the t^{12} mutation in mouse embryos. *Devl Biol.* **89**, 254–60.

—— and Solter, D. (1983). Nuclear transplantation in the mouse embryo by microsurgery and cell fusion. *Science* **220**, 1300–2.

——, —— (1984). Completion of mouse embryogenesis requires both the maternal and paternal genomes. *Cell* **37**, 179–83.

McLaren, A. (1960). New evidence of unbalanced sex-chromosome constitution in the mouse. *Genet. Res.* **1**, 253–61.

—— (1972). Germ cell differentiation in artificial chimaeras of mice. In *The genetics of the spermatozoon* (ed. R. Beatty and S. Gluecksohn-Waelsch) pp. 313–24. Bogtrykkeriet Forum, Edinburgh.

—— (1974). Embryogenesis. In *Physiology and genetics of reproduction*. Part B pp. 297–316. Plenum Press, New York.

—— (1975). Sex chimaerism and germ cell distribution in a series of chimaeric mice. *J. Embryol. exp. Morphol.* **33**, 205–16.

—— (1976a). Gene mutations in early mammalian development (general discussion). *CIBA Foundation Symposium*: *Embryology in mammals*, Vol. 40, p. 231. North Holland, Amsterdam.

—— (1976b). Genetics of early mouse embryos. *A. Rev. Genet.* **10**, 361–88.

—— (1976c). *Mammalian chimaeras*. Cambridge University Press, Cambridge.

—— (1979). The impact of preimplantation events on postfertilization development in mammals. In *Maternal effects in development* (ed. D.R. Newth and M. Balls) pp. 287–320. Cambridge University Press, Cambridge.

—— (1981a). *Germ cells and soma: a new look at an old problem*. Yale University Press, New Haven and London.

—— (1981b). Analysis of maternal effects on development in mammals. *J. Reprod. Fert.* **62**, 591–6.

—— (1982). The embryo. In *Reproduction in mammals* (ed. C.R. Austin and R.V. Short) Vol. 2, pp. 1–25. Cambridge University Press, Cambridge.

—— (1983). Does the chromosomal sex of a mouse germ cell affect its development? In *Current problems in germ cell differentiation* (ed. A. McLaren and C.C. Wylie) pp. 225–40. Cambridge University Press, Cambridge.

—— and Bowman, P. (1969). Mouse chimaeras derived from fusion of embryos differing by nine genetic factors. *Nature* **224**, 238–240.

——, —— (1973). Genetic effects on timing of early development in the mouse. *J. Embryol. exp. Morphol.* **30**, 491–8.

—— and Buehr, M. (1981). GPI expression in female germ cells of the mouse. *Genet. Res.* **37**, 303–9.

—— and Michie, D.L. (1956). Studies on the transfer of fertilized mouse eggs to uterine foster-mothers. *J. exp. Biol.* **33**, 394–416.

——, Chandley, A.G., and Kofman-Alfaro, S. (1972). A study of meiotic germ cells in the gonads of foetal mouse chimaeras. *J. Embryol. exp. Morphol.* **27**, 515–24.

McMahon, A., Fosten, M. and Monk, M. (1981). Random X-chromosome inactivation in female primordial germ cells of the mouse. *J. Embryol. exp. Morphol.* **64**, 251–8.

——, ——, —— (1983). X-chromosome inactivation mosaicism in the three germ layers and the germ line of the mouse embryo. *J. Embryol. exp. Morphol.* **74**, 207–20.

Magnuson, T. (1983). Genetic abnormalities and early mammalian development. In *Development in mammals* (ed. M.H. Johnson) Vol. 5, pp. 209–50. Cambridge University Press, London.

—— and Epstein, C.J. (1981). Genetic control of very early mammalian development. *Biol. Rev.* **56**, 369–408.

——, Smith, S., and Epstein, C. (1982). The development of monosomy 19 embryos. *J. Embryol. exp. Morphol.* **69**, 223–36.

—— and Epstein, C.J. (1984). Oligosyndactyly: a lethal mutation in the mouse that results in mitotic arrest very early in development. *Cell* **38**, 823–33.

Makino, S., Sasaki, M.S., and Takishima, T. (1964). Triploid chromosome constitution in human chorionic lesions. *Lancet* 1273–5.

Manes, C. (1969). Nucleic acid synthesis in preimplantation rabbit embryos. I. Quantitative aspects, relationship to early morphogenesis and protein synthesis. *J. exp. Zool.* **712**, 303–10.

—— (1975). Genetic and biochemical activities in preimplantation embryos. In *Developmental Biology of Reproduction*, 33rd Symposium pp. 133–63. Academic Press, New York.

Mangia, F., Erickson, R.P., and Epstein, C.J. (1976). Synthesis of LDH-1 during mammalian oogenesis and early development. *Devl Biol.* **54**, 146–50.

Mann, J.R. and Lovell-Badge, R.H. (1984). Inviability of parthenogenones is determined by pronuclei not egg cytoplasm. *Nature* **310**, 66–7.

Markert, C.L. (1982). Parthenogenesis, homozygosity and cloning in mammals. *J. Hered.* **73**, 390–7.

—— and Petters, R.M. (1977). Homozygous mouse embryos produced by microsurgery. *J. exp. Zool.* **201**, 295–302.

—— and Seidel, G. (1981). Parthenogenesis, identical twins, and cloning in mammals. In *New techniques in animal breeding* (ed. B.G. Brackett, G.E. Seidel, and S.M. Seidel) pp. 181–200. Academic Press, London.

Maro, B., Johnson, M.H., Pickering, S., and Flach, G. (1984). Changes in actin distribution during fertilization of the mouse egg. *J. Embryol. exp. Morphol.* **81**, 211–37.

Marston, J.H. and Chang, M.C. (1964). The fertilizable life of ova and their morphology following delayed insemination in mature and immature mice. *J. exp. Zool.* **171**, 465–73.

Martin, R.H. (1983). Selection against chromosomally abnormal sperm—fact or fiction? *Hum. Genet.* **63**, 406.

—— (1983). A detailed method for obtaining preparations of human sperm chromosomes. *Cytogenet. Cell. Genet.* **35**, 252–6.

—— (1984). Analysis of human sperm chromosome complements from a male heterozygous for a reciprocal translocation t(11;22) (q23; q11). *Clin. Genet.* **25** (4), 357–61.

——, Balkan, W., and Burns, K. (1983b). Cytogenetic analysis of Q-banded pronuclear chromosomes in fertilized Syrian hamster eggs. *Cytogenet. Cell Genet.* **35**, 41–5.

——, Lin, C.C., Balkan, W., and Burns, K. (1982). Direct chromosomal analysis of human spermatozoa: preliminary results from 18 normal men. *Am. J. Hum. Genet.* **34**, 459–68.

——, Balkan, W., Burns, K., Rademaker, A.W., Lin, C.C., and Rudd, N.L. (1983b). The chromosome constitution of 1000 human spermatozoa. *Hum. Genet.* **63**, 305–9.

Martin-Deleon, P.A., Shaver, E.L., and Gammal, E.B. (1973). Chromosome abnormalities in rabbit blastocysts resulting from spermatozoa aged in the male tract. *Fert. Steril.* **24**, 212–19.

—— and Boice, M.L. (1982). Sperm aging in the male and cytogenetic anomalies. An animal model. *Hum. Genet.* **6**, 70–7.

——, —— (1983). Spontaneous heteroploidy in one-cell mouse embryos. *Cytogenet. Cell Genet.* **35**, 57–63.

Matsunaga, E., Tonomura, A., Oishi, H., and Kikuchi, Y. (1978). Re-examination of paternal age effect in Down's syndrome. *Hum. Genet.* **40**, 259–268.

Matthey, R. (1965). Cytogenetic mechanisms and speciation of mammals. In *The chromosome: structural and functional aspects.* Vol. 1, pp. 1–11. Tissue Culture Association, New York.

Maudlin, I. and Fraser, L.R. (1978). Maternal age and the incidence of aneuploidy in first-cleavage mouse embryos. *J. Reprod. Fert.* **54**, 423–6.

Maxon, S.C., Platt, T., Shenker, P., and Trattner, A. (1982). The influence of Y-chromosome of Rg/1 Bg mice on agonistic behaviors. *Aggress. Behav.* **8**, 285–91.

Mayr, E. (1963). *Animal species and evolution.* Belknap Press, Cambridge, Mass.

Meredith, R. (1969). A simple method for preparing meiotic chromosomes from mammalian testes. *Chromosoma* **26**, 254–8.

Michels, V.V., Medrano, C., Venne, V.L., and Riccardi V.M. (1982). Chromosome translocations in couples with multiple spontaneous abortions. *Am. J. Hum. Genet.* **34**, 507–13.

Mihamo, K. (1968). Intrafollicular overripeness and teratogenic development. *Cytogenetics* **7**, 212–33.

—— (1971). Cytogenetic studies on the possible teratogenecity of contraceptive steroids. Proceedings Seventh International Conference on Fertility and Sterility, Tokyo – Kyoto, pp. 215–18.

—— and Hamaguehi, H. (1975). Chromosomal disorder occurred by preovulatory over-ripeness of oocytes. In *Aging gametes* (ed. R.J. Blandau) pp. 72–97. Karger, Basel.

Mikkelsen, M., Poulsen, H., Grinsted, J., and Lange, A. (1980). Non-disjunction in trisomy 21: study of chromosomal heteromorphisms in 110 families. *Ann. hum. Genet.* **44**, 17–28.

Miller, D.A. and Miller, O.J. (1972). Chromosome mapping in the mouse. *Science* **178**, 949–55.

——, Dev, V.G., Tantravahi, R., Miller, O.J., Schiffman, M.B., Yates, R.A., and Gluecksohn-Waelsch, S. (1974). Cytological detection of the c^{25H} deletion involving the albino (c-) locus on chromosome 7. *Genetics,* **78**, 905–10.

Miller O.J. and Miller D.A. (1975). Cytogenetics of the mouse. *A. Rev. Genet.* **9**, 285–303.

Miller, J.F., Williamson, E., Glue, J., Gordon, Y.B., Grudzinskas, J.G., and Sykes,

A. (1980). Fetal loss after implantation. A prospective study. *Lancet* **ii**, 554–6.

Mintz, B. (1962). Incorporation of nucleic acid and protein precursors by developing mouse eggs. *Am. Zool.* **2**, 432.

—— (1963b). Experimental recombination of cells in the developing mouse egg: normal and lethal mutant genotypes. *Am. Zool.* **2**, 541–2.

—— (1964a). Formation of genetically mosaic mouse embryos, and early development of lethal (t^{12}/t^{12}) normal mosaics. *J. exp. Zool.* **157**, 273–92.

—— (1964b). Gene expression in the morula stage of mouse embryos as observed during development of t^{12}/t^{12} lethal mutants *in vitro*. *J. exp. Zool.* **157**, 267–72.

—— (1964c). Synthetic processes and early development in the mammalian egg. *J. exp. Zool.* **157**, 85–100.

—— (1967). Nucleic acid and protein synthesis in developing mouse embryos. *CIBA Foundation Symposium: Preimplantation stages of pregnancy*, pp. 145–78. Churchill, London.

Mittwoch, U. (1978). Parthenogenesis. *J. med. Genet.* **15**, 165–81.

—— and Delhanty, J.D.A. (1972). Inhibition of mitosis in human triploid cells. *Nature* **238** 11–13.

Miyabara, S., Gropp, A., and Winking, H. (1982). Trisomy 16 in the mouse fetus associated with generalized edema and cardiovascular and urinary tract anomalies. *Teratology* **25**, 369–80.

Modlinski, J.A. (1975). Haploid mouse embryos obtained by microsurgical removal of one pronucleus. *J. Embryol. exp. Morphol.* **33**, 897–905.

—— (1978). Transfer of embryonic nuclei to fertilized mouse eggs and development of tetraploid blastocysts. *Nature* **273**, 466–7.

—— (1980). Preimplantation development of microsurgically obtained haploid and homozygous diploid mouse embryos and effects of pretreatment with Cytochalasin B on enucleated eggs. *J. Embryol. exp. Morphol.* **60**, 153–162.

—— (1981). The fate of inner cell mass and trophectoderm nuclei transplanted to fertilized mouse eggs. *Nature* **292**, 342–3.

Molinaro, M., Siracusa, J., and Monesi, V. (1969). Andamanto del cycle cellulare in vivo e in coltura durante l'embriogenes iniziale del topo. *Biol. Latino-am.* **22**, 227–34.

——, ——, —— (1972). Differential effects of metabolic inhibitors on early development in the mouse embryo at various stages of the cell cycle. *Exp. Cell Res.* **71**, 261–4.

Molls, M., Zamboglon, N., and Streffer, C. (1983). A comparison of the cell kinetics of preimplantation mouse embryos from two different mouse strains. *Cell Tissue Kinet.* **16**, 277–83.

Monesi, V. and Salfri, V. (1967). Macromolecular synthesis during early development in the mouse embryo. *Exp. Cell Res.* **46**, 632–5.

——, Molinaro, M., Spalletta, E., and Davoli C. (1970). Effect of metabolic inhibitors on macromolecular synthesis and early development in the mouse embryo. *Exp. Cell Res.* **59**, 197–206.

Monk, M. (1978). Biochemical studies on mammalian X-chromosome activity. In *Development in mammals* (ed. M.H. Johnson) Vol. III, pp. 189–223. North Holland, Amsterdam.

—— (1981). A stem-line model for cellular and chromosomal differentiation in early mouse development. *Differentiation* **19**, 71–6.

—— and Kathuria, H. (1977). Dosage compensation for an X-linked gene in preimplantation mouse embryos. *Nature* **270**, 599–601.

Monk, M. and Harper, M.I. (1979). Sequential X chromosome inactivation coupled with cellular differentiation in early mouse embryos. *Nature* **281**, 311–13.

—— and McLaren A. (1981). X-chromosome activity in foetal germ cells of the mouse. *J. Embryol. exp. Morphol.* **63**, 75–84.

Morris, T. (1968). The XO and OY chromosome constitution in the mouse. *Genet. Res.* **12**, 125–37.

Muggleton-Harris, A.L. and Johnson, M.H. (1976). The nature and distribution of serologically detectable alloantigens on the preimplantation mouse embryos. *J. Embryol. exp. Morphol.* **35**, 59–72.

——, Whittingham, D.G., and Wilson, L. (1982). Cytoplasmic control of preimplantation development *in vivo* in the mouse. *Nature* **299**, 460–2.

Mukherjee, A.B. (1976). Cell cycle analysis and X-chromosome inactivation in the developing mouse. *Proc. nat. Acad. Sci. USA* **78**, 1608–11.

Mullen, R.J. (1975). Genetic dissection of the CNS with mutant–normal mouse and rat chimeras. *Soc. Neurosci., Symp.* **2**, 47–65.

—— and Herrup, K. (1979). Chimeric analysis of mouse cerebellar mutants. In *Neurogenetic and genetic approaches to the nervous system* (ed. X.O. Bakfield). Elsevier, Amsterdam.

Muramatsu, S. (1974). Frequency of spontaneous translocations in mouse spermatogonia. *Mutat. Res.* **24**, 81–2.

Nadeau, J.H. and Eicher, E.M. (1982). Conserved linkage of soluble aconitase and galactose-1-phosphate uridyl transferase in mouse and man: assignment of these genes to mouse chromosome 4. *Cytogenet. Cell Genet.* **34**, 271–81.

Nadijcka, M.D., Hillman, N., and Gluecksohn-Waelsch, S. (1979). Ultrastructural studies of lethal c^{25H}/c^{25H} mouse embryos. *J. Embryol. exp. Morphol.* **52**, 1–11.

Nash, H.R., Brooker, P.C., and Davis, S.J.M. (1983). The Robertsonian translocation house-mouse populations of North East Scotland: a study of their origin and evolution. *Heredity* **50**, 303–10.

Nejfakh, A.A. (1962). *The problem of nucleus–cytoplasmic relationships in development.* Institute of Animal Morphology, Academy of Science USSR, Moscow (in Russian).

—— (1965a). Gene and Phene. In *General genetics* pp. 232–98. Nauka, Moscow (in Russian).

—— (1965b). Specific inhibitors of nucleic acid metabolism in studies of morphogenetic activity of nuclei during development. In *Cell differentiation and induction mechanisms* pp. 38–68. Nauka, Moscow (in Russian).

—— (1971). Steps of realization of genetic information in early development. *Curr. Top. devl. Biol.* **6**, 45–77.

—— (1974). Molecular aspects of nucleocytoplasmic relationship in embryonic development. In *Neoplasia and cell differentiation* pp. 27–59. Karger, Basel.

—— and Timofeeva M.J. (1977). *Molecular biology of developmental processes.* Nauka, Moscow (in Russian).

—— and Timofeeva, W.S. (1978). *Regulation problems in molecular developmental biology.* Nauka, Moscow (in Russian).

Nesbitt, M.N. (1978). Attempts at locating the site of action of genes affecting behaviour. In *Genetic mosaics and chimaeras in mammals* (ed. L.B. Russell) pp. 51–8. Plenum Press, New York.

—— and Donahue, R.P. (1972). Chromosome banding patterns in preimplantation mouse embryos. *Science* **177**, 805–6.

—— and Francke, U. (1973). A system of nomenclature for banding patterns of mouse chromosomes. *Chromosoma* **41**, 145–58.

Nevo, E., Naftali, G., and Guttman, R. (1975). Aggression patterns and speciation. *Proc. nat. Acad. Sci. USA* **72** (8), 3250–4.

Niebuhr, E. (1974). Triploidy in man. Cytogenetical and clinical aspects. *Humangenetik* **21**, 103–25.

——, Sparrevohn, S., Hinnigsen, K., and Mikkelin, M. (1972). A case of liveborn triploidy (69 XXX). *Acta paediat. scand.* **61**, 203–8.

Nielsen, J.T. and Chapman, V.M. (1977). Electrophoretic variation for X-chromosome linked phosphoglyceratekinase (PGK-1) in the mouse. *Genetics* **87**, 319–25.

Niemierko, A. (1975). Induction of triploidy in the mouse by cytochalasin B. *J. Embryol. exp. Morphol.* **34**, 279–89.

—— (1981). Postimplantation development of CB-induced triploid mouse embryos. *J. Embryol. exp. Morphol.* **66**, 81–9.

—— and Komar, A. (1977). Cytochalasin B induced triploidy in mouse oocytes fertilized *in vitro. J. Reprod. Fert.* **48**, 279–84.

—— and Opas, J. (1978). Manipulation of ploidy in the mouse. In *Methods in mammalian reproduction* (ed. J.C. Daniel Jr.) pp. 50–67. Academic Press, New York.

Nijhoff, J.H. and de Boer, P. (1981). Spontaneous meiotic non-disjunction in mammals. A study evaluating the various experimental approaches. *Genetica* **56**, 99–121.

Nikawa, N. and Tadashi, K. (1974). Triploid human abortuses due to dispermy. *Humangenetik* **24**, 261–4.

Nishimura, N. (1975). Prenatal versus postnatal malformations based on the Japanese experience on induced abortions in human being. In *Aging gametes* (ed. R.J. Blandau) pp. 349–68. Karger, Basel.

—— and Yamamura, H. (1969). Comparison between man and some other mammals of normal and abnormal developmental processes. In *Second International Workshop on Teratology*, pp. 223–40. Sakuoku, Kyoto.

Niwa, K., Araki, M., and Iritani, A. (1980). Fertilization *in vitro* of eggs and first cleavage of embryos in different strains of mice. *Biol. Reprod.* **22**, 1156–89.

Nombela, J.J. and Murcia, C.R. (1978). Spontaneous double Robertsonian translocations Rb(2.3) and Rb(X.3) in the mouse. *Cytogenet. Cell Genet.* **19**, 227–30.

Noniashvili, E.M. (1985). Spontaneous parthenogenesis in different strains of laboratory mice. In *Mechanisms of normal and abnormal embryogenesis in mammals*, pp. 11–16. Nauka, Leningrad (in Russian)

—— and Dyban, A.P. (1985). Parthenogenetic development of mouse ova induced by ethyl alcohol and heat shock both *in situ* and *in vitro*. In *Mechanisms of normal and abnormal embryogenesis in mammals*, pp. 22–30. Nauka, Leningrad (in Russian).

Norberg, H.S., Refsdal, A.O., Garm, O.N., and Nes, N.A. (1976). A case report on X-trisomy in cattle. *Hereditas* **82**, 69–72.

Norby, D.E., Hegreberg, G.A., Thuline, H.C., and Finley, D. (1974). An XO cat. *Cytogenet. Cell Genet.* **13**, 448–53.

Novitsky, E. (1967). Nonrandom disjunction in *drosophila. A. Rev. Genet.* **1**, 71–86.

Nusbacher, J. and Hirschhorn, K. (1968). Autosomal anomalies in man. In *Advances in teratology* (ed. D.U.M. Woollam) Vol. 3, pp. 11–64. Logos Press, London.

Ohama, K. and Tadashi, K. (1972). Monosomy 21 in spontaneous abortus. *Human-genetik* **16**, 267–70.

——, Kajii, T., Okamoto, F., Fukuda, Y., Imaizunni, K., Tsukahara, M., Kobagashi, K., and Hagiwara, K. (1981). Dispermic origin of XY hydatiform moles. *Nature* **292**, 551–2.

Ohno, S. (1967). *Sex chromosomes and sex-linked genes.* Springer, Berlin.

—— (1969). Evolution of sex chromosomes in mammals. *A. Rev. Genet.* **3**, 495–524.

——, Kaplan, W.D., and Kinosita, R. (1969). On the end-to-end association of the X and Y chromosomes of *Mus musculus. Exp. Cell Res.* **18**, 282–90.

—— (1979). *Major sex determining genes.* Springer, New York.

Olds, P.J. (1971). Effect of the T-locus on fertilization in the house mouse. *J. exp. Zool.* **4**, 417–34.

—— Stern, S., and Biggers, J.D. (1973). Chemical estimates of the RNA and DNA contents of early mouse embryos. *J. exp. Zool.* **186**, 39–47.

Opas, J. (1977). Effects of extremely low osmolarity on fertilized mouse eggs. *J. Embryol. exp. Morphol.* **37**, 65–77.

Oprescu, S. and Thibault, C. (1965). Duplication de l'ADN dans les oeufs de lapine après la fécondation. *Ann. Biol. Anim. Biochim. Biophys.* **8**, 151–6.

Oshimura, M. and Takagi, N. (1975). Meiotic disjunction in T(14, 15)6Ca heterozygotes and fate of chromosomally unbalanced gametes in embryonic development. *Cytogenet. Cell Genet.* **15**, 1–16.

—— and Sandberg, A.A. (1978). Sexual differences in meiotic disjunction of murine T (1;13)70 H translocation heterozygotes. *Jap. J. Genet.* **53**, 215–18.

Otis, E.M. and Brent, R. (1954). Equivalent ages in mouse and human embryos. *Anat. Rec.* **120**, 33–63.

Paigen, K. (1979). Genetic factors in developmental regulation. In *Physiological genetics* (ed. J.G. Scandalios). Academic Press, New York.

Papaioannou, V. and Gardner, R.L. (1979). Investigation of lethal yellow A^y/A^y embryos using mouse chimeras. *J. Embryol. exp. Morphol.* **52**, 153–63.

—— and West, J.D. (1981). Relationship between the parental origin of the X chromosomes, embryonic cell lineage and X chromosome expression in mice. *Genet. Res.* **47**, 183–97.

Parington, J.M., West, L.E., and Povey, S. (1984). The origin of ovarian teratomas. *J. med. Genet.* **21** (1), 4–12.

Park, W.W. (1957). The occurrence of sex chromatin in early human and macaque embryos. *J. Anat.* **91**, 369–73.

Patau, K., Therman, E., Smith, D.D., Inhorn, S.L., and Picken, B.F. (1961). Partial trisomy syndrome. I. Sturge-Weber's disease. *Am. J. hum. Genet.* **13**, 287–98.

Paterson, H.F. (1979). *In vivo* and *in vitro* studies on the early embryonic lethal oligosyndactylism (Os) in the mouse. *J. Embryol. exp. Morphol.* **52**, 115–25.

—— (1980). *In vivo* and *in vitro* studies on the early embryonic lethal tail-short (Ts) in the mouse. *J. exp. Zool.* **211**, 247–56.

Patkin, E.L. (1979). Fluorescent analysis of heterochromatin in early embryogenesis of mammals. Ph.D. thesis, Leningrad (in Russian).

—— (1980). A study of structural heterochromatin at the initial embryonic stages in mice. *Ontogenez* **11**, 49–55 (in Russian with English summary).

—— and Sorokin, A.V. (1983). Nucleolus organizer regions of chromosomes in early embryogenesis of the laboratory mouse. *Bull. exp. Biol. Med.* **8**, 92-4 (in Russian with English summary).

Paton, G.R., Silver, M.F., and Allison, A.C. (1974). Comparison of cell cycle time in normal and trisomic cells. *Humangenetik* **23**, 173-82.

Pearson, P.L., Ellis, J.D., and Evans H.J. (1970). A gross reduction in chiasma formation during meiotic prophase and a defective DNA repair mechanism associated with a case of human infertility. *Cytogenetics* **9**, 460-7.

——, Roderick, T.H., Davison, M.T., Lalley, P.A., and O' Brien, S.J. (1982). Report of the Committee on comparative mapping. *Cytogenet. Cell Genet.* **32**, 208-37.

Pedersen, R.A. (1974). Development of lethal yellow (A^y/A^y) mouse embryos *in vitro*. *J. exp. Zool.* **188**, 307-19.

—— and Spindle, A.I. (1976). Genetic effects on mammalian development during and after implantation. *CIBA Foundation Symposium: Embryogenesis in mammals* Vol. 40, pp. 133-45. North Holland, Amsterdam.

Penrose, L.S. (1966). The causes of Down's syndrome. In *Advances in teratology* (ed. D.U.M. Woollam) Vol. 1. Academic Press, London.

—— and Delhanty, J.D.A. (1961). Triploid cell cultures from a macerated foetus. *Lancet* **i**, 1261-2.

Pesonen S. (1946). Abortive egg cells in the mouse. *Hereditas* **32**, 93-6.

Petzoldt, U., Hoppe, P.C., and Illmensee, K. (1980). Protein synthesis in enucleated fertilized and unfertilized mouse eggs. *Wilhelm Roux' Arch. Entwicklungsmech. Org.* **189**, 215-19.

——, Illmensee, G.R., Burki, K., Hoppe, P.C., and Illmensee, K. (1981). Protein synthesis in microsurgically produced androgenetic and gynogenetic mouse embryos. *Mol. gen. Genet.* **184**, 11-16.

——, Bürki, K., Illmensee, G., and Illmensee, K. (1983). Protein synthesis in mouse embryos with experimentally produced asynchrony between chromosome replication and cell growth. *Wilhelm Roux' Arch. Entwicklungsmech. Org.* **192**, 34-42.

Pexieder, T., Miyabara, S., and Gropp, A. (1981). Congenital heart diseases in experimental (fetal) mouse trisomies: incidence. In *Mechanisms of cardiac morphogenesis and teratogenesis* (ed. T. Pexider) pp. 389-99. Raven Press, New York.

Phillips, E. and Boué, J.G. (1970). Placenta et aberrations chromosomiques au cours des avortements spontanés. *Presse méd.* **78**, 641-6.

Phillips, R.I.S. and Kaufman, M.H. (1974). Bare-patches, a new sex-linked gene in the mouse associated with a high production of XO females. II. Investigations into the nature and mechanisms of the XO production. *Genet. Res.* **24**, 27-41.

Pierce, G.B. Jr. (1967). Teratocarcinoma: model for developmental concept of cancer. *Curr. Top. devl. Biol.* **2**, 229-46.

Piko, L. (1958). Etude de la polyspermie chez le rat. *C.r. séanc. Soc. Biol.* **152**, 1356-8.

—— (1970). Synthesis of macromolecules in early mouse embryos cultured *in vitro*: RNA, DNA, and polysaccharide components. *Devl Biol.* **21**, 737-9.

—— and Bomsel-Heimreich, O. (1960). Triploid rat embryos and other chromosomal durations after colchicine treatment and polyspermy. *Nature* **186**, 737-9.

—— and Clegg, K.P. (1982). Quantitative changes in total RNA, total poly (A) and ribosomes in early mouse embryos. *Devl Biol.* **89**, 362-78.

Pincus, G. (1939a). The comparative behaviour of mammalian eggs *in vivo* and *in vitro*. IV. The development of fertilized and artificially activated rabbit eggs. *J. exp. Zool.* **82**, 85–130.

—— (1939b). The breeding of some rabbits produced by recipients of artificially activated ova. *Proc. nat. Acad. Sci. USA* **25**, 557–9.

—— and Waddington, C.H. (1939). The effects of mitosis-inhibiting treatments on normally fertilized pre-cleaved rabbit eggs. *J. Hered.* **30**, 515–18.

——, —— (1936). The comparative behaviour of mammalian eggs *in vivo* and *in vitro*. II. Activation of tubal eggs of the rabbit. *J. exp. Zool.* **73**, 195–206.

—— and Enzmann, E.V. (1937). The growth, maturation and atresia of ovarian eggs in the rabbit. *J. Morphol.* **61**, 351–76.

—— and Shapiro, H. (1940). Further studies on the parthenogenetic activation of rabbit eggs. *Proc. nat. Acad. Sci USA* **26**, 163–5.

Pluznik, D.H., Herbst, E.W., Lenz, R., Sellin, D., Hertzog, Ch. F., and Gropp, A. (1981). Controlled production of trisomic hematopoietic stem cells: an experimental tool in hematology and immunology. In *Experimental hematology today*, pp. 3–11. S. Karger, Basel.

Polani, P.E. (1969). Abnormal sex chromosomes and mental disorder. *Nature* **223**, 680–6.

—— (1972). Centromere localization at meiosis and the position of chiasmata in the male and female mouse. *Chromosoma* **36**, 343–74.

—— (1977). Abnormal sex-chromosomes, behaviour and mental disorder. In *Developments in psychiatric research* (ed. J.M. Tanner) pp. 89–128. Hodder and Stoughton, London.

—— and Jagiello, G.M. (1976). Chiasmata, meiotic univalents and age in relation to aneuploid imbalance in mice. *Cytogenet. Cell Genet.* **16**, 505–29.

Poletaeva, I.I. and Romanova, L.G. (1977). Genetical differences in complex behaviour in rodents. *Act. nerv. super.* **19** (3), 246–7.

Popescu, C.P. and Boscher, J. (1982). Cytogenetics of preimplantation embryos produced by pigs heterozygous for the reciprocal translocation (4q$^+$; 14q$^-$). *Cytogenet. Cell Genet.* **34**, 119–23.

Posnakhirkina, N.A., Serov, O.L., and Korochkin, L.I. (1975). A study of lactate dehydrogenase isozymes in rat ova. *Biochem. Genet.* **13**, 65–72.

Poulson, D.F. (1940). The effect of certain X-chromosome deficiencies on the embryonic development of *Drosophila melanogaster*. *J. exp. Zool.* **83**, 271–326.

—— (1945). Chromosomal control of embryogenesis in *Drosophila*. *Am. Nat.* **79**, 340–63.

Pratt, H.P.M., Bolton, V.N., and Gudgeon, K.A. (1983). The legacy from the oocyte and its role in controlling early development of mouse embryo. In *The molecular biology of egg maturation. CIBA Symposium* Vol. 98, pp. 197–227. Pitman Press, New York.

Pratt, R.T. (1967). *The genetics of neurological disorders*. Oxford University Press, Oxford.

Prescott, D.M. (1976). *Reproduction of eukaryotic cells*. Academic Press, New York.

Prokofieva-Belgovskaya, A.A. (1946). Asynchrony of paternal chromosome transformations. *Dokl. Acad. nauk. USSR* **54** (2), 169–172 (in Russian with English summary).

—— (1947). Asynchrony of cell nucleus system. *Zh. Obshch. Biol.* **8** (2), 247–80 (in Russian).

—— (1969). Hereditary substance. In *Basic studies in human cytogenetics* pp. 64–75, Meditsina, Moscow (in Russian).

——, Grinberg, K.N., Revazov, A.A., Mikelsaar, A.V.N., and Kuliev, A.M. (1973). Phenotype–karyotype interrelationships in chromosomal diseases. *Vestn. Acad. Med. nauk. USSR* 37–44 (in Russian with English summary).

Putz, B., Krause, G., Garde, T., and Gropp, A. (1980). A comparison between trisomy 12 and vitamin A-induced exencephaly and associated malformations in mouse embryos. *Virchows Arch.* **368**, 65–80.

—— and Morris-Kay, G. (1981). Abnormal neural fold development in trisomy 12 and trisomy 14 mouse embryos. I. Scanning electron microscopy. *J. Embryol. exp. Morphol.* **66**, 141–58.

Race, R.R. and Sanger R. (1969). The origin of sex chromosome imbalance in man. *Br. med. Bull.* **25**, 99.

Raff, E.C., Brothers, A.I., and Kaff, R.A. (1976). Microtubule assembly mutant. *Nature* **260**, 615–17.

O'Rahilly, R. (1979). Early human development and the chief source of information on staged human embryos. *Eur. J. Obstet. Gynecol. reprod. Biol.* **9** (4), 273–80.

Rao, P.N., Wilson, B.A., and Sunkara, P. (1978). Inducers of DNA synthesis present during mitosis of mammalian cells lacking G_1 and G_2 phases. *Proc. nat. Acad. Sci. USA* **75**, 5043–7.

Rastan, S. (1982). Timing of X-chromosome inactivation in postimplantation mouse embryos. *J. Embryol. exp. Morphol.* **71**, 11–24.

Rathenberg, K. and Müller, D. (1973). X and Y chromosome pairing and disjunction in a male mouse with an XYY sex-chromosome constitution. *Cytogenet. Cell Genet.* **12**, 87–92.

Reeve, W.J. and Ziomek, C.A. (1981). Distribution of microvilli on dissociated blastomeres from mouse embryos: evidence for surface polarization at compaction. *J. Embryol. exp. Morphol.* **62**, 338–50.

Reichert, W., Hansmann, I., and Röhrborn, G. (1975). Chromosome anomalies in mouse oocytes after irradiation. *Humangenetik* **28**, 25–38.

Rieck, G.W., Höhn, H., and Herzog, A. (1970). X-Trisomie beim Rind mit Anzeichen familiärer Disposition for Meioses-Störungen. *Cytogenetics* **9**, 401–9.

Robertson, G.G. (1942). An analysis of the development of homozygous yellow mouse embryos. *J. exp. Zool.* **89**, 197–231.

Röhrborn, G. (1972). Frequencies of spontaneous nondisjunction in metaphase II oocytes of mice. *Humangenetik* **16**, 123–4.

Roux, C. (1970). Etude morphologique des embryons humains atteints d'aberrations chromosomiques. *Presse méd.* **78**, 647–562.

——, Emerit, J., and Taillemite, J.L. (1971). Chromosomal breakage and teratogenesis. *Teratology* **4**, 303–16.

Rudak, E., Jacobs, P.A., and Yanagimachi, R. (1978). Direct analysis of chromosome constitution of human spermatozoa. *Nature* **274**, 911–13.

Ruddle, F. (1981). Linkage studies employing mouse–man somatic cell hybrids. *Fed. Proc.* **30**, 921–5.

Russell, L.B. (1961). Variegated-type position effects in the mouse. *Genetics* **46**, 509–25.

—— (1962). Chromosome aberrations in experimental mammals. In *Progress in medical genetics* Vol. 2, pp. 230–94. Academic Press, New York.

—— (1964). Experimental studies on mammalian chromosome aberrations. In *Mammalian cytogenetics and related problems in radiobiology* pp. 61–86. Pergamon Press, Oxford.

—— (1965). Death and chromosome damage from irradiation of preimplantation stages. *CIBA Foundation Symposium; Preimplantation stages of pregnancy* pp. 217–45. Churchill, London.

—— and Saylors, C.L. (1960). Factors causing a high frequency of mice having the XO sex-chromosome constitution. *Science* **131**, 1321–2.

——, —— (1961). Spontaneous and induced abnormal sex-chromosome number in the mouse. *Genetics* **46**, 894.

——, —— (1962). Induction of paternal sex-chromosome losses by irradiation of mouse spermatozoa. *Genetics* **47**, 7–10.

—— and Woodiel, F.N. (1966). A spontaneous mouse chimera formed from separate fertilization of two meiotic products of oogenesis. *Cytogenetics* **5**, 106–19.

——, Russell, W.L., Cacheiro, N.L.A., Vaughan, C.M., Popp, B.A., and Jacobson, K.B. (1975). A tandem duplication in the mouse. *Genetics* **80** (Suppl.), 71.

Russell, W.L., Russell, L.B., and Gower, J.B. (1959). Exceptional inheritance of a sex-linked gene in the mouse explained on the basis that the X/O sex-chromosome constitution in female. *Proc. nat. Acad. Sci. USA* **45**, 554–60.

Salonen, K., Paranko, J., and Parvinen, M. (1982). A colcemid-sensitive mechanism involved in regulation of chromosome movements during meiotic pairing. *Chromosoma* **85**, 611–18.

Samarina, O.P., Lukandin, E.M., and Georgiev, G.R. (1973). Ribonucleoprotein particles containing RNA and pre-mRNA. In *Protein synthesis in reproductive tissue*. Sixth Karolinska Symposium, Stockholm, pp. 130–67.

Samoshkina, N.A. (1965). Incorporation of ^{3}H-thymidine by embryonic cell nuclei during preimplantation development in mice. *Dokl. Acad. nauk. USSR* **161** (6), 1467–70 (in Russian).

—— (1968). DNA-synthesis patterns in the cleaving eggs of mice (*in vitro* studies). *Tsitologiya* **10** (7), 856–64 (in Russian with English summary).

—— (1970). DNA-synthesis pattern in early mammalian embryos (autoradiographical studies). In *Cell nucleus metabolism and nucleus–cytoplasmic relationships* pp.247–8. Naukova Dumka, Kiev (in Russian).

—— (1972). DNA synthesis and nucleus functions in early mammalian embryos. In *Mechanisms of cell nucleus regulation*, pp.102–3. Mintsniereba, Tbilisi (in Russian).

—— (1978). Histoautoradiographical analysis of nucleic acid synthesis and mitotic cycle peculiarities during early development of mammalian embryos. DSc thesis, Leningrad (in Russian).

—— and Khozhai, L.I. (1975). RNA-synthesis patterns at initial stages of embryogenesis in mice (histoautoradiographical studies). Fifth All-Union Conference of Embryologists, Leningrad, p.156.

Sankaranarayanan, K. (1979). The role of non-disjunction in aneuploidy in man. An overview. *Mutat. Res.* **61**, 1–28.

Sawicki, J.A., Magnuson, T., and Epstein, C.J. (1981). Evidence for expression of the paternal genome in the two-cell mouse embryo. *Nature* **294**, 450–1.

Sawicki, W., Abramczuk, J., and Bolton, O. (1978). DNA synthesis in the second and third cell cycles of mouse preimplantation development. *Exp. Cell. Res.* **112**, 199–205.

Schindler, J. and Sherman, M.I. (1981). Effects of α-amanitin on programming of mouse blastocyst development. *Devl. Biol.* **84**, 332–40.

Schleiermacher, E. (1966). Uber den Einfluss von Trinemon und Endoxan auf die

Meiose der Mänlichen Maus. I. Methode der Präparation und Analyse Meiotischer Teilungen. *Humangenetik* 3, 127–33.

—— (1972). Frequency of spontaneous chromosome mutations in spermatogenesis of mice. *Humangenetik* 16, 115–17.

Schneider, B.F. and Norton, S. (1979). Equivalent stages in rat, mouse and chick embryos. *Teratology* 19, 273–8.

Schneider, E.L. and Epstein, C. J. (1972). Replication rate and lifespan of cultured fibroblasts in Down's syndrome. *Proc. Soc. exp. Biol. Med.* 141, 1092–4.

Schultz, G., Manes, C., and Hahn, W. (1973a). Synthesis of RNA containing polyadenylic acid sequences in preimplantation rabbit embryos. *Devl Biol.* 30, 418–26.

——, Manes, C., and Hahn, W.E. (1973b). Estimation of the diversity of the transcription in early rabbit embryos. *Biochem. Gen.* 9, 247–59.

Schwarzacher, H.G. (1976). *Chromosomes in mitoses and interphase* pp. 234–6. Springer, Berlin.

Searle, A.G. (1974). Nature and consequences of induced chromosome damage in mammals. *Genetics* 78, 173–86.

—— (1982). Chromosomal variants. In *Genetic variants and strains of the laboratory mouse* (ed. M.C. Green) pp. 324–53. Gustav Fischer, Stuttgart.

—— and Beechey, C.V. (1978). Complementation studies with mouse translocations. *Cytogenet. Cell Genet.* 20, 282–303.

——, —— (1982). The use of Robertsonian translocations in the mouse for studies of non-disjunction. *Cytogenet. Cell Genet.* 33, 81–7.

——, Beechey, C.V., and Evans, E.P. (1978). Meiotic effects in chromosomally derived male sterility of mice. *Ann. Biol. anim. Biochem. Biophys.* 18 (2B), 391–8.

——, Berry, R.J., and Beechey, C.V. (1970). Cytogenetic radiosensitivity and chiasma frequency in wild living male mice. *Mutat. Res.* 9, 137–40.

——, Evans, E.P., Ford, C.E., and West, B.J. (1968). Studies on the induction of translocations in mouse spermatogonia. I. The effect of dose-rate. *Mutat. Res.* 6, 427–36.

——, Ford, C.E., and Beechey, C.V. (1971b). Meiotic disjunction in mouse translocations and the determination of centromere position. *Genet. Res.* 18, 215–35.

——, Beechey, C.V., Evans E.P., Ford, C.E., and Papworth, D.G. (1971a). Studies on the induction of translocations in mouse spermatogonia. IV. Effects of acute gamma-irradiation. *Mutat. Res.* 12, 411–16.

Searle, R.F., Sellens, M.H., Elson, J., Jenkinson, E.J., and Billington W.D. (1976). Detection of alloantigens during preimplantation development and early differentiation. *J. exp. Med.* 143, 348–59.

Seidel, G.E. (1983). Mammalian oocytes and preimplantation embryos as methodological components. *Biol. Reprod.* 28, 36–49.

Sellin, D., Schlizio, E., and Herbst, E.W. (1982). Immunkapazitat trisomer Lymphozyten Vergleichende Untersuchungen an Lymphozyten der trisomie 12 und trisomie 19 der Maus (Strahlenchimaran) mit Stimulierung *in vitro* sowie Sensibilisierung *in vivo. Verh. Dtsch. Ges. Pathol.* 66, 261–4.

Semyonova-Tian-Shanskaya A.G. and Patkin, E.L. (1978). Changes in gonocyte nuclei at different stages of their differentiation in early human female embryos. *Arkh. Anat. Histol. Embriol.* 74, 91–7 (in Russian with English summary).

——, —— (1977). Relationship between X-chromosome heterochromatization in the gonocytes of early human embryos and their position in the germ anlage epithelium. *Arkh. Anat. Histol. Embriol.* **73** (9), 124.

Serebrowski, A.S. (1935). *Animal hybridization.* Medgiz, Moscow (in Russian).

Serov, O.L. and Manchenko, G.P. (1974). Genetical control of electrophoretically distinct isozymes of 6-phosphogluconatedehydrogenase in wild and in laboratory rats. *Genetika* **10** (3), 40-3 (in Russian with English summary).

Severova, E.L. (1985). Prevital studies of mouse embryos with monosomy and nullisomy for the autosomes 15 and 17. In *Mechanisms of normal and abnormal embryogenesis in mammals.* pp. 51-4. Nauka, Leningrad (in Russian).

—— and Dyban, A.P. (1984). The live selection of early mouse embryos by their sex and karyotype peculiarities. *Ontogenez* **15** (6), 585-91 (in Russian with English summary).

Shaver, E.L. and Carr, D.H. (1967). Chromosome abnormalities in rabbit blastocysts following delayed fertilization. *J. Reprod. Fert.* **14**, 415-20.

——, —— (1969). The chromosome complement of rabbit blastocysts in relation to the time of mating and replication. *Can. J. Genet. Cytol.* **11**, 287-98.

—— and Martin-Deleon, P.A. (1975). Effects of aging of sperm in the female and male reproductive tracts before fertilization on the chromosome complement of the blastocysts. In *Aging gametes* (ed. R.J. Blandau) pp. 151-65. Karger, Basel.

Shelton, J.A. and Goldberg, E.H. (1984). Male-restricted expression of H-Y antigen on preimplantation mouse embryos. *Transplantation* **37** (1), 7-8.

Sherman, M.I. (1979). Developmental biochemistry of preimplantation mammalian embryos. *A. Rev. Biochem.* **48**, 443-70.

—— and Wudl, L.R. (1977). The mouse T/t complex. In *Concepts in Mammalian Embryogenesis* (ed. M.I. Sherman) pp. 136-234. MIT Press, Cambridge, Mass.

Shin, H.S., Stavnezer, J., Artzt, K., and Bennett, D. (1982). Genetic structure and origin of t-haplotypes of mice, analyzed with H-2 cDNA probes. *Cell* **29**, 969-76.

Shire, I.G.M. and Whitten, E.K. (1980a). Genetic variation in the timing of first cleavage in mice: effect of the paternal genotype. *Biol. Reprod.* **23**, 363-8.

——, —— (1980b). Genetic variation in the timing of first cleavage in mice: effect of maternal genotype. *Biol. Reprod.* **23**, 369-76.

Silver, L.M. (1982). Genomic analysis of the H-2 complex region associated with mouse t-haplotypes. *Cell* **29**, 961-8.

—— and White, M. (1982). A gene product of the mouse t-complex with chemical properties of a cell surface-associated component of the extracellular matrix. *Devl Biol.* **91**, 423-30.

Singh R.P. and Carr, D.H. (1966). The anatomy and histology of XO human embryos and fetuses. *Anat. Rec.* **155**, 369-84.

Sinow, R.M. and Luthardt, F.W. (1978). Chromosome analysis of 4-cell embryos from XO mice. *Am. J. hum. Genet.* **30**, 94A.

Siracusa, G., Colletta, M., and Monesi, V. (1975). Duplication of DNA during the first cell cycle in the mouse embryo. *Devl Biol.* **42**, 355-89.

Sirlin, J.L. and Edwards, R.G. (1950). Timing of DNA synthesis in ovarian oocyte nuclei and pronuclei of the mouse. *Exp. Cell Res* **18**, 190-4.

Skakkebaek, N.E., Zeuthen, E., Nielsen, J., and Yde, H. (1973). Abnormal spermatogenesis in XYY males, a report on 4 cases ascertained through a population study. *Fert. Steril.* **24**, 390-5.

Skalko, R.G. and Morse J.M. (1969). The differential response of the early mouse embryo to actinomycin D treatment *in vitro. Teratology* **2**, 47-54.

Slye, M., Holmes, H.F., and Wells, H.G. (1920). Primary spontaneous tumours of the ovary in mice. Studies on the incidence and inheritability of spontaneous tumours in mice. *J. Cancer Res.* **5**, 205-26.

Smith, L.J. (1956). A morphological and histochemical investigation of a preimplantation lethal (t^{12}) in the house mouse. *J. exp. Zool.* **132**, 51-83.

Smith, R. and McLaren, A. (1977). Factors affecting the time of formation of the mouse blastocoele. *J. Embryol. exp. Morphol.* **41**, 73-92.

Snell, G.D. (1941). Linkage studies with induced translocations in mice. *Genetics* **26**, 169.

—— (1946). An analysis of translocation in the mouse. *Genetics* **31**, 157-80.

—— (1968). The H-2 locus of the mouse. Observations and speculations concerning its comparative genetics and polymorphism. *Fol. biol.* **14**, 335-58.

—— and Stevens, L.C. (1966). Early embryology. In *Biology of the laboratory mouse* pp.205-45. McGraw-Hill, Toronto.

——, Bodemann, E., and Hollander, W. (1934). A translocation in the house mouse and its effect on development. *J. exp. Zool.* **67**, 93-104.

Snow, M.H.L. (1973). Tetraploid mouse embryos produced by cytochalasin B during cleavage. *Nature* **244**, 513-15.

—— (1975). Embryonic development of tetraploid mice during the second half of gestation. *J. Embryol. exp. Morphol.* **34**, 707-21.

—— (1976). The immediate postimplantation development of tetraploid mouse blastocysts. *J. Embryol. exp. Morphol.* **35**, 81-6.

——, Aitken, J., and Ansell, I.D. (1976). Role of the inner cell mass in controlling implantation in the mouse. *J. Reprod. Fert.* **48**, 403-4.

Sokolov, N.P. (1959). *Nucleus–cytoplasmic interactions in interspecific animal hybridization.* Academy of Science, Moscow (in Russian).

Sonta, S. (1980). X-ray-induced aneuploid Chinese hamsters trisomic for chromosome 10. *CIS* **28**, 17-18.

Sorokin, A.V. (1984). Metaphase chromosome studies at initial stages of mouse embryo cleavage. Ph.D. thesis, Moscow (in Russian).

Speed, R.M. (1977). The effects of aging on the meiotic chromosomes of male and female mice. *Chromosoma* **64**, 241-54.

Stene, J., Stene, E., Stengel-Rutkowski, S., and Murken, J.D. (1981). Paternal age and incidence of chromosomal aberrations. Prenatal diagnosis data. Sixth International Congress of Human Genetics, Jerusalem, pp.2-12.

Stevens, L.C. (1967). The biology of teratomas. *Adv. Morphol.* **6**, 1-31.

—— (1975). Teratocarcinogenesis and spontaneous parthenogenesis in mice. 33rd Symposium on Developmental Biology. *The developmental biology of reproduction*, pp.93-106. Academic Press, New York.

—— (1978). Totipotent cells of parthenogenetic origin in a chimaeric mouse. *Nature* **276**, 266-7.

—— and Varum, D.S. (1974). The development of teratomas from parthenogenetically activated ovarian mouse eggs. *Devl Biol.* **37**, 367-80.

——, ——, and Eicher, E.M. (1977). Viable chimaeras produced from normal and parthenogenetic mouse embryos. *Nature* **269**, 515-17.

Stolla, R. and Gropp, A. (1974). Variation of the DNA content of morphologically normal and abnormal spermatozoa in mice susceptible to irregular meiotic segregation. *J. Reprod. Fert.* **38**, 335-46.

Streffer, C., van Beuningenen, D., Molls, M., Zamboglon, N., and Schultz, G.A.

(1980). Kinetics of cell proliferation in the preimplanted mouse embryo *in vivo* and *in vitro*. *Cell Tiss. Kinet.* **13**, 135–43.

Sugawara, S. and Mikamo, K. (1983). Absence of correlation between univalent formation and meiotic non-disjunction in aged female Chinese hamster. *Cytogenet. Cell Genet.* **35**, 34–40.

Sumner, A.T. (1971). Frequency of polyploid spermatozoa in men. *Nature* **231**, 49.

——, Evans, H.J., and Buckland R.A. (1971). New technique for distinguishing between human chromosomes. *Nature* **232**, 31–2.

Surani, M.A. and Barton, S.C. (1983). Development of gynogenetic eggs in the mouse: implications for parthenogenetic embryos. *Science* **222**, 1034–6.

——, Azim, H., and Kaufman, M.H. (1977). Influence of extracellular Ca^{2+} and Mg^{2+} ions on the second meiotic division of mouse oocytes: relevance to obtaining haploid and diploid parthenogenetic embryos. *Devl Biol.* **59**, 86–90.

——, Barton, S.C., and Kaufman, M.H. (1977). Development of chimeras between diploid parthenogenetic and fertilised embryos. *Nature* **270**, 601–3.

——, ——, and Norris, M.L. (1984). Development of reconstituted mouse eggs suggests imprinting of the genome during gametogenesis. *Nature* **308**, 548–50.

Szemere, G. and Chandley, A.C. (1975). Trisomy and triploidy induced by X-irradiation of mouse spermatocytes. *Mutat. Res.* **33**, 229–38.

Szollosi, D. (1966). Time and duration of DNA synthesis in rabbit eggs after sperm penetration. *Anat. Rec.* **154**, 209–12.

—— (1973). Mammalian eggs aging in the fallopian tubes. In *Aging gametes* (ed. R.J. Blandau) pp. 98–121. Karger, Basel.

Szulman, A.E. and Surti, U. (1978). The syndromes of hydatiform mole. I. Cytogenetic and morphologic correlations. *Am. J. Obstet. Gynecol.* **131**, 665–71.

——, —— (1984). The syndromes of partial and complete molar gestation. *Clin. Obstet. Gynecol.* **27**, 172–80.

Tabor, A., Andersen, O., Lundsteen, C., Neibuhr, E., and Sardemann, H. (1983). Interstitial deletion in the 'critical region' of the long arm of the X-chromosome in a mentally retarded boy and his normal mother. *Hum. Genet.* **64** (2), 196–9.

Tadashi, K., Koso, O., Niikawa, N., Ariane, F., and Sugandhi, A. (1973). Banding analysis of abnormal karyotypes in spontaneous abortion. *Am. J. hum. Genet.* **25**, 539–47.

Takagi, N. (1974). Superovulation and chromosomal fetal karyotype in mice. *Jap. J. Genet.* **46**, 443 (in Japanese).

—— (1974). Differentiation of X-chromosomes in early female mouse embryos. *Exp. Cell Res.* **86**, 127–35.

—— (1978). Preferential inactivation of the paternally derived X chromosome in mice. In *Basic life sciences, Vol. 12, Genetic mosaics and chimeras in mammals* (ed. L.B. Russell) pp. 341–60. Plenum Press, New York.

—— and Sasaki, M. (1975). Preferential inactivation of the paternally derived X-chromosome in the extraembryonic membranes of the mouse. *Nature* **250**, 640–2.

——, Wake, N., and Sasaki, M. (1978). Cytological evidence for preferential inactivation of the paternally derived X chromosome in XX mouse blastocysts. *Cytogenet. Cell Genet.* **20**, 240–8.

Tarkowski, A.K. (1966). An air-drying method for chromosome preparations from mouse eggs. *Cytogenetics* **5**, 394–400.

—— (1970). The nature of blastomere differentiation in early development of mammals: epigenesis or preformism? In *Inter-cellular interactions in growth and*

differentiation pp. 117–25. Nauka, Moscow (in Russian).
—— (1971). Recent studies on parthenogenesis in the mouse. *J. Reprod. Fert.* **14** (Suppl.), 31–9.
—— (1975). Induced parthenogenesis in the mouse. In *33rd Symposium of the Society of Developmental Biology: Developmental Biology of Reproduction* (ed. C.L. Markert) pp. 107–29. Academic Press, New York.
—— (1977). *In vitro* development of haploid mouse embryos produced by bisection of one-cell fertilized eggs. *J. Embryol. exp. Morphol.* **38**, 187–202.
—— (1982). Nucleo-cytoplasmic interaction in oogenesis and early embryogenesis in the mouse. In *Embryonic development, part A. Genetic aspects* (ed. M.M. Burger and R. Neber). Academic Press, New York.
—— and Balakier, H. (1980). Nucleo-cytoplasmic interactions in cell hybrids between oocytes, blastomeres and somatic cells in the mouse. *J. Embryol. exp. Morphol.* **55**, 319–30.
—— and Rossant, J. (1976). Haploid mouse blastocysts developed from bisected zygotes. *Nature* **259**, 663–5.
—— and Wroblewska, J. (1967). Development of blastomeres of mouse eggs isolated at the 4 and 8-cell stage. *J. Embryol. exp. Morphol.* **18**, 155–80.
——, Witkowska, A., and Nowicka, J. (1970). Experimental parthenogenesis in the mouse. *Nature* **266**, 162–5.
——, Witkowska, A., and Opas, I. (1977). Development of cytochalasin B-induced tetraploid and diploid/tetraploid mosaic mouse embryos. *J. Embryol. exp. Morphol.* **41**, 47–64.
Tasca, R.J. and Hillman, N. (1970). Effects of actinomycin D and cycloheximide on RNA and protein synthesis in cleavage stage mouse embryos. *Nature* **225**, 1022–5.
Tease, C. (1982). Similar dose related chromosome non-disjunction in young and old female mice after X-irradiation. *Mutat. Res.* **95**, 287–96.
Tettenborn U. and Gropp A. (1970). Meiotic nondisjunction in mice and mouse hybrids. *Cytogenetics* **9**, 272–83.
——, Schwinger, E., and Gropp, A. (1970). Zur Hodenfunktion bei Männern mit XYY-Konstitution. Meiose und morphologische Befunde. *Dtsch. med. Wochenschr.* **95**, 158–61.
Therkelsen, A.J., Jensen, O.M., Jonassen, J., Lamm, L.U., Lauritsen, J.G., Lindsten, J., and Petersen, B. (1973). Studies on spontaneous abortions. In *Chromosome identification* (ed. I. Caspersson and L. Zech) pp. 251–7. Academic Press, New York.
Thibault, C. (1959). Analyse de la fécondation de l'oeuf de la truie après accouplement ou insemination artificielle. *Ann. zootechn.* (Suppl.), 165–177.
—— (1977). Are follicular maturation and oocyte maturation independent processes? *J. Reprod. Fert.* **51**, 1–15.
Thompson, H., Meinyk, J., and Hecht, F. (1967). Reproduction and meiosis in XYY. *Lancet* **ii**, 831.
—— and Biggers, J.D. (1966). Effect of inhibitors of protein synthesis on the development of preimplantation mouse embryos. *Exp. Cell Res.* **41**, 411–27.
Tiepolo, L., Zuffardi, O., and Rodewald, A. (1977). Nullisomy for the distal portion of Xp in a male child with an X/Y translocation. *Hum. Genet.* **39**, 277–81.
Titenko, N.Y. (1977). Preimplantation development of mouse embryos in homo-and heterogeneous crosses. *Ontogenez* **8**, 27–33 (in Russian with English summary).

Uchida, I.A. and Freeman, C.P.V. (1977). Radiation induced non-disjunction in oocytes of aged mice. *Nature* **265**, 186–7.

—— and Lee, C.P.V. (1972). Radiation induced non-disjunction in mouse oocytes. *Nature* **250**, 601.

—— and Lin, C.C. (1972). Identification of triploid genome by fluorescence microscopy. *Science* **176**, 304–5.

Udalova, L.D. (1969). Chromosome analysis at early stages of embryogenesis in normal rats and after teratogen application. Ph.D. thesis, Leningrad (in Russian).

—— (1971). Method of bone-marrow biopsy in karyotype studies of laboratory animals. *Arkh. Anat. Histol. Embriol.* **9**, 87–88 (in Russian with English summary).

Uebele-Kallhardt, B.M. (1978). *Human oocytes and their chromosomes. An atlas.* Springer, Berlin.

Unaki, W. and Hsu, T.C. (1972). The C- and G-banding patterns of *Rattus norvegicus* chromosomes. *J. nat. Cancer Inst.* **49**, 1425–31.

Van Blerkom, J. (1975). Qualitative aspects of protein synthesis during the preimplantation stages of pregnancy. *Res. Reprod.* **7**, 2–3.

—— (1981). Intrinsic and extrinsic patterns of molecular differentiation during oogenesis, embryogenesis and organogenesis in mammals. In *Cellular and molecular aspects of implantation* (ed. S.R. Glasser and D.W. Bullock) pp. 155–76. Plenum Press, New York.

—— and Brockway, G.O. (1975a). Qualitative patterns of protein synthesis in the preimplantation mouse embryo. I. Normal pregnancy. *Devl Biol.* **44**, 148–157.

——, —— (1975b) Qualitative patterns of protein synthesis in the preimplantation mouse embryo. II. During release from facultative delayed implantation. *Devl Biol.* **46**, 446–51.

—— and Mance, C. (1974). Development of preimplantation rabbit embryos *in vivo* and *in vitro*. II. Comparison of quantitative aspects of protein synthesis. *Devl Biol.* **40**, 41–51.

—— and Runner, M.N. (1984). Mitochondrial reorganization during resumption of arrested meiosis in the mouse oocyte. *Am. J. Anat.* **171**, 335–55.

——, Barton, S.C., and Johnson, M.H. (1976). Molecular differentiation in the preimplantation mouse embryo. *Nature* **259**, 319–21.

Van Valen, P. (1966). Oligosyndactylism, an early embryonic lethal in the mouse. *J. Embryol. exp. Morphol.* **15**, 119–24.

Vassilakos, P., Riotton, G., and Kajii, T. (1977). Hydatiform mole: two entities. A morphological and cytogenetic study with some clinical considerations. *Am. J. Obstet. Gynecol.* **127**, 167–70.

Verbanac, K.M. and Warner, C.M. (1981). Role of the major histocompatibility complex in the timing of early mammalian development. In *Cellular and molecular aspects of implantation* (ed. S. Glasser and D. Bullock) pp. 467–70. Plenum Press, New York.

Vickers, A.D. (1969). Delayed fertilization and chromosomal anomalies in mouse embryos. *J. Reprod. Fert.* **20**, 69–76.

Vogel, F. (1973). Genotype and phenotype in human chromosome aberrations and in the minute mutants of *Drosophila melanogaster. Humangenetik* **19**, 41–56.

—— and Motulsky, A.G. (1979). *Human genetics. Problems and approaches.* Springer, Berlin.

Vorontzov, N.N. (1966). Karyotype evolution. In *Handbook of cytology* (ed. A. Troshin) Vol. 2, pp. 359–89. Nauka, Leningrad (in Russian).

Wabl, M. and Steinberg, C. (1982). A theory of allelic and isotypic exclusion for immunoglobulin genes. *Proc. nat. Acad. Sci. USA* **79**, 6976–83.

Waisman, B.L., Dyban, A. P., and Udalova, L.D. (1981). Gene dosage effects on cytoplasmic glutamate–oxaloacetate transaminase in mouse embryos with trisomy for autosome 19. *Bull. exp. Biol. med.* **5**, 598–9 (in Russian with English summary).

Wakasugi, N. (1973). Studies on fertility of DDK mice: reciprocal crosses between DDK and C57BL/6J strains and experimental transplantation of the ovary. *J. Reprod. Fert.* **33**, 283–91.

—— (1974). A genetically determined incompatibility system between spermatozoa and eggs, leading to embryonic death in mice. *J. Reprod. Fert.* **41**, 85–96.

—— and Morita, M. (1977). Studies on the development of F_1 embryos from inter-strain crosses involving DDK mice. *J. Embryol. exp. Morphol.* **38**, 211–16.

——, Tomita, T., and Kondo, K. (1967). Differences of fertility in reciprocal crosses between inbred strains of mice: DDK, KK and NC. *J. Reprod. Fert.* **13**, 41–50.

Wake, N., Takagi, N., and Sasaki, M. (1976). Non-random inactivation of the X-chromosome in the rat yolk sac. *Nature* **262**, 580–1.

——, ——, —— (1978). Androgenesis as a cause of hydatiform mole. *J. nat. Cancer Inst.* **60**, 51–7.

Waksmundska, M., Krysiak, E., Karasiewicz, J., Czolowska, R., and Tarkowski, A. (1984). Autonomous cortical activity in mouse eggs controlled by a cytoplasmic clock. *J. Embryol. exp. Morphol.* **79**, 77–96.

Walbaum, R., Dehaene, P., Breynaert, R., and Fontaine, G. (1971). Les transloca-tions Robertsoniennes 21-D. Etude de quatre familles. *Lille med.* **16**, 1398–409.

Wallace D.C., Surti U., Adams C.W., and Szulman A.E. (1982). Complete moles have paternal chromosomes but maternal mitochondrial DNA. *Hum. Genet.* **61**, 145–7.

Warner, C.J. and Vestergh, L.R. (1974). *In vivo* and *in vitro* effect of α-amanitin on preimplantation embryo RNA polymerase. *Nature* **248**, 678–80.

—— and Spannaus, D.J. (1984). Demonstration of H-2 antigens on preimplantation mouse embryos using conventional antisera and monoclonal antibody. *J. exp. Zool.* **230**, 37–52.

Washburn, L. and Eicher, E. (1975). *Mouse News Letter* **56**, 43.

Wassarman, P.M., Schultz, R.M., and Letourneau, G. (1978). Protein synthesis dur-ing meiotic maturation of mouse oocytes *in vitro*. Synthesis and phosphorylation of a protein localized in the germinal vesicle. *Devl Biol.* **69**, 94–107.

Watanabe, T. and Endo, A. (1979). Cytogenetic analysis of mammalian oocytes as a tool for chromosomal mutagen screening. *Congenital Anomalies (Senten Ijo)* **19**, 323–35.

Weil, J. and Epstein, C.J. (1979). The effect of trisomy 21 on the patterns of poly-peptide synthesis in human fibroblasts. *Am. J. hum. Genet.* **31**, 478–88.

Weiss, G., Weick, R., Knobil, E., Wolman, S.R., and Gorstein, F. (1973). An XO anomaly and ovarian dysgenesis in a rhesus monkey. *Fol. primatol.* **19**, 24–7.

Weitlauf, H.M. and Greenwald, G.S. (1967). A comparison of the *in vivo* incorporation of S^{35} methionine by two-celled mouse eggs and blastocysts. *Anat. Rec.* **159**, 249–54.

Welshons, W.J. and Russell, L.B. (1959). The Y chromosome as the bearer of male determining factors in the mouse. *Proc. nat. Acad. Sci. USA* **45**, 560–6.

West, J.D. and Green, J.F. (1983). The transition from oocyte-coded to embryo-coded glucose phosphate isomerase in the early mouse embryo. *J. Embryol. exp. Morphol.* **78**, 127–40.

White, B.J. and Tjio, J.H. (1968). A mouse translocation with 38 and 39 chromosomes but normal N.F. *Hereditas* **58**, 284–9.

——, ——, Water, L.C., and Crandell C. (1972a). Studies of mice with balanced complement of 36 chromosomes derived from F_1 hybrids of T_1Wh and T_1ALD translocation homozygotes. *Proc. Nat. Acad. Sci. USA.* **69**, 2757–61.

——, ——, ——, —— (1972). Trisomy of the smallest autosome of the mouse and identification of the T_1 Wh translocation chromosome. *Cytogenetics* **11**, 363–78.

——, ——, ——, —— (1974a). Trisomy 19 in the laboratory mouse. I. Frequency in different crosses at specific developmental stages and relationship of trisomy to cleft palate. *Cytogenet. Cell Genet.* **13**, 217–31.

——, ——, ——, —— (1974b). Trisomy 19 in the laboratory mouse. II. Intra-uterine growth and histological studies of trisomies and their normal littermates. *Cytogenet. Cell Genet.* **13**, 232–45.

White, M.J.D. (1969). Chromosomal rearrangements and speciation in animals. *Rev. Genet.* **3**, 75–98.

—— (1973). *Animal cytology and evolution*. Cambridge University Press, London.

—— (1975). Chromosomal repatterning—regularities and restrictions. Symposium on Evolution, XIII International Congress of Genetics. *Genetics* **79** (Suppl), 63–72.

—— (1978). *Modes of speciation*. W.H. Freeman, San Francisco.

Whitten, W.K. and Dagg, C.P. (1962). Influence of spermatozoa on the cleavage rate of mouse eggs. *J. Exp. Zool.* **148**, 173–83.

Whittingham, D.G. (1975). Fertilization, early development and storage of mammalian ova *in vitro*. In *The early development of mammals* (ed. M. Balls and A.E. Wild) pp. 1–24. Cambridge University Press, Cambridge.

—— (1980). Parthenogenesis in mammals. In *Oxford reviews of reproductive biology* (ed. C.A. Finn) Vol. 2, pp. 205–31. Oxford University Press, Oxford.

Widmeyer, M.A. and Shaver, E.L. (1972). Estrogen, progesteron and chromosome abnormalities in rabbit blastocyst. *Teratology* **6**, 207–14.

Willadsen, S.M. (1979). A method for culture of micromanipulated sheep embryos and its use to produce monozygotic twins. *Nature* **277**, 298–300.

—— (1980). The viability of early cleavage stages containing half the normal number of blastomeres in the sheep. *J. Reprod. Fert.* **59**, 357–62.

——, Lehn-Jensen, H., Fehilly, C.B., and Newcomb, R. (1981). The production of monozygotic twins of preselected parentage by micromanipulation of non-surgically collected cow embryos. *Theriogenol.* **15**, 23–9.

Witkowska, A. (1973a). Parthenogenetic development of mouse embryos *in vivo*. I. Preimplantation development. *J. Embryol. exp. Morphol.* **30**, 519–45.

—— (1973b). Parthenogenetic development of mouse embryos *in vivo*. II. Postimplantation development. *J. Embryol. exp. Morphol.* **30**, 547–60.

—— (1981). Pronuclear development and the first cleavage division in polyspermic mouse eggs. *J. Reprod. Fert.* **62**, 493–8.

Wolf, U. and Engel, W. (1972). Gene activation during early development of mammals. *Humangenetik* **15**, 99–118.

—— and Zenezes, M.T. (1979). Gonadendifferenzierung und H-Y-Antigenen. *Verh. Anat. Ges.* **73**, 379–84.

Womack, J.E. (1978). Biochemical loci:mouse. In *Inbred and genetically defined strains of laboratory animals. Part I. Mouse and rat* (ed. P. Altman and D.D. Katz) pp. 96–100. Federation of Animal Society for Experimental Biology, Bethesda, Md.

Woodland, H.R. and Graham, C. (1969). RNA synthesis during early development of the mouse. *Nature* **221**, 327–32.

Wroblewska, I.J. (1971). Developmental anomaly in the mouse, associated with triploidy. *Cytogenetics* **10**, 199–207.

—— (1978). Spontaneous triploidy and tetraploidy in inbred strains of mice and their crosses. Ph. D. thesis, Warsaw.

—— and Dyban, A.P. (1969). Chromosome preparations from mouse embryos during early organogenesis. Dissociation after fixation followed by air-drying. *Stain Technol.* **44**, 147–50.

Wudl, L. and Chapman, V. (1976). The expression of β-glucuronidase during preimplantation development of mouse embryos. *Devl Biol.* **48**, 104–9.

—— and Sherman, M.J. (1976). *In vitro* studies on mouse embryos bearing mutations at the T-locus: t^{w5} and t^{12}. *Cell* **9**, 523–31.

Yadov, J.S. and Yadov, A.S. (1981). Spontaneous centric fusion and karyotypic variations in *Mus musculus. Genet. pol.* **22**, 217–21.

Yamamoto, M. and Ingalls, T.H. (1972). Delayed fertilization and chromosome anomalies in the hamster embryo. *Science* **176**, 518.

——, Endo, A., and Watanabe, G. (1971). Chromosomal aneuploids and polyploids in embryos of diabetic mice. *Arch. envir. Health* **22**, 462–75.

——, ——, —— (1973). Maternal age dependence of chromosome anomalies. *Nature* **241**, 141–2.

——, Fujimori, H., Ito, T., Kamimura, K., and Watanabe, G. (1975). Chromosome studies in 500 induced abortions. *Humangenetik* **29**, 9–14.

Yanagisawa, K., Bennett, D., Boyse, E.A., Dunn, L.C., and Dimeo, A. (1974a). Serological identification of sperm antigens specified by lethal t-alleles in the mouse. *Immunogenetics* **1**, 57–67.

—— Pollard, D.R., Bennett, D., Dann, L.C., and Boyse, E. A. (1974b). Transmission ratio distortion at the T-locus: serological identification of two sperm populations in t-heterozygotes. *Immunogenetics* **1**, 91–6.

Yongbloet, P.H. (1975). The effects of preovulatory overripeness of human eggs on development: seasonality of birth. In *Aging gametes* (ed. R.J. Blandau), pp. 300–29. Karger, Basel.

Yosida, T.H. (1978). An XXY male appeared in the F_2 hybrids between Oceanian and Ceylonese type black rats. *Proc. Jap. Acad.* **B54** (3), 121–4.

——, Tsuchiya, K., and Moriwaki, K. (1971). Karyotypic differences of black rats *Rattus rattus*, collected in various localities of East and South East Asia and Oceania. *Chromosoma* **33**, 252–67.

Young, R.J., Sweeney, K., and Bedford, J.M. (1978). Uridine and guanosine incorporation by mouse one-cell embryos. *J. Embryol. exp. Morphol.* **44**, 133–48.

Zartman, D.E., Hinesley, L.L., and Gnathowski, M.W. (1983). A 53, X female sheep *(Ovis aries). Cytogenet. Cell Genet.* **30** (1), 54–8.

Zavarzin, A.A. (1967). *DNA synthesis and cell population kinetics in mammalian ontogenesis.* Nauka, Moscow (in Russian).

——, Samoshkina, N.A., and Dondua, A.K. (1966). DNA-synthesis and cell kinetics in early mouse embryos. *Zh. Obsch. Biol.* **27**, 697–709 (in Russian with English summary).

Zech L., Evans E.P., Ford E., and Gropp A. (1972). Banding patterns in mitotic chromosomes of tobacco mouse. *Exp. Cell Res.* **70**, 263–8.

Ziomek, C.A., Pratt, H.P.M., and Johnson, M.H. (1982). The origin of cell diversity in the early mouse embryo. In *The integration of cell in animal tissues* (ed. M.E. Finbow and J.E. Pitts) pp. 149–65. Cambridge University Press, Cambridge.

Index

blastomeres (*cont.*)
  fusion producing tetraploid embryos 61, 63
  totipotency of 281
  see also two-cell embryos; four-cell embryos
blastulation 229, 277, 283
  cell numbers required for 64, 284
  effect of α-amanitin on 249, 250, 251
  in parthenogenetic embryos 34–5, 37
  protein synthesis during 251
brain
  changes in partially trisomic mice 216
  retarded development in trisomy 1 mice 110, 141

c^{25H} (albino) deletions 170, 264–5, 279, 280, 289
calcium (Ca^{2+})
  changes in maturing oocytes 270
  deficient medium for oocyte activation 21, 23, 32
  role in compaction 282
cardiovascular malformations in trisomic embryos 119, 121, 125, 131, 136, 141, 240, 293–4
cats, numerical sex chromosome aberrations in 172, 173, 178, 182
cattle, numerical chromosome aberrations in 75, 172
caudal malformations in trisomy 17 mouse embryos 132
cell cycles
  in early embryos 242–3, 274–5, 277–8
  genetic control of 278, 280–1
  in triploid cells 52, 277
  in trisomic cells 286–7
  see also meiosis; mitosis; mitotic activity
cell membrane proteins 253, 285
  in t^{12} and t^{w32} homozygotes 260–1
cell surface protein p63/6, 9a 261
'cell syndrome' of aneuploid cells 286
chiasma formation, factors affecting frequency of 79, 86
chimaeras 291, 296
  behavioural abnormalities in 224–5
  monosomy 19-diploid 162
  parthenogenetic-diploid 37–8
  secondary trisomic, *see* haemopoietic stem cells, trisomic
  spontaneous 61
  tetraploid–diploid 66, 67
  trisomic-diploid aggregated embryos 132–4, 142, 162, 239, 287
  trisomy 16-diploid 131–2, 240
  see also mosaics
chimpanzees, trisomy in 107
chromosomal aberrations

genetic background and expression of 297
  in mice heterozygous for different translocations 94–5
  numerical
    mechanisms of origin 68–91
    of sex chromosomes 172–84
    see also aneuploidy; haploid embryos; monosomy; nullisomy; tetraploidy; triploidy; trisomy
  spontaneous incidence 15–17, 232–3
  structural, *see* structural chromosomal aberrations
chromosomes
  analytical methods 2–3
  comparisons of mouse and human 239–40
  functions during post-implantation development 291–4
  interactions in early embryos 271–3, 287, 288–91, 292, 297, 298
  non-disjunction, *see* non-disjunction
  see also individual human chromosomes and mouse chromosomes
cleavage 229
  block leading to tetraploidy 61, 62
  cell cycles during 243
  cytoplasmic factors controlling 271, 273–4, 281
  in embryos with chromosomal aberrations 277–81
  genetic control of 170–1, 242–66, 274–7
  in haploid embryos 33–4, 35, 37, 277
  in monosomic embryos 148, 153, 160–2, 278
  non-disjunction during 75–7, 91, 104
  'proteins' 277–8
  in tetraploid embryos 64–5, 277, 278
  in triploid embryos 49–50, 277, 278
  in XO mice 175
  see also blastocysts; blastomere numbers; four-cell embryos; two-cell embryos
cleft palate 294
  in trisomic embryos 125, 131, 141, 237
  in trisomy 13 121
  in trisomy 18 134
  in trisomy 19 136, 137
clustering of 'early' genes 278, 285, 292, 297
colcemid 9–10, 48
colchicine 48, 50, 62
compaction 281–2
  abnormalities in t^{12} and t^{w32} homozygotes 260–1
  effect of α-amanitin on 249, 251, 282
  in haploid parthenogenetic embryos 34–5, 37, 282
conditioned reflexes in mice with reciprocal translocations 213, 214–19
congenital malformations, *see* malformations